Monika J. A. Schröder

Food Quality and Consumer Value

Springer

Berlin
Heidelberg
New York
Hong Kong
London
Milan
Paris
Tokyo

Monika J. A. Schröder

Food Quality and Consumer Value

Delivering Food that Satisfies

With 14 Figures and 36 Tables

 Springer

Dr. Monika J. A. Schröder

Faculty of Business and Arts
Queen Margaret University College
Edinburgh EH12 8 TS,
UK
mschroder@qmuc.ac.uk

ISBN 3-540-43914-5 Springer-Verlag Berlin Heidelberg New York

Cataloging-in-Publication Data applied for

Bibliographic information published by Die Deutsche Bibliothek
Die Deutsche Bibliohek lists this publication in the Deutsche Nationalbibliografie; detailed bibliographic data
is available in the Internet at <http://dnb.ddb.de>.

Springer-Verlag Berlin Heidelberg New York
a member of BertelsmannSpringer Science+Business Media GmbH

http://www.springer.de

© Springer-Verlag Berlin Heidelberg 2003
Printed in Germany

The use of general descriptive names, registered names, trademarks, etc. in this publication does not imply, even in the absence of a specific statement, that such names are exempt from the relevant protective
laws and regulations and therefore free for general use.

Product liability: The publisher cannot guarantee the accuracy of any information about dosage and application contained in this book. In every individual case the user must check such information by consulting
the relevant literature.

Typesetting: Fotosatz-Service Köhler GmbH, Würzburg
Coverdesign: design & production, Heidelberg

Printed on acid-free paper   52/3020 ra – 5 4 3 2 1 0

For John

Preface

Consumer markets for foods and beverages in developed countries are well supplied and highly fragmented. Yet despite the ever-increasing number of product lines on offer to consumers in such countries, observers' views diverge about whether this means genuine choice for an individual. The question being asked is how close food retailers actually come to fulfilling their customers' requirements. The global nature of the modern food market provides access for ordinary consumers to unfamiliar products from around the world. For example, a considerable range of tropical fruit is now readily available to them. Today authentic ingredients for the preparation of all sorts of ethnic and foreign dishes, from Japanese sushi to Thai curries, are relatively easy to get hold of throughout the UK. The nature of the market also offers continuity of supply throughout the year of produce traditionally thought of as seasonal, e.g., strawberries and green beans. At the same time, the availability of a varied supply of locally grown, fresh produce is widely perceived as having been severely reduced in recent years. For example, whilst the UK National Collection of apples contains several thousand varieties, supermarkets may only stock around ten varieties of apple, and the offer will be similar throughout a retailer's national branch network. Yet consumer perception of local foods tends to associate them with fuller flavour and greater purity than produce grown in distant parts of the world.

Through food research, many food components with potential, specific health benefits for humans are being discovered, especially components naturally present in edible plant tissues. Based on such knowledge, technological innovations have allowed functional attributes to be specifically designed into modern foods, adding value both for consumers and producers. Apart from health-related attributes, such added features may relate to the convenience in use of a food. Extended shelf life and ready-to-consume technology are especially important here, but there are many other aspects, e.g., portion control and control of energy intake, possibly in conjunction with a suggested dietary plan. Recently, health-related food quality attributes have been attracting increasing interest for incorporation into various foods, the market in so-called functional foods representing a clear manifestation of this trend. The perceived downside of food technology tends to be identified with a certain loss of control on the part of modern food consumers. This is felt more strongly in some sectors of the market than in others, and is related to the level of an individual's emotional involvement in diet and food choice decisions. Modern food supply chains are complex in structure, and as a result, communication between the producer and the consumer of a food may suffer. Food labels must include ingredients list by law. However, food labelling by itself should not be taken as the be-all and end-all of communication with the consumer; and in practice, it tends not to be employed to their best advantage. As an example, food ingredients lists typically use very small print size, and are barely legible even by the privileged few who possess perfect vision. Few consumers have a clear understanding any longer of the true nature of manufactured food products, e.g., the role of each of the ingredients, and the processes involved. This applies not only to high-technology procedures, but also to so-called traditional food technologies. As an example, the use of curing salts in organic bacon has been criticised by green campaigners, suggesting lack of knowledge of what bacon is. Further confusion may result from the misinterpretation of legitimate marketing claims and, in certain cases, from dishonest claims being made.

Irrespective of whether individual products in some product category are differentiated in a manner that is meaningful in terms of individual requirements, consumers must choose between them. Food choice also presents the consumer with a paradox: it is the very prototype of a routinised behaviour, but at the same time, it is potentially highly involving. Depending on the individual, and their particular condition and consumption context, very little attention may be paid to the activity of eating and drinking, or a very great deal. Some consumers, for whom choice represents a mere chore, may develop strategies to reduce the need to actively choose. One such rule might be for a person to select only from a particular category, e.g., organic foods, or brand, e.g., an own-label value range. Product differentiation, combined with the impossibility of direct communication between the food chain players, creates the need for food producers and distributors to signal the nature of products to consumers, focusing in particular on key differences with the product's close competitors. Thus the term quality, as well as other terms with similar meanings, has become ubiquitous in modern marketing communications about food. Quality, as this book will demonstrate, is an operationally useful concept, both in product design and in routine product delivery. However, in the context of food labelling and advertising, the word quality *per se* is relatively meaningless, a mere signal. It manages to draw attention, but needs to be further qualified. The prospective purchaser, or consumer, of a food with the quality label is urged to evaluate further the specific properties of the product which may be of value to them. Assuming that he or she responds to the challenge, a more meaningful context may emerge, whether in terms of attractive flavour or presentation, nutritional value or ethical production; in fact anything pertinent to individual consumer requirements. It is inevitable, and logical, that food manufacturer and retailers will use positive language and imagery to promote their products.

Certain individuals and groups, who are known for their vociferous disapproval of modern foods and food production systems might disagree. These might wish to attribute to food supply chains some moral duty to work actively for the optimum health of individual consumers. In their worldview, modern food consumers are the passive, more or less defenceless victims of an overpowering opponent; and worst of all, one with irresistible powers to seduce. From these quarters originate familiar expressions such as junk foods, as a general label for a somehow undesirable food, and Frankenstein foods, for genetically modified foods. However, markets do not work efficiently if consumers do not play an active role in them. The concept of consumer value is one of the main pillars underpinning the theory of market differentiation. In the ideal market, competing products are fully described in terms of their key attributes. This allows prospective customers to evaluate each product and product attribute, and enables optimal matches of product attributes with consumer requirements, i.e., rational consumer choice. The current UK government acknowledges the critical role that consumers play in the effective functioning of modern markets and modern society. In the realm of food, mechanisms for consumer representation and education have begun to be explored with the creation in 2000 of a Food Standards Agency. Similar moves are afoot at the level of the European Union. An important aspect of the public debate about food quality is the question whether consumers are being misled by the use of non-specific, quality-related terminology in advertising, and whether there is a need for statutory intervention. Formal quality management plays an increasingly important role in the UK food supply chain. Formal hazard assessment and management is a legal requirement for all food businesses, and there are many commercial quality assurance schemes with focus on specific product attributes. In recent years, food from organic production systems has consistently and significantly increased in popularity, and standards are now protected through statutory controls.

What then are the attributes of a food that contribute significantly to its quality? What are the influences that shape different individuals' attitudes towards these attributes? How do we define good foods and bad foods, and in whose terms? In order to supply consumer satisfaction via the supply of targeted goods, both product knowledge and an understanding of the nature of consumer behaviour and what motivates consumption choices are needed. Because such motives can be very complex, consumption preferences and choices must be attributed accurately. In common with consumer goods generally, food and drink are not consumed simply for functional purposes, such as satiation, thirst quenching, nutrition, convenience and so forth. They are also used, or chosen above alternative products, to reaffirm an individual's self-perception or how they wish to be perceived by others. Considerations of this nature clearly play an important role in the recent trend towards ethical food consumption, in particular, the rapid growth in the market for organic foods. Consumption of high-status food and drink serves to make similar, person centred statements.

In this book, I take an interdisciplinary approach to the analysis of satisfaction in relation to the consumption of food, with both food science and consumer science playing central parts. Each area is discussed in detail, using the appropriate technical terminology, but keeping the text accessible to readers from both academic traditions, as well as to non-specialist readers. As far as I am aware, this is a novel approach to the subject of consumer value in food, especially in so far as it is tackled as a single-author monograph. Food scientists who choose to work for food manufacturers invariably find that they need to expand their horizons away from simple product focus. My own professional career has included roles as a researcher in a dairy research institute, quality management in the flavourings industry, and latterly, teaching and research activity in food and consumer studies. I became interested in consumer and marketing theory whilst employed in the industry, and cemented my theoretical understanding of these disciplines after returning to academia. These combined experiences have allowed me to experience many aspects of food quality and value, including the provision of customer service, and participation in policy and regulatory consultative processes. My aim in this book has been to allow readers to examine the specific issues in the topical debate about good and bad foods in an orderly fashion, and from different perspectives. I hope that the structure I have adopted will be effective in supporting this goal. Given the breadth of coverage, not every quality issue that has been raised here could be debated in depth. However, suitable literature references will allow interested readers to further pursue specific topics. Throughout the book, I seek to identify the links between the food as the object of consumption (whose properties can be measured in a value free manner), those properties that contribute to value (which therefore should be controlled), and the nature of the value derived by consumers.

I offer the book as my contribution to a better understanding of what is important about foods and of what motivates food choice. I hope it will serve to help to bridge perceptional differences that sometimes exist between the different disciplines contributing to the functioning of modern food systems. The book is aimed at anyone with an interest in food quality, in particular those who, like me, have been frustrated with the polarity of the debate about what is good and bad in food. Food scientists and technologists, whose training often denies them exposure to social scientific thought, will benefit from reading it, as will any food industry personnel with either technical or commercial roles and responsibilities. It should also provide a resource for those involved in the communication, teaching and researching of food issues, of consumer and marketing issues, and to students in each of these areas.

Monika Schröder
September 2002

Acknowledgements

I thank all those who have collaborated with me on research in recent years. They have been instrumental in formulating my current understanding of food quality and consumer issues. Their names appear in literature references throughout the book. Food preferences, cooking and eating out are popular topics of everyday discourse, and there are a number of people who, as the book was taking shape, provided useful impulses for it. I am especially grateful to John Bower and John Beath, each of whom read, and provided me with valuable feedback on, sections of the manuscript.

Table of Contents

Part 2
Food Quality Attributes 97

Part 3
Understanding the Food Consumer 211

Abbreviations

AA	Arachidonic acid
ACE	(Vitamins) A, C and E
ACh	Acetylcholine
ACNFP	Advisory Committee on Novel Foods and Processes
ANS	Autonomic nervous system
APA	(UK) Association of Public Analysts
ASP	Amnesic shellfish poisoning
ATP	Adenosine triphosphate
BHA	Butylated hydroxyanisole
BHT	Butylated hydroxytoluene
BRC	British Retail Consortium
BSE	Bovine spongiform encephalopathy
CAMRA	Campaign for Real Ale
CAP	Common Agricultural Policy
CCK	Cholecystokinin
CNS	Central nervous system
COMA	Committee on Medical Aspects of Food Policy
DFD	Dark, firm, dry (meat)
DGLA	Dihomo-gamma-linolenic acid
DHA	Docosahexaenoic acid
DSP	Diarrhetic shellfish poisoning
ECR	Efficient Consumer Response
EFA	Essential fatty acid
EFSIS	European Food Safety Inspection Service
EPA	Eicosapentaenoic acid
EPEC	Enteropathogenic *E. coli*
EU	European Union
FACE	(Vitamins) folate, A, C and E
FEMA	(US) Flavor and Extract Manufacturers' Association
GABA	Gamma-butyric acid
GLA	Gamma-linolenic acid
GM	Genetically modified
GMP	Good Manufacturing Practice
GMP	Guanosine monophosphate
GRAS	Generally recognised as safe
HACCP	Hazard Analysis Critical Control Point
HDL	High-density lipoprotein
HM	High methoxyl- (pectins)
HTST	High-temperature-short-time

HVP	Hydrolysed vegetable protein
IAA	Indispensable amino acid
IFST	Institute of Food Science and Technology (UK)
IMF	Intermediate-moisture foods
IMP	Inosine monophosphate
JND	Just noticeable difference
kJ	KiloJoule
LDL	Low-density lipoprotein
LM	Low methoxyl- (pectins)
MAFF	Ministry of Agriculture, Fisheries and Food
MAP	Modified atmosphere packaging
3-MCPD	3-Monochloropropane-1,2-diol
MRL	Maximum residue level
MSG	Monosodium glutamate
MUFA	Monounsaturated fatty acid
NE	Norepinephrine
NSP	Neurotoxic shellfish poisoning
NSP	Non-starch polysaccharides
OA	Ochratoxin A
o/w	Oil-in-water (emulsion)
PAH	Polycyclic aromatic hydrocarbons
PCB	Polychlorinated biphenyl
PKU	Phenylketonuria
PNS	Peripheral nervous system
Ppb	Parts per billion
PROP	6-n-propylthiouracil
PSD	Pesticides Safety Directorate
PSE	Pale, soft, exudative (meat)
PSP	Paralytic shellfish poisoning
PUFA	Polyunsaturated fatty acid
rBST	Recombinant bovine somatotropin
QDA	Quantitative descriptive analysis
QFD	Quality function deployment
SaFA	Saturated fatty acid
SDA	Stearidonic acid
SME	Small and medium sized enterprise
SNS	Somatic nervous system
SSS	Sensory specific satiety
TCDD	2,3,7,8-tetrachlorodibenzo-p-dioxin
TFA	*Trans*-unsaturated fatty acid
TPB	Theory of Planned Behaviour
TRA	Theory of Reasoned Action
TVP	Texturised vegetable protein
UHT	Ultra-high-temperature
vCJD	New variant Creutzfeldt-Jacob disease
WHC	Water holding capacity
w/o	Water-in-oil (emulsion)

Introduction

"Quality", "fine(st)", "best", "prime", "original", "real", "traditional", "whole (some)", "health(y)" – these are just some of the familiar adjectives used by marketers to draw consumers' attention to specific foods. They do this because such descriptions undoubtedly create expectations about the desirability of the product. However, both the marketer and the consumer know that further enquiry by the latter, e.g., perusal of the food label, is necessary for them to establish the exact nature of the food attributes being signalled. Similarly, terms such as "new" and "improved" are designed to prompt consumers, current users in particular, to reset their "expected utility barometer" for the product being advertised. Defective quality often means some deviation from an ideal or an expectation, e.g., the presence of some defect. In some areas of the food market, it is the special role of expert tasters and graders to maintain and transmit quality standards, familiar examples being wine, tea and certain cheeses. However, in an environment containing thousands of individual foods and in which there are high rates of new product development, this traditional approach to the management of the sensory quality of foods has become infeasible for most foods and beverages. Instead, in the modern food industry, this type of work is usually carried out by a formal taste panel. As in any highly competitive market, food producers need to offer products that consumers will not only like, but for which they are also motivated to modify current purchasing behaviours. New introductions therefore need to identify some kind of added consumption value. To be successful, food product designers, producers and marketers need to ask themselves, and find answers to, certain questions, including how best to research and define the types of consumer value required of a product; how to translate requirements into product attributes; how to consistently deliver these attributes in routine production; and how to signal the presence of these attributes to the consumer. However, it may be futile to ask people what tangible new features they would like to see in some particular kind of food. People often find it difficult to imagine improved products, especially where improvements can be achieved through the application of new technologies. A more promising approach is to enquire into the perceived benefits consumers value in a food. For example, consumers might prefer a more natural product, a product with an extended shelf life or a snacking product with improved portability, allowing the food to be eaten away from home. Each type of consumer value, once uncovered, can then be translated into technical specifications for production processes, product formulations and food and packaging technologies.

Use of the word "quality", or of related terms, in the marketing of foods in order to signal a degree of relative excellence is not necessarily harmful or misleading to consumers. For examples, fine Belgian chocolates and fine wine are each targeted towards a dedicated market sector, whose members will be adept at identifying any products that do not deliver on their promise, and the product will suffer the consequences of this in the market place. This is an example of stable, self-regulating systems, which work because producer and consumer converge strongly in their perceptions of the product, and because there is little benefit to be gained from producer dishonesty. This does not mean that examples of fine wines and fine chocolates, as perceived by experts, will not be a source of disappointment to some people who try them, especially when tasting them for the first time. This is not only likely but, in fact, inevitable; with food especially, the truism applies that people tend to prefer what they are used to. Food producers and marketers need to keep themselves constantly aware of the characteristics and expectations of their target market. For example, they need to understand the effects of how label information is framed on consumer attitudes towards the product. Sometimes, quality signalling and framing may be found to be not only misleading, but positively puzzling. Milk from Ayrshire cows, which is naturally relatively high in fat, tends to be valued by its users as a particularly rich, creamy milk, making it especially suitable perhaps for pouring over certain foods, e.g., breakfast cereals. However, the Ayrshire brand is marketed as "96% fat free milk" [1]. The framing of claims affects their evaluation, and in this example, the emphasis is on the low fat content. Yet relative to milk generally, Ayrshire milk is a high fat product. In this type of example, one wonders whether the efficacy of the claim was ever evaluated in terms of the target sector, which one presumes would be relatively affluent and well educated. One might expect these consumers to consider the claim as rather ill judged, and therefore, reflecting badly on the company. Here marketing appears to be following blind a currently popular trend of negative framing of the fat contents of foods. Framing is an important issue in consumer choice. Simple words can have considerable powers of persuasion, and food categories often become shorthand for a whole host of either benefits or problems. Thus, as a side effect of the rise of the organic food market, and by being seen as the opposite of organic farming, all of conventional farming has, in many people's eyes, acquired the stigma of food which is somehow contaminated.

In recent years, food has assumed increasing prominence as a marker of lifestyle. In fact, loyalty to the organic food market is viewed, in some quarters, as aspirational purchasing or consumption. In the UK media, there is widespread coverage of food and diet issues. TV cookery series introducing viewers to ethnic cuisines from around the word, which have been popular for many years, have been joined more recently by various food and cookery shows. These aim to provide a blend of information and entertainment, serving as much as a platform to showcase so-called celebrities as they do to present food. Such programmes may or may not encourage viewers to be more experimental in their cookery; they are however likely to raise the profile of food as well as awareness of available ingredients. Consumers may be encouraged to experiment more when choosing food, even if it means choosing one of the many ready meals, for which there is a particularly buoyant and innovative market at present [2]. Food related journalism has become a near-compulsory feature of newspapers, especially the weekend editions, and there are a number of magazines focusing on foods and diets. Whilst the recipe writers tend to be concerned about the quality and range of ingredients, there is a fashion among many food journalist of wholesale rejection of anything mass produced and of new food technologies. Instead artisanal, small-scale food production systems are promoted to a largely affluent audience or readership. The mixture of information and entertainment can be detrimental to consumer education, as in the instance when a TV chef contributed to the debate about genetically modified foods by asking people to check food labels for modified starch – no relation to genetic modification. The need for authoritative information about food is clearly widespread and does not merely concern consumers lacking formal education in food science. Those representing food manufacturing may be frustrated by what they refer to as the junk food fallacy, claiming that there are no bad foods, only bad diets [3]. But are they right, or are they too guilty of refusing to debate the relevant issues in a dispassionate manner? The Food Standards Agency was set up in April 2000 [4], largely as a consequence of the low consumer confidence in food control that resulted from high profile food scares [5]. The agency exists to represent the consumer interest, but in addition, has set itself the task to educate consumers about the nature of modern foods and the modern food system. This represents an opportunity for a fresh start both in terms of food control and public education about food. There is increasing uptake of formal quality management systems within all sectors of the food supply chain. A major motivating force since the 1990s has been the Food Safety Act 1990 [6]. This is due to the fact that the Act requires food businesses to exercise due diligence and to document the measures that enable them to do so. Specifically, they must implement formal hazard management systems along the lines of the technique widely known as Hazard Analysis Critical Control Point (HACCP) [7]. Quality management systems with broader scope, in particular BS EN ISO 9001 [8], are being adopted for improved competitiveness and profitability of food businesses. Accreditation to ISO 9001 signals a firm's commitment to quality and may be a condition of becoming an approved supplier. Effective implementation of ISO 9001 leads to streamlined processes including production, product development and marketing. Quality failures in the UK food supply chain involving issues of food safety have been well publicised. The most serious of these has been the emergence of a new variant of Creutzfeldt Jakob Disease (vCJD) in humans which is thought to have originated from beef carrying bovine spongiform encephalopathy (BSE) [9]. Less visible is the poor success rate of new product launches [10]. This does however bear witness to the difficulties inherent in designing products representing and communicating genuine consumer vale, where the value that has been added is sufficient to cause consumers to change existing behaviours. ISO 9001 provides a framework for effective product development. However, it is only a framework and does require the added ingredients of motivation and competence. It is interesting that the Business Excellence Model is founded on three outcomes, namely, customer satisfaction, employee satisfaction and impact on society

[11]. Impact on society is an important issue, yet is frequently overlooked. Food production systems can impact non-consumers, for example in the areas of animal welfare and new food technologies, e.g., food irradiation and genetic modification. Added value to consumers may be related to cost or product function, or it may involve less tangible benefits, such as feelings about the product. Even perceived corporate values, where these are consonant with those of consumers, can be a source of added value to consumers. Examples are companies that stand for fair trade, environmental care and animal welfare.

People differ in the foods they like or dislike and in what they like or dislike about specific foods, whether as such or in relation to the context of consumption. A food that is highly valued by one consumer may be rejected outright by another. Statutory consumer protection, like commercial practice, does not always demonstrate awareness of, or concern about consumer needs. The egg is a case in point. Whilst superficially, the egg may appear as one of the least complicated foods, deeper enquiry reveals that this is far from the truth. In fact, consumers forced to rely on the retail market for their egg supplies may find themselves in something of a quandary. Statutory labelling information relates solely to the shelf life and presumably, safety aspects of shelf life. This denies the fact that cooking performance depends on the actual age of the egg, successful poaching, in particular, depending on a high degree of freshness. Yet a few years ago, a UK egg producer was prosecuted for revealing the laying date of his hens' eggs to his customers. However, perhaps the logic of this story is less confused than it first appears. After all, ever since large-scale infection of British eggs with salmonellae was revealed, government advice has discouraged people from consuming eggs in which the yolk remains semi-liquid. In fact, the sale of old eggs favours those consumers wishing to hard-boil them, because an old egg is much easier to shell than a new one. This is an example of the issues covered in this book. The book is seen as a contribution to the bridging of communication gaps, especially that between food science as a natural science and consumer science as a social science. Foods can be described in terms of many different attributes, and consumers can be described in terms of the motivations that cause them to purchase, eat or drink a particular product, and the factors affecting these motivations. Quality arises when product attributes are identified as significant in terms of their contribution to consumer satisfaction. It is the consumer's perceptions and attitudes that determine which subset of the totality of product attributes is critical to consumer value and therefore, critical to the likely commercial success of the product. In addition, intangible quality attributes, although not directly verifiable in terms of product analysis, may be as crucial in terms of consumer value. They may relate to production systems, in particular, moral issues. Much of the confusion in the public debate about food quality appears to derive from a lack of analysis of the type of consumer value that particular food items represent. In the junk food debate in particular, a key point that often seems to be forgotten is that any product for which there is repeat demand is, by definition, a quality product. However, whilst such a product clearly satisfies requirements, they are not necessarily everybody's requirements, and in the case of so-called junk foods, they are certainly not the requirements of the health campaigner.

The book is divided into three parts, each of which contains three chapters. Part 1 sets the scene in terms of the key theories and concepts that form the foundation of the book, focusing on the nature of food and, at the same time, the nature and origins of consumer behaviour. Definitions are given for product quality and consumer value, and both concepts are illustrated. The key academic disciplines on which the book draws are introduced. They include business theory, in particular, quality and marketing theory; food science; and the behavioural disciplines of economics, psychology and sociology. At the core of Part 1 lies the question of what motivates consumption in general terms, and what are the specific quality issues in food consumption. Academic authors tend to concur with the popular view that quality is difficult to define. Inter-disciplinary debates about quality present particular difficulties as each discipline has its own focus and language. Grunert, an academic and industry consultant, is concerned about what he perceives as the inability of food technologists and behavioural scientists employed in the food industry to communicate effectively about food quality [12]. Grunert attributes at least some of the blame for the high failure rate of product introductions to this. Another area of conflict exists between economics and psychology, where some take the view that the main stream economic analysis of consumer behaviour is psychologically naive [13]. Part 2 of the book deals with the natural science of foods and beverages. However, unlike standard food science texts, the organisation reflects the book's focus on quality. Hence, for example, carbohydrates are discussed twice, i.e., as structural elements of foods (impacting texture) and as nutritional fuels. The first chapter describes foods from the point of view of their physico-chemical nature. This is followed by a discussion of the main performance attributes of foods and beverages, in particular, their sensory attributes. The concepts of keeping quality and fitness for purpose are also discussed. The sensory attributes of foods are regarded as objective attributes, i.e., attributes belonging to the foods. This is despite the fact that their evaluation involves the human senses. The sensory attributes of foods clearly play a crucial role in whether a food is desirable to consumers. The final chapter of Part 2 deals with matter that is added to food, both deliberately and as undesired contamination, as food additives and contaminants are major consumer concerns in the present market. In summary, Part 2 of the book discusses food from the perspective of any objectively definable and measurable attributes. It provides the scientific, and largely value-neutral, background for the analysis of specific types of consumer value derived from product attributes, which is the main focus of Part 3. Part 3 is concerned with consumers, their food choices and what motivates these choices. The role of individual physiological and psychological make up in the perception of quality in foods is addressed together with socio-economic and other influences. The analysis of good and bad foods draws strongly on issues raised in the public debate about food quality mainly over the last decade. Links between the food object (whose properties can be measured in value free manner), those properties that influence value (and therefore should be measured), and the nature of the value derived by consumers are highlighted throughout the book.

References

1. Morrison I. Milking the Ayrshires. *The Scotsman*, 16th March 1998
2. Griffiths J (ed). *Key Note. UK Food Market. 1999 Market Review.* 11th edition. Hampton, Middlesex: Key Note Ltd, 1999
3. Woollen A. Junk Food Hysteria. *Food Processing* 2001; January: 8
4. Food Standards Agency. The Board. Retrieved on 6th April 2000 from http://www.food-standards.gov.uk/the_board.htm
5. Davies S, Todd S. The need for an independent food agency. *Consumer Policy Review* 1996; 6: 82–86
6. HMSO. *Food Safety Act 1990.* London: HMSO, 1990
7. Shapton N. Implementing a food safety programme. *Food Manufacture* 1989; August: 47–50
8. ISO. *ISO 9001: Quality management systems – Requirements.* Geneva: International Standards Organization, 2000
9. Connor S. Portrait of a nation fed on reassurances. *The Independent*, 27th October 2000
10. Sloan E. Why New Products Fail. *Food Technology* 1994; January: 36–37
11. BQF. *Links to the Business Excellence Model.* London: British Quality Foundation, 1998
12. Grunert KG. Food Quality: A Means-End Perspective. *Food Quality and Preference* 1995; 6: 171–176
13. Lewin SB. Economics and Psychology: Lessons For Our Own Day From the Early Twentieth Century. *Journal of Economic Literature* 1996: XXXIV: 1293–1323

Part 1

Defining Quality, Value, Food and the Consumer

Why does a shopper or a diner choose one food over another that is similar? Why do other people choose differently in similar circumstances? What causes a particular food to be prized in one culture and rejected in another? Which attributes are critical in the acceptance or rejection of a food, which attributes are less important? Why are individual food habits so difficult to unlearn, even when the individual displaying them perceives them to be harmful? These are just some of the questions, relevant to food quality and consumer value, which are addressed in this part of the book. This is done by examining the relevant theories, mainly deriving from the social sciences. Key issues included here are various conceptual approaches to quality and value, the biological and cultural meaning of food and consumer motivations in food choices. Marketers tend to subscribe to the view that perception is everything, i.e, that product perception is often more important in achieving consumer satisfaction than the so-called reality of the product. This chimes with the commonplace that one man's food is another man's poison. However, as will be demonstrated, perceptions, and therefore behaviours, can be changed and even manipulated.

Conceptually, "quality" and "value" lie at the interface between the consumer's mind processes and the objects of the external world. Both are concerned with a consumer and an object, and with interactions taking place between them. Food value is slanted more towards the consumer and food quality towards the food, but there is an overlap. This interface, although critical, is only one of several that are relevant to an understanding of consumer behaviour in the area of food quality. "Value" and the associated concept of "values", and "consumer behaviour" draw on several academic disciplines, both in the social and the natural sciences, for their theoretical underpinnings; yet there is no unified view of food quality and the consumer. In fact, there is often a distinct lack of understanding, which tends to prevent effective communication between food professionals raised in the different academic traditions, and which can lead to difficulties in the business environment. One outcome of this may be poor success rates in new product development.

Few observers would deny that all food companies require a thorough understanding of the characteristics and behaviours of the materials they use, as well as of the capabilities of the processes they employ. But this is not enough. They must also understand the mental processes that will cause consumers to develop perceptions of the finished product. Such perceptions engage with consumers' motivational systems and therefore directly affect choices and other food related behaviours. As for marketing and communication strategies for novel foods and technologies, it is crucial that proper account is taken of the beliefs and attitudes of both consumers and the wider community. This is a lesson that consumers have recently taught the industry when they actively rejected the introduction of genetically modified foods into the market [1]. Another group of professionals who require in-depth knowledge of consumer behaviour theory are those working in the area of healthy eating promotion. As Part 2 of the book is dedicated almost entirely to natural science related aspects of food, the focus in this part is mostly on the social sciences and on the management sciences derived from these.

Consumers are the final link of food supply chains, i.e., they are the end users. This does not mean that, in each case, the food shopper is the person who will consume the food in the sense of eating it. For example, foods may be bought for other family or household members and as gifts. Consumer-related terminology is also sometimes used even more loosely, i.e., to denote the public at large. This can be problematic. For example, following lobbying by consumers, the UK government recently introduced a law prohibiting the tethering of sows in the production of pork. In fact, those participating in the lobbying were, by definition, committed to actively supporting animal welfare initiatives. As such, they would have been unlikely to be purchasers of meat originating from production systems perceived to be cruel to animals. In the event, following the introduction of the law, demand from UK consumers for pork from tethered sows did not decline. This was because the actual customers for this (cheaper) meat switched to imported products [2].

One of the uses of language is to assign individual (food) products to established, or new, product categories. In this way, the naming of products and the framing of product information influence the way in which these foods are perceived. A child announcing a dislike of vegetables condemns a vast range of fundamentally different foods, even though their attitude may originate in exposure to just a few members of the category "vegetables". Furthermore, these vegetables may not have been in prime condition, or they may have been inexpertly prepared. For example, a dislike of tomatoes may be caused by exposure to barely ripe, watery specimens before one has had the chance to sample a freshly picked, properly ripened tomato. The same argument may be applied to some of the tropical fruit currently gaining popularity with both UK food retailers and consumers. Returning to the issue of novel foods, it has been suggested that part of the problem with the consumer rejection of genetically modified foods may have been the decision to class all of these together under that particular label [3]. In an alternative approach, distinctions might have been made between wide transfer (gene transfer between quite different organisms, e.g., Arctic fish into strawberries), close transfer (e.g., from wild to domestic plant forms) and tweaking (e.g., altering the activity of an existing gene). Poorly defined concepts and categories may put up barriers against consumers' ability (and motivation) to learn about them. A current example of this are the so-called functional foods discussed in Chapter 6. Both food law and commercial marketing activities are to a large extent about categorisation and semantics.

Chapter 1 starts with an examination of the theoretical bases of the quality and value concepts and discusses what these have in common and where they differ. Food examples are used to illustrate the general theory and to try to answer the question: What types of consumer requirements can be satisfied through consumption of foods? Having defined food quality and consumer value, the discussion moves on to take stock of the role of these issues in contemporary food supply chains. It explains why the food supply chain needs to be responsive, through new product development, to ever changing consumer requirements and tastes. It also highlights the difficulties that derive from the growing distance in food supply chains between consumers and producers, in particular, primary producers. There are some fundamental differences in how quality management is applied in the contexts of new and existing products. Quality management of existing products is predominantly the responsibility of food technologists insofar as it is related to assuring batch-to-batch consistency (quality of conformity). Quality of design needs to be managed jointly by food scientists and marketers and other managers in food companies. In new products, it is not only the food product per se that requires attention. During product development, the requirements of the target consumer need to be constantly reviewed, and this must be done in the context of the profitability and/or competitive advantage the product is likely to generate for the company.

Chapter 2 provides a view of how food quality attributes may be classified and weighted. However, it starts by asking what it is that makes something a food. This introduces biological, individual and cultural arguments and contexts. As well as examining tangible and use attributes that are specific to foods, it introduces concepts such as positioning (relative quality) and signalling. Whilst signalling is usually associated with advertising, a freshly picked bunch of carrots can be made to signal that very freshness by selling it with the green tops left on. Inasmuch as the chapter looks at tangible food attributes, it provides a link with Part 2 of the book (Food Quality Attributes).

In common with the two preceding chapters, Chapter 3 is designed to provide more of the foundations required for the detailed arguments about consumer value in the contemporary food market, most of which is conducted in Part 3 of the book. The chapter provides a précis of the social sciences insofar as they contribute to consumer behaviour theories, in particular, theories underpinning food choice. Special emphasis is placed on information processing and motivation. Microeconomics studies consumer behaviour quite literally, i.e., it is concerned primarily with demand and some basic rules governing it. There is less interest in the precise reasons for this demand and none in what consumers think or say they might do in certain situations. Psychologists agree that actions alone are observable; however, psychologists are keen to infer the causes of particular behaviours, e.g., perceptions, beliefs, attitudes and motivations, and they use a range of theoretical frames to help them in this task. This can help explain why (food) choices may often appear as irrational. Social aspects may play a major role in this, and the chapter therefore ends with an examination of the anthropology and sociology of food choice.

1 Quality and Value

1.1
Introduction

In this chapter, several models and definitions relating to quality and value are presented and evaluated. The types of quality (attributes) and consumer value identified are explored and illustrated, with particular reference to food. The boundary and overlaps between the two concepts are examined in some detail. Both tangible and intangible quality attributes of foods are discussed, with special attention being paid to the nature of quality of conformity and quality of design and the differences between them.

Food safety is of paramount importance both for suppliers and consumers of food. Food safety assurance is therefore the primary objective for all quality assurance schemes, both statutory and voluntary, consuming considerable company resources. The Food Safety Act 1990 [4] is designed to protect consumers from unsafe food as well as from food fraud, but food safety and other food standards cannot be separated completely. Mislabelled products can constitute safety hazards as in the case of sheep's milk (declared) yoghurt containing a proportion of (undeclared) cow's milk. This may threaten the life of a consumer with a serious cow's milk allergy [5]. Generally, in the eyes of consumers, food safety is an implicit quality attribute, i.e., they would not specifically demand "safe" food in a shop or restaurant. However, where a consumer does not have trust in their supplier, they will try to take charge of food safety assurance themselves. A customer at a butcher's may decide not to purchase cooked meats there, e.g., if the shop fails to demonstrate effective procedures for the prevention of cross contamination of those meats via raw meats.

Aspects of food quality besides safety are largely determined by individual preferences. The range and diversity of foods available to consumers in the UK today is considerable. In fact, there are few foods that cannot be obtained by the individual with the time and money to pursue and acquire them. This means that, in principle, each consumer can be matched with his or her ideal foods. Whilst most would not go to such lengths, most people have their own individual mental lists of products that they would avoid under all circumstances. As far as individual supermarkets are concerned, some might question exactly what level of product differentiation there really is. For example, does an aisle's length of different branded flavoured yoghurts represent choice when it can be difficult to locate a straightforward live yoghurt capable of serving as a starter for domestic yoghurt making? On the other hand, it may be difficult for consumers with strongly held ethical values to choose food accordingly, e.g., meat from what they would perceive as cruelty-free production systems. In this case, the difficulty lies with the poor state of the consumer information and advice systems that exist in the food arena, and with the associated issue of food labelling. From these examples, the initial impression of choice becomes less certain. If therefore consumers end up by buying sub-optimally, can the economist's assertion that demand reflects preferences be maintained? And, if consumers choose a product repeatedly, does that mean that it meets their requirements for quality?

To address these issues, one first needs to establish the meaning of quality and value because, despite extensive debate over recent years, these concepts remain somewhat elusive, with a lack of a common understanding among those with a professional interest in food as well as among consumers. Broach the subject of added-value food products with a group of nutritionists and the focus is likely to be on nutritional aspects of quality. In sharp contrast, the commercial perspective on the added-value concept sees product differentiation in all its forms, and added profit as the ultimate goal. Food technologists may be uncomfortable with consumer issues; they usually lack the academic training to think these through and discuss them. Scientists generally are inclined to presume that consumers are ignorant of all things scientific, and that they must be "educated", i.e., told by them what things are really like. What is usually missing here is an appreciation that communication is at least a two-way process, and that learning too relies on exchanges of information. Grunert, who acts as a consultant to food companies, especially in Denmark, laments the fact that too often quality is discussed purely relative to products and processes, neglecting user perceptions – even in product development [6]. He suggests that food technologists and marketers do not know how to communicate effectively about food quality, which may lead to product failures in the marketplace. Cardello is concerned about the extent of product attribute measurements undertaken to the detriment of measurements and evaluations from the consumer and use perspectives [7].

Technologists and consumers may not share the same value system regarding food production. Two examples come to mind. In the first, a dairy technologist considered that yoghurt was the output of a highly skilled, highly controlled, almost beautiful, process which linked him, the maker, with a long line of traditional craftsmen in history. How could a product of this nature be allowed to be "tampered with", its texture and subtle flavour destroyed, by stirring it and crudely flavouring it, its bioactivity lost through heating? In the second example, a cheesemaker had been recruited in the 1970s by a UK dairy company to take charge of their mozzarella production but returned to Italy as soon as he learned what the process involved. Of course today, flavoured yoghurts and the new types of mozzarella have both become very popular; so, who has the last word here? The technologist's professional pride in the product, where it exists, may be akin to the disdain of certain food writers for the mass consumer's "appalling" taste in food. Consumers, their champions, food journalists and others in the media who have a professional interest in food typically express their own particular viewpoints about food quality, reflecting their lifestyles, value systems and personal or professional agendas. However, there are many different consumer requirements and many different products that can be, and are, used by consumers, in different combinations, to meet these requirements.

1.2
General Perspectives on Quality and Value

Most foods, especially those based on whole plant and animal tissues, have a highly complex structure and composition. Chemical, physical and/or microbiological processes take place in foods during storage, which may alter some of these attributes. Food structure and composition translate into sensory attributes and other performance aspects for the consumer. What all this means is that, theoretically at least, hundreds of different attributes could be defined for individual foods. Clearly, to specify food quality in those terms is impractical; more importantly, it is inappropriate because quality does not refer to the totality of attributes a product possesses. Instead, the quality concept introduces a filtering device through which attributes are weighted in terms of their contribution to user satisfaction with the product. Therefore, the definition of product quality incorporates subjective elements. Table 1.1 shows some of the quality-related terms recently defined by the British Standards Institution [8].

According to Table 1.1, customer satisfaction means the customer's perception of requirements having been fulfilled [8]. Other models of quality include, in addition, the requirements of the other stakeholders in a process or product. For example, the Business Excellence Model is based on the premise that, whilst customer satisfaction is critical, the requirements of employees and impact on society must be similarly satisfied [9]. This approach is particularly relevant to food supply chains. The recent, highly negative, public response to the genetic modification of foods [1] has taught scientists that, in the introduction of this type of new technology, costs and benefits must be evaluated in respect of all stakeholders. The types of value consumers perceive in, and the overall satisfaction they may gain from, a given food varies between individuals and with the

Table 1.1. Terms relating to quality as defined in BS EN ISO 9000:2000 [8]

Quality	Degree to which a set of inherent characteristics fulfils requirements (e.g., poor, good or excellent)
Inherent	Existing in something (especially permanent characteristics) – as opposed to assigned (e.g., price)
Characteristic	Distinguishing feature (qualitative or quantitative)
Requirement	A need or expectation that is stated (specified), generally implied (customary) or obligatory; e.g., product requirements, customer requirements ...
Specification	Document stating requirements
Conformity	Fulfilment of a requirement
Customer satisfaction	Customer's perception of the degree to which his or her requirements have been fulfilled

context of use. Furthermore, perceptions may change with time. A consumer may have a preference for a certain inexpensive type of wine. When receiving a gift of wine, they will expect this to be from the top of the preferred range, but a glass drunk at night to relax with might be from the cheaper end. Over a period of time, as the person's earnings increase, they may join a wine club to gain more experience of the different wines available. Gradually, they will explore more expensive products, which may cause their preferences to change as their knowledge base increases. The wine club will track their evolving tastes, catering to them for mutual benefit. This example illustrates the dependence of value (perception) on the context, i.e., who uses what, when, where, how and why. As products have many different attributes, not all of these will have an equal impact on consumer satisfaction. It is therefore important for suppliers to identify the true quality attributes and to assess each in terms of their contribution to consumer value.

The value and the satisfaction consumers derive from a food product is therefore closely related to its quality. In fact, both terms appear throughout the academic literature, although their use by different authors is not consistent. Grunert, representing one one end of this spectrum, favours "quality" centred terminology. He distinguishing between objective and subjective quality, where the former is concerned with product attributes, in value-neutral terms, and the latter with perceptions about products [6]. Grunert assigns the management of objective quality to the sphere of food science and technology and considers management of subjective quality to be the responsibility of psychologists and marketers. Objective quality is concerned with meeting product and process specifications to achieve consistency between product batches; subjective quality focuses on product design and the creation of product specifications that accurately reflect consumer requirements. One of the key merits of Grunert's model of quality is that it points towards important practical consequences for the quality management of foods. It highlights the fact that natural and social scientists employed in the food industry have different, but equally important, roles and responsibilities in the design and reliable delivery of quality products. Cardello's approach is strongly focused on the consumer. Nevertheless, he too chooses "quality" based terminology. In his view, quality is not something belonging to a food; rather, he defines quality as a "consumer-based perceptual/evaluative construct that is relative to person, place and time" [7]. Cardello emphasises the point that quality is achieved only when the right products are provided to the right consumers for use in the right contexts.

Porter uses both terms, selecting "quality" when it is the product that is to be emphasised and "value" when discussing the impact of the product on the buyer, i.e., when emphasising product benefits [10]. It is Porter's convention that has been adopted for the purposes of the present book. To illustrate this approach, a fat-free biscuit may be perceived as a "slimming" biscuit; an instant chocolate powder may become a convenient snack, and organically grown produce may be seen as a means for expressing and affirming a particular ethical stance. Typically in food product development programmes, initial concepts are generated with focus on the needs of a possible target market. Detailed points of recipe, formulation and process are considered at a later stage. When it is thought ad-

Table 1.2. The nature of consumer value as defined by Holbrook [12]

"Consumer value is an interactive relativistic preference experience."	
Interactive	Consumer value entails an interaction between some subject and some object.
Relativistic	Consumer value is comparative, personal and situational.
Preferential	Consumer value embodies a preference judgement.
An experience	Consumer value resides in the consumption experience(s) derived from products, brands and possessions.

vantageous to reposition an existing product for marketing purposes, the attributes of the product will be re-examined in terms of the types of value they may represent to possible target markets. In this manner, the traditional recovery drink Lucozade, for many years offered to those convalescing from sickness, was successfully repositioned as an "energy/sports" drink [11]. A related scenario is that where an existing product is being introduced to new, e.g., export, markets. This too requires a re-examination of product benefits specifically in terms of the target market. A similar approach, from quality towards value, may be taken where a new technology suddenly becomes widely available, as was the case with the introduction of microwave ovens. These examples suggest that there are many different types of consumer value, something that has been investigated in some detail by Holbrook [12]. Table 1.2 gives his definition of the nature of consumer value.

The interactive nature of Holbrook's consumer value suggests that food value does not reside in foods in an objective, context-free manner. Rather, value is generated in the interaction with a consumer and mediated through that consumer's perceptions and attitudes. Returning to the example of the fat-free biscuit, it is easy to see that this is not a product that would appeal to all consumers equally. The relativistic aspect relates to consumer value involving preferences among objects and variations across people, and to its being specific to the context. Individual differences in the evaluation of foods may be due to personality, specific physiological needs and social reference groups, whilst context may relate to the time of day, other foods and drinks consumed at the same time, presence of eating companions, whether a person feels hungry, and so on. Finally, consumer value is based on preference judgements being made, i.e., it is the outcome of an evaluation. The general concept of preference embraces a wide variety of value-related terms including affect (pleasing vs. displeasing), attitude (like vs. dislike), evaluation (good vs bad), and predisposition (favourable vs. unfavourable) [12]. Attitude is one of the key concepts in psychology and its role in consumer behaviour will be examined in detail in Chapter 3.

Some authors highlight a distinction between so-called goods, e.g., a car, and services, e.g., an insurance policy or a medical examination. In BS EN ISO 9000:2000, definitions are given for neither; instead, both concepts are incorporated in that of the product. Here a product is defined as the "result of a set of interrelated or interacting activities which transform inputs into outputs" [8]. When examined in the context of Holbrook's view of the nature of consumer value, the idea of goods as repositories of value does appear somewhat redundant. If consumer value resides in the consumption experience rather than the possession of goods, then it might be said that there are elements of service inherent in all goods that consumers wish to acquire. On the other hand, the value of goods changes according to context; for example, like many articles of daily living, cars are simultaneously "bads" (e.g., when there is no parking space available) and "goods" (e.g., ease of transportation, comfort, reliability, prestige). Similar examples can readily be found in all sorts of contexts, including foods. For example, an ice cream eaten whilst stopping off at a café in the middle of a shopping trip is an indulgence; the same ice cream offered for breakfast or with a glass of beer is off-putting. Many prepared and processed foods and their packaging explicitly, sometimes by definition, incorporate various service elements, and consumer demand for convenience is a major driving force in the food market. Retail foods and restaurant meals were once considered prototypes of goods and services, respectively. Today the dividing line between them is often rather thin, with ready meals available in supermarkets and fast food operations resembling manufacturing processes more than they do traditional restaurant operations.

In order to set accurate quality specifications for foods, the types of consumer value expected must be discovered and prioritised. Price is obviously an important aspect of this, and value for money is a familiar concept. The economist Adam Smith called the notion of price "value in exchange", and contrasted it to the notion of "value in use" [13]. He used the example of water to illustrate the point that the things that have the greatest value in use often have little or no value in exchange. What something is worth at any particular time and in any particular place depends instead on scarcity, or supply and demand. The pricing of foods plays an important role in how their value is perceived. It is a signaller of quality, e.g., of the types and grades of raw material used in a particular recipe. Although the value-in-exchange concept may be seen to underlie certain aspects of Holbrook's typology of consumer value [12], the typology comes about from quite different considerations. The eight distinctive types of consumer value emerge from a three-dimensional grid based on certain key dimensions of consumer value, i.e., extrinsic vs. intrinsic, self-oriented vs. other-oriented, and active vs. reactive (Table 1.3).

Extrinsic value relates to consumption when it is valued as a means to an end, e.g., a hammer to drive in a nail or money to make some purchase [12]. In the food arena one might think of someone who is ravenously hungry and for whom any food will do to still that hunger. Another example is microwaveable food packaging. This serves as a means to a quick, mess-free meal but does not otherwise contribute to the consumption experience. By contrast, intrinsic value occurs when some consumption experience is appreciated for its own sake, i.e., a day on the beach or listening to a piece of music [12]. There are many food-related examples of intrinsic value, e.g., indulgence foods. The ice cream eaten while spending a day on the beach exemplifies this. Value is self-oriented when consumption is for one's own sake, and it is other-oriented when it looks beyond the self to someone or something else. The other(s) in question can range from the micro level (family, friends, colleagues) to an intermediate level (community,

Table 1.3. Morris B. Holbrook's Typology of Consumer Value. Reproduced with permission [12]

		Extrinsic	Intrinsic
Self-oriented	Active	EFFICIENCY (O/I, Convenience)	PLAY (Fun)
	Reactive	EXCELLENCE (Quality)	AESTHETICS (Beauty)
Other-oriented	Active	STATUS (Success, Impression Management)	ETHICS (Virtue, Justice, Morality)
	Reactive	ESTEEM (Reputation, Materialism, Possessions)	SPIRITUALITY (Faith, Ecstasy, Sacredness, Magic)

country, world) to the macro level (the Cosmos, Mother Nature, the Deity). At the most macro level of all, and typical of certain Eastern religions as well as Freudian psychoanalysis, the "other" could relate to some inaccessible inner self [12]. Someone might practice meditation and adopt a yogic diet to communicate with their inner self. A couple might serve caviar at a dinner party to both indulge and impress their guests. Yet another person might stop purchasing prepackaged foods as their contribution to protecting the environment, or they might turn vegetarian because of concerns over animal welfare. Finally, value is active when it involves something being done to or with a product as part of some consumption experience, and it is reactive when the object acts upon the consumer [12]. Food generally provides consumers with three fundamental kinds of value, which can be summarised as nutrition, hedonistic experiences and cultural belonging. These basically explain why people eat. However, consumer value derived from food is more differentiated than suggested by this scheme, and Holbrook's typology provides a useful framework to examine consumer value in relation to food in more depth.

Efficiency involves extrinsic value that results from the active use of a product or consumption experience to achieve some self-oriented purpose and is often measured as a ratio of outputs to inputs (O/I) [12]. For example, one might assess the value of a meal as the number of calories consumed per money expenditure. Convenience is another key example and one that is particularly relevant to the food market. This is usually taken to mean output per unit of time, but can also relate to other inputs, e.g., food preparation skills. A convenience product such as a prepared mixed salad can also incorporate other benefits, e.g. variety (of types of lettuce). Packages with easy-opening mechanisms dispense with the need for additional kitchen gadgets, whilst re-sealable packaging allows food to be kept without having to transfer it into special storage containers. Excellence involves a reactive appreciation of some the ability of some object or consumption experience to serve as an extrinsic means to some personal self-oriented end [12]. Excellence is experienced when positive expectations are met effectively, resulting in a high degree of satisfaction, and in turn, demonstrating high product quality [8]. This engenders feelings of admiration and delight, and these may be independent of whether the product will be of direct practical benefit to the observer. Dishes that are beautifully presented, perhaps displaying a high level of food craft skills on the part of the maker, are typical of this category; such foods may "look too good to eat". On the other hand, any food or dish in peak condition and free from defects illustrates the concept of excellence.

Ostentation, or conspicuous consumption, means some people's desire to provide prominent visible evidence of their ability to afford luxury goods [14]. Status-as-value refers to the active manipulation of one's own consumption behaviour in order to influence others. Status-seeking may be understood as a response to a perceived gap between a present state and some ideal state. This leads to a motivational state which is strongest among social groups chronically affected by status anxiety. Dining in a celebrity chef's restaurant may be motivated as much by the wish to be seen at a fashionable venue as about the food. Choosing consumption experiences to project the sort of image one wants to create is part of impression management [12]. In attribution theory, the causes of behaviour can be either dispositional or situational. Attribution theory is discussed in the psychology section of Chapter 3, but in these terms clearly, status-related consumption is situationally motivated. Food and drink choices are a means of signalling, and re-enforcing, one's status on a more or less ongoing basis, both within the family and in a wider context. Indeed, a cursory scan of the shopping trolleys of customers in a supermarket usually allows wide ranging inferences to be made about their status in society. Status displays include what one does not consume as well as what one does, to the extent that avoidance products may become more important to status construction than the products one does consume [14]. Individuals may avoid certain supermarkets or manufacturers' brands of food, or they may boycott foods from some country in order to make a political point. They may reject products from intensively reared animals on the grounds of animal cruelty, and so on. The motivations for specific consumption behaviours can change with time, so that a situationally motivated behaviour can eventually become dispositionally motivated. When this happens, consumption value ceases to be other-oriented and derives instead from the re-enforcement of self-image. For example, an individual may have become conditioned to enjoying excellence in consumer products and may have started to identify with the lifestyle associated with them. Holbrook defines esteem as a type of other-oriented value. However, the very existence of the term self-esteem suggests that esteem has a self-oriented dimension. Specifically, Holbrook sees esteem as the reactive counterpart to status whereby an individual appreciates their consumption and lifestyle as a potential extrinsic means to enhancing their public image [12].

Play is an end in itself and implies the consumer having fun [12]. Play allows for experimentation and the discovery of novelty. Children enjoy playing with their food, but adults may order them not to. Some commercial snack foods currently targeted for inclusion in children's lunch boxes encourage play, e.g., by allowing biscuits and shaped slices of cheese and ham to be assembled in different ways. A stringy cheese stick is also available that a child can pull at, shredding the cheese while eating it. Eating shellfish, especially crustaceans such as crabs and lobsters which have not been dressed, provides a socially acceptable opportunity for adults to play with their food. The same might be said about eating spaghetti and cheese fondue; after all, these do not represent the most efficient way of eating pasta and melted cheese, respectively. Playing means improvising, and improvisation encourages learning. Such learning can lead to increased involvement of a consumer with a particular product. Consumers may pursue playful value by challenging the rules of the food marketer; on the other hand, marketers can sanction and encourage play, e.g., by suggesting that consumers customise dishes, such as adding fresh herbs to ready-made soups.

Aesthetics refers to an appreciation of some consumption experience valued intrinsically as a self-oriented aim in itself [12]. As stated previously (excellence), food can be beautiful to look at. However, a more interesting illustration of the aesthetic principle is given through its link with palatability. Palatable foods are pleasant to eat, but this pleasantness extends beyond the purely sensory attributes of foods. News that abattoir wastes, including sewage, were being rendered down for incorporation into French animal feeds caused revulsion and much newspaper coverage in the UK [15]. Yet this was neither a food safety issue nor a nutritional one, nor were the sensory attributes of the meat affected. Instead, the aesthetic view of meat production, which tends to picture healthy, happy animals foraging on fresh, wholesome crops, had been disturbed.

Although personal values affect the perception of consumer value obtained from products, they are not the same conceptually. Values are culturally defined standards by which people assess desirability, goodness and beauty, and which serve as broad guidelines for social living [16]. Values underlie both aesthetic and moral beliefs. The pursuit of ethics involves doing something for the sake of others – that is, with a concern for how it will affect them or how they will react to it – where such consumption experiences are valued for their own sake [12]. Ethical food choices result from the desire to sacrifice some of one's consumption benefits out of consideration for other stakeholders; they derive from altruistic concerns for other living things, both people (e.g., fair trade) and animals (e.g., animal welfare). Meat consumption presents consumers with a particular moral dilemma, namely the fact that the life of another sentient being is sacrificed for the benefit of the consumer [17]. Purchasing only organic foods may be an expression of a consumer's ethical stance on a range of food related issues, including animal welfare and environmental protection. Boycotts too can be moral acts.

Spirituality entails a dispositional "acceptance, adoption, appreciation, admiration, or adoration of an Other where the "Other" may constitute some Divine Power, some Cosmic Force, some Mystical Entity, or some otherwise inaccessible Inner Being" and where such an experience is prized for its own sake [12]. This is a type of consumer value that will not bring to mind too many food related examples, except perhaps in a religious or quasi-religious context. In the context of the Yogic diet, it is said that "he who practices Yoga without moderation of diet, incurs various diseases, and obtains no success" [18]. Foods are placed into three classes associated with corresponding states of consciousness: gross, intermediate and spiritual for tamasic (impure), rajasic (stimulating), and sattvic (pure) foods, respectively; examples of sattvic foods are milk, butter, fruit, vegetables and grains. Another example of spiritual consumption value in foods is associated with Japanese vegetarian shōjin cooking. This is a discipline meant to improve one's training in, and practice of, the Buddhist faith through the consumption of only the simplest foods [19]. Holbrook's typology of value will be revisited extensively in Part 3 of the book.

Although Grunert's quality-related terminology is not to be adopted in the book, the model he proposes offers a highly visual, and practically useful, characterisation of the components of quality and/or value [6]. Conformity means the fulfilment of requirements (Table 1.1). Both the requirements of the product and those of the consumer need to be fulfilled. Figure 1.1, which is based on Grunert's model and on Porter's [10] terminology, presents a graphic representation of this.

The left-hand side of Fig. 1.1 (FOOD QUALITY) relates to the quality attributes of the food and to the processe(s) for making that food. Processes will have been designed to consistently achieve the required attributes within the parameters of the product specification. The assurance of product conformity to specification is one of the most important tasks of the technical function within food companies. The right-hand side of the scheme (CONSUMER VALUE) relates to the design of the food, i.e., the types of value the consumer expects or desires from the product. It is for the marketing function to ascertain both the needs of the consumer and the context the product is to be used in Fig. 1.1 highlights the close relationship between food quality and consumer value and the need for the technical and marketing functions to co-operate in the realisation of new products and in the management of existing products. Both functions must be able to understand the nature of the consumer requirements as well as process capabilities and achievable product specifications. Supply problems, technological progress, changes in food law and changing consumer attitudes are examples of situations where it may be necessary to substitute existing food ingredients in successful product lines. Marketing input into both new and modified product specifications is imperative for successful change management.

In recent years, consumers have become increasingly concerned with process attributes, even where these are not reflected in tangible product attributes. Palatability has already been mentioned in the context of animals being fed ma-

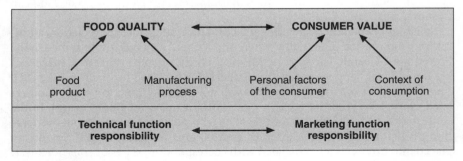

Fig. 1.1. Responsibility for conformity to product, process and consumer requirements in foods

terials containing sewage. Consumer responses to that issue had already been sensitised by the history of BSE, when many consumers had been shocked to learn of the practice of feeding animal carcasses back to herbivores [20]. Processes relevant to food quality therefore go far beyond processes used in manufacturing. Indeed, concerns about provenance often run deep, and may encompass perceived producer motivations and integrity as well as production systems per se. In the UK, the demand for organic food has risen sharply and consistently in recent years [21], one of the main motivations for being the desire to avoid foods contaminated with pesticides [22]. The UK House of Commons Select Committee on Agriculture in its Report on Organic Farming [21] stated that the term organic refers to a process, not the final product. This means that the entitlement to label vegetables, meat or any other foodstuff as organic depends on the way in which it was produced and the procedures involved in processing, rather than in intrinsic, testable quality in the food itself. The position of the UK Food Standards Agency is to dispute the validity of the widely held belief that the pesticide residues on conventional crops make these more hazardous than the corresponding organic crops [23]. Meanwhile, committed consumers of organic food are unlikely to worry about risk assessments involving population studies that lead to reassuring results. They are concerned with their personal health and with the health of those close to them, and for them trust in the process (and the process owner) leads to trust in the product. Strong demand for any type of food is surely the clearest possible sign that consumers are capable of making value judgements about it.

In the practical marketing context, a frequently used term related to quality and value is added value. Added value allows for product differentiation within specific product categories. For the producer, added value tends to be defined ultimately in monetary, or profit, terms. However, added value for the consumer has various origins, as demonstrated in Holbrook's typology. Although a consumer may have to spend more money on an added-value product, other types of cost may be reduced, e.g., time related costs (convenience foods). Alternatively, an added value product may satisfy key consumer requirements more effectively, or it may satisfy a greater number of requirements, compared with the standard product. Commercial food processing may be said to add value to more basic foods as there is consumer demand for them. Value may be added by adding certain ingredients, by adding a process step, by using a particular process, or by removing an ingredient (so-called lite products). Value may also be added to a product in strict marketing terms, for example by revealing certain types of intrinsic value. For example, there are currently many brands and varieties of apple juice available in UK supermarkets. There is considerable potential for characterising these juices in terms of sensory attributes and use contexts. As it is, although superficially the market for apple juice appears to be a differentiated, the actual differences between products are not currently highlighted effectively to consumers [24].

The competitive advantage of firms depends on the effectiveness of their value chains and on the impact of these on consumer value chains [10]. Value chains are used to disaggregate the activities performed by firms to design, produce, market, deliver and support their product. Each of these activities can then be analysed in terms of costs and of value created. At the level of consumers, households and food, value chains are focused on the way in which foods are used. An important consideration in determining the overall value of a food is the way in which it fits into existing dietary and other household patterns. Relevant issues include the habits and preferences of individual household members (e.g., gatekeepers, household size, presence of vegetarians or children), as well as general lifestyles of household members, both individually and as a group. A new product is likely to be purchased only if, on the one hand, it is seen to fit well into a household structure and, on the other, if it can readily displace an existing, but less preferred competitor. The value chains for representative households can provide an important tool for differentiation analysis. The more direct the positive impact a product has on a household's value chain, the greater the level of achievable differentiation [10].

The value created for the consumer must be conveyed to them accurately [10]. If consumers do not understand the benefits of a product, or if they use it incorrectly, the product may fail. According to Porter, purchase criteria can be divided into use criteria, e.g., product features, and signalling criteria, e.g., advertising [10]. Whilst use criteria are specific measures of what creates buyer value, signalling criteria are measures of how buyers perceive the presence of value. Use criteria include intangibles such as style, prestige, perceived status and brand connotation. A particularly important use criterion for branded foods is batch-to-batch conformity with the product specification. For example, in the case of a fast food dish such as a McDonald's hamburger and chips, consistency over time and across locations may be as important as the exact taste and portion size. This is because a non-standard item will be perceived as faulty and thus upset the trust the consumer had invested in the brand. Product conformity and reliability add value to the consumer's value chain because the risk of disappointment with the purchase is low. Signalling criteria typically stem from marketing activities and in the case of foods, the main vehicle for this is food labelling information and claims and packaging design, although other common tools in the marketing and advertising armoury are also used [25].

It is often difficult to assess the value of a food before eating it (e.g., flavour), and it may even be impossible for consumers to verify some of the claims made for a food (e.g., vitamin content). This is why price is often used as a heuristic, or rule of thumb, to ascertain value. Perception of a firm and its products can be as important as the offer itself. Quality marks and endorsements on food labels are also likely to signal value, despite the fact that consumers are often unable to assess the meaning of such labels [26]. The sudden, unexplained withdrawal of a product that does not generate value for a retailer directly can however generate a high degree of dissatisfaction in the customers for that product which may affect their perception of, and future behaviour towards, the retailer. Neatly attired, clean and knowledgeable staff, working at the point of sale, can also add value to foods for consumers. As use criteria measure the source of buyer value and are likely to determine signalling criteria, they should be identified first.

Consumers' lack of understanding of what might be valuable to them presents an opportunity for a firm's differentiation strategy, since they may be able to adopt new products or processes pre-emptively and educate the buyer to value them [10]. Brands are created by augmenting a core product with distinctive values [27]. The core benefits of brands derive from the core product, which must achieve the basic functional requirements expected of it. Branding involves developing a distinctive name, packaging and design. It must be accompanied by faultless attention to detail in post-launch quality management. This is because one of the key functions of brands is quality certification for the consumer. McDonald's global success has been based on creating added value for its customers which is based not only on the food products it sells but on the complete delivery system that goes to make up a fast-food restaurant. It aims to set high standards in what is called QSCV – Quality, Service, Cleanliness and Value [27]. Customer value can therefore be derived from many aspects of what the company delivers to customers – not just the basic product.

1.3
Role of Food Quality and Consumer Value in the Food Supply Chain

There are two major motivating forces for the implementation of quality management in the food industry, i.e., the need for statutory compliance and that for competitive advantage. Food, by its very nature, carries the potential to inflict harm on those who consume it. Consequently, many aspects of the food supply are subject to government regulation, especially those related to food safety. This is mostly initiated by the European Union (EU) in order to ensure consistent standards and free trade. In one type of approach, typified by milk pasteurisation, the process is prescribed; in another, maximum levels are set for potentially harmful substances, e.g., pesticides. More generally, the Food Safety Act 1990 requires food companies to exercise "due diligence" in the way in which they handle food [4]. This approach demands a general attitude of care as well as the practical implementation of processing and quality control steps designed to pre-empt any safety failure of food production. The requirement for practical measures implies the adoption by companies of formal quality management systems. Competitive pressures too encourage formal quality planning approaches in the food industry. Here the aims are to control the cost of quality and to ensure that research and development activities are effective. As the food market is in constant flux, with continuously changing demands by consumers, the level of product development in food companies is high; unfortunately, so are failure rates [28]. Product development is a very complex activity, requiring many different kinds of input. Careful planning and review at each stage are therefore critical to a successful outcome. Developers must ensure that products conform to statutory food standards, including the product name and labelling. Apart from this, they have considerable creative freedom in terms of the product design, but need to be constantly aware of, and responsive to, consumer requirements as well as process and supply chain capabilities.

The UK food chain comprises the agriculture, horticulture, fisheries and aquaculture, food and drink manufacturing, food and drink wholesaling, food and drink retailing and catering industries [29]. Some industries also harvest foods from the wild. Many other important industries are associated with some, several or all of these sectors. Important manufacturing industries associated with food include the chemical, pharmaceutical, engineering and packaging industries. Service industries associated with food include analytical and legal services, market research and marketing, and product development. Retail prices have to cover the cost of processing and distributing food. There may be limited demand for parts of the primary product and, with livestock especially, waste may actually incur a disposal cost. One area of value creation in meat supply chains is product development and innovation aimed at the improved usage of the less popular cuts of meat, and the upgrading of "waste" [29]. Although value is added at each stage of food supply chains, the different ends of these chains operate in different economic and regulatory frameworks. For example, primary production is heavily influenced by the Common Agricultural Policy (CAP) which may send economic signals that are not totally aligned with those of the market [29]. The CAP is often held responsible for sub-optimal competitiveness, diversity and flexibility in the farming industry, and for making it unresponsive to changing market opportunities. As individual food supply chains vary considerably, Fig. 1.2 is a stylised representation of such chains, with the finished (retail) food product at the centre.

The quality of a finished food product may be compromised at any stage, so that it is necessary to integrate quality management throughout supply chains. Quality assurance schemes in particular are effective only if quality is correctly defined and implemented at every stage. An example of this is the supply chain for Specially Selected Scotch Beef, where livestock breeders, rearers and finishers, abattoirs and cutting plants, manufacturers and retailers all participate in the relevant quality assurance scheme [30]. During the last thirty years, there have been many changes in European food supply chains, e.g., the increase in

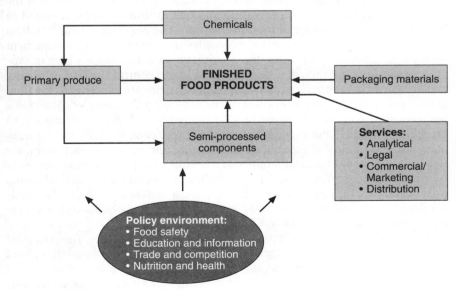

Fig. 1.2. The food supply chain

market power of a decreasing number of retailers, who operate large numbers of outlets. Coupled with the drive by manufacturers to achieve economies of scale, this has led to more international sourcing [31]. There have also been changes in the nature and length of distribution channels as large orders enable the multiple retailers to deal directly with manufacturers, circumventing traditional wholesalers. These retailers exert control over products by becoming involved in specification and control at an early stage [31]. They do this both when sourcing primary produce, e.g., animal carcasses for fresh meat sales, and in the development of own-label manufactured foods for contract manufacture.

Because consumers spend a decreasing proportion of their incomes on food as incomes increase, the food industry cannot increase its prosperity by producing more, but by providing better products, with all kinds of added value designed into them. Providing better quality food means achieving better matches between individual products and the requirements of individual market sectors. However, UK food processors typically produce relatively large quantities of standard product [32]. Growing consumer affluence has lead to demand for wider ranges of specialist food products, offering opportunities for smaller food companies [31]. As the food market fragments, more flexible processes, as in ingredient sourcing and manufacturing, need to be considered for a larger number of products. In contrast to the increasing concentration in food processing and distribution, the UK still has a very large number of farm businesses which have little market power and which have tended to supply commodities markets as price takers. Whilst elsewhere in Europe farmers have sought to overcome this through co-operative buying and marketing ventures, UK farmers have tended to maintain their independence, with some exceptions, e.g., horticulture [29]. One way in which value can be added to primary products is through the wide variety of regional and speciality foods. Though niche markets, demand is growing rapidly, resulting in innovative marketing structures, e.g., farmers' markets [29]. The current focus on organic production in particular may be seen as part of a wider movement to strengthen the place of farmers in food supply chains by highlighting quality [33].

Firms throughout the world face slower growth as well as greater domestic and global competition, and competitive advantage grows out of the value companies create for their customers that exceeds the cost of creating it [10]. The higher price received per unit of differentiated or added-value product reflects the greater use or enjoyment that it provides. Long term, sustainable growth in the UK food market depends on product differentiation, development and innovation. Government initiatives are aimed at facilitating more effective UK food chains in order to close the performance gap with competitors in terms of productivity (efficiency) and the ability to innovate ahead of the market (effectiveness). In the recent report "Working Together for the Food Chain", three quality management tools and processes are highlighted, i.e., benchmarking, efficient consumer response (ECR) and quality assurance schemes [29]. For businesses to thrive in increasingly global, competitive markets, they need to regularly benchmark their performance against that of their competitors. Some multiple food retailers' rolling programmes of product assessment against competitors' products exemplify the benchmarking process. ECR is a total food and grocery supply chain initiative which enables companies to work together to fulfil consumer wishes better, faster, and at a lower cost [29]. The enabling technology for this is sales information captured at the till to drive ordering systems and to provide a detailed understanding of consumer demand. Among ECR initiatives are Efficient Assortment (items offered in each product category), Efficient Replenishment (continuous replenishment inventory system), Efficient Promotion and Efficient New Product introductions (to curb product over-proliferation) [34]. Finally, quality assurance schemes can help to promote confidence in the product and process quality. Such schemes have proliferated in the farming sector since the start of the BSE crisis [29].

Market research conducted by the Institute of Grocery Distribution shows UK food consumers to be generally satisfied (81%), with product range attracting the highest approval rate (31%) [35]. Also highlighted were quality/freshness (20%), convenience (15%) and availability (13%), with cost the main cause of dissatisfaction. Most multiple retailers aim to provide groceries covering the full range of consumer needs. This is evidenced by the presence of economy ranges being sold side-by-side with luxury lines, and by the stocking of a comprehensive range of organic foods. In-store bakeries, butchers and fishmongers are other examples. Encouraged by the multiples themselves, many consumers have settled into one-stop grocery shopping routines, with one major shopping trip a week [36]. This, combined with the use of shopping lists, has led to a high level of expectation with regard to product availability. Therefore the non-availability of a product in the store when required is a major source of consumer complaint to retailers. Related to availability is the issue of withdrawal of new products from stores if these do not prove themselves profitable quickly enough. ECR together with use of the information available to retailers from loyalty card schemes may be used to minimise this problem. Once-a-week shopping routines, often involving car journeys of considerable lengths, have also led to increased emphasis on shelf life extension and control in retailers' product management.

Another aspect of the food system that consumers are generally dissatisfied with concerns the difficulty of accessing accurate product information, exacerbated by the prevalence of more or less misleading packaging claims and labels, some of them officially sanctioned. Even the "Scotch Beef" half of the Specially Selected Scotch Beef brand, although constituting a statutory designation does not mean that the animal was born and raised in Scotland, but only that it was finished there for a minimum of three months [37]. This is contrary to consumer expectation [38]. Government asserts that it wants to strengthen consumer confidence through improved product labelling, addressing in particular the volume of information to be conveyed and the multiplicity of labels that currently exist for British foods [29].

Market-driven businesses know by which choice criteria their products are being evaluated against the competition, and so attempt to create customer value in order to attract and retain customers [10]. By employing both technical and commercial resources, companies can open up latent markets for possible products. New product development should be based on sound interfacing between perceived customer needs and technological research. The marketing function must understand the types and nature of barriers to acceptance and how to overcome them. Each company bases its marketing mix on an understanding of its customers' needs, taking into account four key decision areas, i.e., product, price, promotion and place (the 4 Ps) [27]. The product decision addresses the selection of products that should be offered to a market segment. It also involves choices regarding brand names, packaging and any associated services. Place deals with issues such as the choice and management of distribution channels, location of outlets, methods of distribution, and inventory levels. In niche, added-value markets for foods consumers exhibit greater emotional involvement in their buying decision than is normally the case [17]. This can be linked to higher expectations about all aspects of quality. Luxury ranges are, by definition, expensive and therefore raise expectations particularly in terms of eating quality, i.e., types and grades of raw materials used. The perceptible taste or declared content of onion powder or garlic oil in a budget beef burger may be expected and acceptable; however, their presence in a high-price, premium beef burger would be a different matter. Where ethical issues are involved, consumer responses to expectations being violated are likely to be even more pronounced.

Social trends underlie consumer value perceptions and individual food requirements. The UK has an ageing population [39]. This implies opportunities for health-focused, added-value healthy as well as single-portion and portion-controlled ready meals for the elderly living alone. In fact, almost three in ten households in Great Britain comprise a person living alone, and about one person in 15 is from an ethnic minority group [39]. Such trends clearly have an impact on the demand for specific foods. Results from the National Food Survey provide some evidence that people in Great Britain are eating more healthily. In particular, the amount of fresh fruit eaten at home has risen steadily since the mid-1980s [39]. The survey also shows consumption of more reduced-fat milks and low-fat spreads at the expense of wholemilk and butter.

Food processing is fundamentally concerned with various types of value addition to edible biological tissues. This may simply mean the scaling up of domestic processes for greater efficiency. Consumers share in the value created in this type of process rationalisation. However, it remains possible for individuals to produce their own flour, bread, fermented milks, beers and wines, jams and pickles and preserved meats domestically. Those that do so derive other types of value and satisfaction from these activities. If raw materials are cheap and abundant, and time is not calculated as a cost, home food production can be cheaper. Alternatively, such an activity may be enjoyed for its own sake, perhaps as a hobby. The eating and/or nutritional quality of home produced foods and beverages may also be perceived as superior to those of industrially produced foods because the maker has both knowledge of and control over the ingredients used. However, industrial processing plant tends to be more effective than domestic appliances, leading to better product quality. For example, whilst consumers are able to freeze certain vegetables at home, these will not mirror the quality resulting from very fast rates of freezing, e.g., using liquid nitrogen or carbon dioxide as the refrigerant [40]. Despite the high-technology nature of processes such as cryogenic freezing, they are perceived to be similar to the equivalent home based processes. This avoids problems of consumer acceptability. Cryogenic freezing provides for the one factor that is crucial in all mass production and processing, i.e., control. This in turn provides for repeatability. Much of the UK's manufactured bread is made using high-speed processes adapted from traditional methods, the most widely known of these being the Chorleywood process. However, bread is an example in which manufacturing efficiency is accompanied by a loss of perceived product quality [41]. Amongst other differences, this type of bread is blander than bread obtained from traditional processes where a more extensive dough fermentation takes place. Current high-technology approaches to innovation in food science and technology are highly targeted to the generation of specific value and functionality. The most obvious example of this is research into genetically modified (GM) foods. Processes that are considered truly novel, e.g., GM foods, are scrutinised by the Advisory Committee on Novel Foods and Processes (ACNFP) which reports to the UK Food Standards Agency. The agency's web site provides access to detailed information on the activities of this and other government advisory committees [42].

For any food company embarking on the adoption of a formal quality management system there are a number of avenues down which they might proceed. The most obvious start is the introduction of a Hazard Analysis Critical Control Point (HACCP) system. In fact, there is an expectation in food law for HACCP systems to be operational in all food companies. HACCP is an internationally recognised system of food control used to identify, evaluate and control any food hazards inherent in a particular product or process [43]. The essence of HACCP analysis and implementation is to prevent any inherent food hazards from turning into tangible risks to consumers. Documentation and record keeping, verification of products and processes and the allocation of responsibilities and authority are key features of any HACCP plan. However, small and medium sized enterprises (SME), in particular, have found it more difficult to implement HACCP than had been anticipated, highlighting a need for stricter regulatory control of certain food businesses. A start has been made on this with the licensing requirement for UK butchers' shops [44]. HACCP is an important aspect of Good Manufacturing Practice (GMP). But it has been estimated that of the over 15,000 UK food manufacturers there are fewer than 3000 with enough technical resources to be able to operate formal systems of GMP [45]. HACCP can also be used to control hazards other than those related to food safety, e.g., financial losses that would result from process breakdown.

The international standard for quality management systems, BS EN ISO 9001: 2000, has recently been radically revised [46]. Its predecessors have been implemented within the UK food industry for well over a decade. The standard provides a comprehensive, but generic, framework for quality management. It requires each individual organisation to adapt it to its own particular processes, both technical and commercial. ISO 9001 is particularly appropriate for food businesses, because the due diligence requirement [4] can only be met if processes and practices are documented fully and records kept accurately, both key requirements of the standard. However, over the years the standard has had its detractors [47] as well as its supporters [48, 49]. One of the loopholes closed during the recent review of the standard is the fact that organisations will no longer be able to choose freely among different levels of the standard. Instead, there is now only one document. Under BS EN ISO 9002: 1996 predecessors it had been possible, and not uncommon, for companies that part of the standard that did not require them to control their p velopment activities, even if such activities did take place. However, many uct faults that materialise after launch can be traced back to deficiencies in design process. Not surprisingly, a system with such a critical flaw can deliver only if it is not abused. By controlling their design and development activities, food companies ensure that the marketing function is fully integrated in the quality assurance process. Accurate information about the nature and requirements of the target market, and about projected sales volumes, is vital to the technical function as product development takes place. Of late, UK food processors have started moving away from ISO 9001 towards more industry-specific, third-party accredited standards [51]. Such standards more closely match the requirements of the industry. In addition, they address the need to rationalise the various retailer-specific compliance schemes. Prominent among these are the European Food Safety Inspection Service (EFSIS) standard and the British Retail Consortium (BRC) standard. With a shift in the power base in food supply chains, whereby retailers' own-branded products now account for over 50% of food sold in the UK, there are clear incentives for companies to adopt a standard that incorporates BRC requirements [51]. Controls identified through HACCP can be developed to form one of the bases of more comprehensive quality management systems such as ISO 9001 and the EFSIS and BRC standards. A series of recent food scares, affecting meat production in particular, has generated a certain desperation among some food producers. In one instance, this has led to the misguided view that one should introduce "objective measures of everything that can be measured" [52]. This attitude betrays a lack of understanding not only of what quality means but, in addition, risks a dangerous waste of resources. This is the opposite of HACCP. The approach that HACCP embodies ensures focus on issues and controls that are truly critical to food quality, in particular, safety.

Communication between producer and consumer is therefore a key issue in quality. Whilst direct communication is ideal, this is rarely possible in increasingly complex food supply chains. Food production for most is now remote from consumption, with few people having detailed knowledge even of farm procedures. Product branding is a proxy for direct communication and it is a very important issue for the food industry. The value of brands lies primarily in the trust and confidence they generate both for producers and consumers. From the business point of view, brand value often exceeds, in monetary terms, the value of physical assets such as buildings, manufacturing plants or stores. For consumers, the purchase or consumption of branded foods and beverages minimises the gap between expectations and delivery, and may also add a dimension of emotional satisfaction. Quality management of brands is about conformity to production specification, as expectations of batch-to-batch conformity will be very high and quality failure costly. Brand management also monitors the market in which a product is sold, allowing any necessary adjustments to be made in a controlled manner. Consumer perceptions of a food company and its products can be as important as the reality of what the firm offers.

In common with other spheres of life, the provision of, and knowledge about, foods has been delegated to professionals. However, when experts are perceived as failing those on whose behalf they act, the delicate balance between trust and the need for control is disturbed. Many of the new food technologies are difficult to comprehend by consumers who cannot relate them to home based processes and who have not been educated in the new technologies. More needs to be done to allow interested consumers access to all food quality related information, an important challenge for the Food Standards Agency. This educational need is also very apparent with regard to many food journalists and food experts with high media profiles, who have a secondary role in consumer education about food. For example, one ought to be able to rely on a celebrity chef, speaking on TV, not to confuse modified starches with genetic modification. In recent years, food and cooking have become integral parts of the UK lifestyle and entertainment industries and the term quality is in frequent use. What this usually means is an attack on all mass-manufactured products in favour of crafted or hand-made ones. For example, standard supermarket fare (in this case, cake) would be rejected as "pre-packaged, over-processed junk", compared with a hand made cake costing four times as much. Compared with this type of approach, which merely expresses a particular individual's attitudes, well argued, informative programmes about foods and the chains which produces them can at present rarely be seen or heard.

1.4
Quality of Conformity and Quality of Design

Delivering quality through a product is about meeting the customer's requirements, as well as the requirements of other interested parties. Requirements may be formally expressed in the form of written specifications, as is typical of relationships and contracts between industrial suppliers and their customers. Alternatively, requirements are either obligatory or implied [8]. The most obvious example of an obligatory requirement in relation to a commercially offered food is the compliance of that food with all applicable statutory standards. Individual consumers' assumptions and expectations about the nature of a food product or process generate implied requirements, which it is the role of suppliers to define and fulfil. Assumptions tend to originate in an understanding of what is commonly done, in advertising and in food packaging and labelling design. It is important for food suppliers to ascertain through consumer research that communication processes, e.g., advertising and labelling, deliver messages accurately. Consumers' beliefs and attitudes constitute vital impulses for the design of such processes. The belief systems underlying individual segments of the food market may be complex and not immediately obvious to suppliers. For example, soy tends to play a large role in vegetarian diets whilst, at the same time, much of the global soy crop is now genetically modified. In attempting to elicit vegetarians' attitudes towards the genetic modification of foods, a company may discover a prevalence of negative attitudes towards that particular technology. In this way, by discovering a link between vegetarianism and opposition to genetic modification, the company will understand its target market's expectations of the type of soy used in food products offered to them.

Table 1.4. Three steps towards quality of design [53]

- Identification of what constitutes fitness for purpose for the user
- Choice of a product concept which will correspond to the identified needs of the consumer
- Translation of the chosen product concept into a detailed set of product and process specifications

Dale and Oakland distinguish between what they call quality of conformance and quality of design, both being aspects of another quality related concept, i.e., fitness for purpose [53]. The term conformance is synonymous with conformity, but is now deprecated; conformity is defined as the fulfilment of a requirement [8]. The two types of requirement to which Dale and Oakland draw attention are related to Grunert's concepts of objective and subjective quality [6]. The three main steps leading to quality of design are shown in Table 1.4.

Quality of design requires the correct identification, at the outset, of critical product quality attributes. Quality attributes must be selected carefully in order to prevent irrelevant attributes from being specified. A company specifying "as many" quality attributes "as possible" for a product [52] reveals poor understanding of its customers' needs, and of the meaning of quality generally. Equally ineffective and misleading are specifications based on product attributes that are either easy to measure or for which there are existing analytical resources within an organisation. Such approaches are wasteful, engender a false sense of security and are likely to store up problems for the future. To direct the focus from food qualities, or attributes, towards quality, and thence value, it is therefore necessary to select critical product attributes from amongst the complete spectrum of attributes, ignoring those that may be merely convenient to measure. An appreciation of the need to select critical product attributes and critical process control points has received a boost with the recent, widespread introduction of HACCP in the food industry [43]. HACCP analysis during the design stage also represents an opportunity for designing hazardous attributes out of products at the outset. Where a market has been found for a product and consumers having adopted the product subsequently find themselves misled about its nature, the issue is one of nonconformity in terms of the design specification. An example of this is many consumers' negative responses to the realisation that "Scotch Beef" does not mean, as they had assumed, that a particular animal had been raised in Scotland [38].

It is possible for an initially well-designed product to be manufactured in such a way that its quality of conformity is low. This means that the product will vary somewhat from batch to batch. Quality of conformity focuses on products which have gone into production and for which a market has been established. Machines and processes, and their operators, all have inherent variability and the extent of this variation is referred to as the process capability [53]. Quality of conformity demands that a product can be made, time and again, such that it meets its specification. Dale and Oakland's definition of quality of conformity is shown in Table 1.5.

Table 1.5. Quality of conformity [53]

The extent to which the product, once it has been generated, conforms to the design, specification or requirement

Specifications must accommodate given process capabilities, i.e., likely variations from the stated dimensions. If specified tolerances for a particular attribute are too tight, inevitably, there will be frequent production failures. Product specifications must be able to be verified, and acceptable ranges for specific attributes must therefore be set and documented with reference to a suitable analytical procedure. This is also helpful when there are differences between the analytical results obtained by the supplier and the customer, preventing damaging disagreements from arising. Simple and rapid methods that are generally accepted and available industry-wide are especially suitable. A food ingredient that is used in a commercial recipe may, for whatever reason, become unavailable; or changes may have to be made to processing plant. Recipe, process and other changes required for established products need to be strictly controlled so as not to impact the quality of conformity.

It is possible in any organisation to identify a quality function. This means that, in order to achieve fitness for purpose, a number of activities need to be performed in a logical sequence [53]. Quality function deployment (QFD) allows the activities of employees from different departments or functions to be co-ordinated into a unified effort. It is a planning and analysis system and a means of translating the customer requirements into the appropriate technical requirements for each stage of marketing, product planning and design, production, sales and service, and so on [53]. The starting point is a customer wants list. Requirements are expressed in customers' original words and translated into technical language. Application of QFD is particularly useful in product development as it enforces a shared focus on both user requirements (voice of the customer) and producer capabilities (voice of the process). The process terminates with the setting of value weights and targets for the different design requirements. A planning matrix, the so-called house of quality, serves as a visual re-enforcement and tracking system of stages and issues addressed or waiting to be addressed. Figure 1.3 illustrates the conceptual map that is the house of quality.

Hofmeister provides a detailed account of how QFD might work in practice, using as an example the development of a chocolate cake mix [54]. The house of quality is divided into various rooms and segments. The first room is known as the WHATs, and it contains the customer requirements. Another room, related to the WHATs, contains important control items, e.g., statutory and company requirements. Product requirements derived from the WHATs are known as the HOWs. They are measurable attributes that describe the product in the language of the engineer. As some of the HOWs affect more than one WHAT, an additional matrix may be created to show the relationships between the WHATs and the HOWs. The fourth key element of any QFD chart is the HOW MUCH section. The correlation matrix is a triangular table, or roof-like structure, which com-

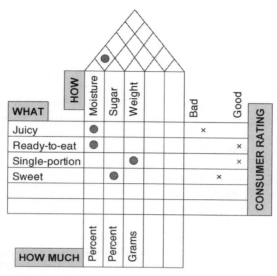

Fig. 1.3. House of quality for a prune snack

pares each of the HOWs to all the others on a pair-wise basis. When consumers ask for certain types of consumer value to be present in a food product, e.g., convenience in use, they do not specify particular product features. QFD assists in translating the one into the other in a controlled manner.

The link between consumer value, or buying motive, and concrete product attributes may also be examined using a technique known as laddering, which is an example of a means-end chain analysis. A means-end chain is a model of consumers' cognitive structures, which reveals the way in which product characteristics are linked to self-relevant consequences and, ultimately, life values [6]. For example, semi-skimmed milk contains less fat, and therefore fewer calories, than wholemilk. In laddering, these product attributes may become linked to the consequences of a slimmer body shape (functional) and, beyond that, social acceptance (psycho-social). This in turn implies self-confidence (instrumental) and, finally, self-esteem (terminal value) [6]. This completes the means-end chain. Alternatively, the means-end chain for semi-skimmed milk drawn up for different categories of consumer may arrive at the terminal values of longevity or good health. Clearly, an in-depth understanding of buying motives is of major importance in achieving quality of design. Quality of design is not necessarily just linked to food product development. For example, laddering is a useful tool for determining product targeting and positioning. This does not necessarily involve new products, but can mean the repositioning (re-targeting) of familiar food products, which may be approaching the end of their life cycle [11].

It is clearly necessary for food companies both to control the attributes of a product through control of production and/or processing and to monitor user requirements of the product in the context of social and market developments. Once again, Grunert's model of quality provides a useful conceptual framework in that it highlights the fact that both technical and marketing functions have a continuing involvement with product specifications throughout a product's life cycle [6]. It therefore serves as an illustration of the role of quality of conformity and of that of design as defined by Dale and Oakland [53]. In particular, it flags up the reality that user requirements of a product are not static. It thereby implies the need for effective change management.

Quality of design is closely associated with the quality of product communications. Finding the product that is the most fit for a particular purpose is possible only where alternatives can be weighed up in an informed manner. Quality of design is therefore an issue even for straightforward selling operations, i.e., those that involve neither product development nor advertising. A combination of product labelling and staff expertise enables a customer to choose correctly, in terms of his or her expressed requirements. An example would be choice of the correct grade of flour for either bread making or cake making. If a consumer chooses a more expensive grade of a particular food than required, money is wasted, and even where this is not an issue, performance may be sub-optimal. For example, shin of beef contains more connective tissue and is therefore cheaper than stewing beef. However used correctly, shin of beef actually performs better in stews as the meat is held together better and a smooth jellied texture is added to the sauce [55].

In the case of man-made foods with many ingredients, quality of design implies that the specifications for individual components are chosen to reflect the standard of quality of the finished product. For example, it is not efficient to build longer-life, and therefore more expensive, components into short shelf-life food products. This would increase the price of the finished product without any associated beneficial effect for the buyer. On the other hand, if a top-of-the range meat product is being developed using top grade raw materials, product packaging needs to reflect this. In this case, formulation will also avoid ingredients and additives which consumers are likely to associate with lower-grade foods [56]. The processed organic foods sector is perceived increasingly as representing not only morally acceptable production methods, but in addition, luxury. This may explain why organic foods are often said to taste better than their conventional counterparts. As they are in any case perceived to be expensive, and the market is willing to pay, higher-grade ingredients tend to be used throughout products and product ranges, e.g., chocolate, fats in flour confectionery, and fruit concentrates in yoghurts. The full spectrum of stimuli emanating from these products can thus be integrated harmoniously in a consumer's perceptions. In contrast, poor packaging or declaration of perceived low-quality ingredient, or of a large number of additives, in an expensive product is likely to produce dissonance in the mind of the consumer as would the over-packaging and over-promotion of economy lines [56].

To achieve quality of design, consumer requirements for products must be explored and formally defined by the marketing function so that they can be incorporated into product development or product marketing briefs. Consumer research provides the opportunity to identify perceptions about new foods being developed and/or marketed. Qualitative and quantitative approaches may be used for this, and often a combination of both is appropriate [57]. This enables key attributes to be pinpointed and weighted. Both function-in-use, including sensory performance, and general perceptions need to be explored. Among qualitative approaches, focus groups are particularly popular. Involving potential users in product development also presents an opportunity to explore their implied needs. Implied needs are especially important in terms of product formulation. Recently, the global fast food retailer McDonald's posted the following message on its web site: "Because it is our policy to communicate to customers, we regret if customers felt that the information we provided was not complete enough to meet their needs" [58]. This was an apology to US customers who had assumed that the company's chips were suitable for vegetarians. The chips actually contained flavourings based on beef fat, and this caused particular outrage among some Hindus who had innocently eaten the McDonald's chips. Implied standards may also relate to food presentation, in particular, the food buying or consumption environment, issues of cleanliness, and staff appearance and conduct.

The most important aspect of the quality of conformity relates to compliance with statutory requirements, i.e., food standards regulated under the Food Safety Act 1990 [4]. A definition of food standards [59] is provided in Table 1.6. The statutory protection of food standards has been based traditionally on so-called recipe law combined with prescriptive designations. However, in the context of the modern food industry, there are major disadvantages to this approach. Firstly it inhibits innovation and secondly, it constitutes a significant barrier to the trade in foods. Designations tend to relate to national standards that can differ considerably between countries. Perhaps the best known example is the case of Cassis de Dijon, a French drink with a 15% alcoholic content. This made legal history when the European Court of Justice ruled that it could not be barred from importation into Germany even though German rules at the time specified that alcoholic drinks must contain a minimum of 32% alcohol [60]. Recipe law is also known as vertical legislation. Following the Court ruling, efforts to harmonise vertical legislation within the European Union began to recede in favour of horizontal legislation. Horizontal measures apply across categories of different foods, for example, permitted additives and, in particular, labelling. Despite increasing emphasis on labelling to protect and inform consumers, national food standards remain in certain areas. For certain basic, high risk, dietary items, standards are especially stringent. For example, the statutory controls over the production of pasteurised milk are so exacting and so tightly specified that one might view these controls as a fully fledged HACCP scheme [61]. However, in terms of the food trade within the Single European Market, the principle of reciprocity now applies. This means that any food legally produced within any member state may be legally sold in any other. Food standards in the global arena are covered by the Codex Alimentarius [62]. Another example of

Table 1.6. Food standards [59]

Legal requirements covering the quality, composition, labelling, presentation and advertising of food and of materials or articles in contact with food the statutory aspect of quality of conformity is the statutory grading of fruit and vegetable, which are covered by EU marketing standards. This is somewhat arbitrary as it singles out one of the many aspects of quality, i.e., appearance, whilst ignoring flavour which, arguably, is more important. In the case of garlic, classification by appearance takes into account absence/presence of blemishes and the degree of regularity of the shape of the bulb [63]. The exclusion of key quality attributes such as flavour from the grading scheme is an example of inappropriate specification of quality and hence, poor quality of design.

Quality of conformity is therefore concerned with product specifications being met. Large-scale industrial and automated food production and manufacturing generally require tighter specifications and greater accuracy in meeting them than do small-scale or craft-based processes. Although in the latter there is more scope for skilled intervention and process adjustment, restaurant kitchens too need to be able to perform comfortably within certain error margins. Whilst there is less pressure in domestic kitchens, increased accuracy is required at certain times, e.g., when entertaining guests. In order to achieve the necessary level of control in large-scale food manufacturing operations, traditional food ingredients, such as flour, may be substituted with ingredients, such as modified starches, that are more defined and less variable in terms of their functionality. Consumer expectations of batch-to-batch conformity are less pronounced for craft based foods than for mass produced ones. Indeed, a degree of variability may be desired as this would suggest that the craft based product resembled its home made counterpart. Food colours and flavourings have an important role in many mass produced foods in that they are able to promote batch-to-batch conformity. Where there is natural and expected variation within a product, for example potato crisps, variation perceivable by consumers through their senses should be minimised. For crisps, a 10-point scale has been developed ranging from 1–2 (match) through 3–5 (acceptable) and 6–8 (unacceptable) to 9–10 (reject) [64]. The design of foods that need to be subjected to further processing by the consumer will take into account the potential for product abuse. For example, the instructions for making up instant snack soups or pot noodles typically specify the use of boiling water. However, agglomerated (instantised) sauce powders may be used in the manufacture of such products in order to ensure satisfactory dispersal even if label instructions are not followed correctly.

British Standard BS 7373:1991 provides guidance for the preparation of specifications [65]. As specifications must convey information, requirements need to be clearly stated so that conformity can be readily checked. Descriptive specifications describe a product or process in such as way as to allow the potential user to establish fitness for their particular purpose. These tend to originate from technical and customer service departments of food companies and typically refer to composition, product applications and legislative compliance in various markets. They also tend to include pack weights and dimensions and storage and distribution criteria. Requirement specifications are statements of attributes of a product or process in a form that provides for objective determination of conformity. Requirement specifications for foods typically include sensory and physico-chemical attributes, microbiological criteria and nutri-

Table 1.7. Achieving quality of conformity through tea grading [66]

"The bushes are picked more frequently if a higher quality is required, and less frequently if greater quantity is the aim. The first spring pick makes the best tea. Frequency of picking depends on weather, fertilization of the soil, pruning and so forth. The tea leaves should be picked dry, but this is not always possible, and large quantities of low-grade material results from 'rain teas'".

tional breakdown. Several specifications may be held for a product, reflecting different user requirements. Among these, the manufacturing specification needs to be the most tightly defined as it must satisfy the requirements of all customer specifications. Manufacturing specifications are also likely to refer to the general manufacturing environment, along HACCP principles, taking into account any possible impact by or upon other products being produced in the facility. Examples are the specification of ingredients and process controls to prevent cross contamination, e.g., from meat, peanuts and other high-risk materials to low risk ones. Agricultural and horticultural produce is inherently variable. Here, commercial quality of conformity is achieved through grading. This procedure is illustrated by Hobhouse using the example of tea grading [66] (Table 1.7).

Quality attributes in foods, and overall quality, evoke three basic types of response in consumers. On one level, the customer will be satisfied with the product. At the bottom end of the range, the product is found faulty whilst at the top, the customer is delighted with the product. However, exceeding customers' expectations, as is frequently recommended, may be unwise unless one is absolutely certain about the requirements. For example, whilst late delivery of a product will most definitely produce a negative response, early delivery may have the same effect if the recipient is not ready to accept it.

2 An Exploration of Food, Food Quality and Food Qualities

2.1
Introduction

If, as the New Oxford Dictionary of English suggests, food means "any nutritious substance that people or animals eat or drink or that plants absorb in order to maintain life and growth" [67], how do we classify dietary fibre? More to the point, how do we classify those modern food ingredients designed specifically not to carry calorific value? The food industry critic Derek Cooper has already provided a possible answer to that question in creating the concept of unfoods, which he applies specifically to the fat substitute Olestra [68]. The introduction of so-called novel foods into European Union food markets is controlled by regulation [69], but fulfilment of what criteria allows a potential food to transform itself into an actual one?

In this chapter, the essential nature of what it is exactly that turns some material or substance into a food suitable for human consumption is put under the spotlight, and is examined from both biological and cultural perspectives. Clearly, there are a number of different types of intrinsic quality attribute that can be applied to food in general. However, few foods, if any, will incorporate all of these. For example, a food product may offer high nutritional value but little hedonic pleasure, or vice versa. From the cultural perspective, the same food that provides a high level of pleasure when offered to one group of people may be rejected equally vehemently by another, for whom it may even constitute a taboo food. Having come to some understanding about what constitutes a food, critical (quality) attributes applicable to foods are examined and categorised. This second part of Chapter 2 thereby sets up a strong link with Part 2 of the book. However, the scope of the discussion is not limited to the kinds of intrinsic, tangible quality attributes that find their origin in physical realities. An equally strong focus is provided on assigned quality attributes, e.g., price, on signalling attributes and on any other relevant categories of quality attributes.

2.2
Foods for Human Consumption

Some of the essential characteristics that, individually or in combination, identify and designate materials as foods suitable for human consumption are: harmlessness, edibility (acceptable flavour, ease of chewing and swallowing), nutritiousness and ease of digestion, and social acceptability. Palatability, or pleasantness, draws both on factors related to edibility and on aspects of social acceptability. For example, a piece of fish may cease to be enjoyed if a parasitic worm is discovered in it. Similarly, the palatability of meat may be reduced if it is learned that the animal from which it derived was likely to have been raised on feed incorporating sewage [15]. Consumers' perceptions of the fundamental nature of food generally, and of good or proper food in particular, are embedded in the wider belief and value systems of individuals which, in turn, are largely culturally determined. The production of energy from food molecules is fundamental to life. However, as well as supplying fuels and building blocks for the body, food and meals serve as vehicles for both hedonistic and cultural consumption. Eating food is therefore not merely a matter of feeding the body. It is well known that many edible and nutritious materials are strongly rejected on cultural grounds. The most extreme manifestation of this is the taboo against cannibalism. Nevertheless, ever since men began sailing the oceans, famished sailors have sustained themselves on the remains of dead shipmates [70]. Taboos forbidding the consumption of the flesh of certain types of animal exist in many cultures and religions. Societies without formal food taboos also have social norms and conventions in this respect.

Hunter-gatherer societies typically have a wider range of food sources available to them than do settled farming communities, whilst urbanisation brings added restrictions. There are very few poisonous plants and most of these are uncommon. It is estimated that only one per cent of the known 500000 plant species are deadly poisonous [71]. The prominent UK naturalist Richard Mabey still includes many hundreds of wild foods in his diet [72]. Australian aborigines traditionally ate a wide variety of plants (fruits, roots, tubers, leaves, flowers), insects (witchetty grubs, bogong moths), small reptiles (snakes, lizards, goannas), as well as large game [73]. They had developed techniques of dealing with potentially harmful foods such as cycad seeds which, when eaten by some of the early explorers and settlers, caused violent vomiting and diarrhoea. One of the possible aims of both traditional and modern food preparation and processing is to transform potential foods into actual ones.

First and foremost, to qualify as a food for human consumption, a material or substances must be safe to eat. However, food safety is both a relative and a subjective concept, as any negative outcome of an eating episode is likely to depend on the amount eaten and on who is doing the eating. Whether an individual becomes sick is often a function of the dose of the poisonous material received. To take this argument to its extreme conclusion, the truism applies that anything consumed in excess is harmful, and this includes water. Food safety therefore needs to be defined in the context of the normal conditions of use of that food. From the subjective perspective, whilst all common food components are, again by definition, harmless to the population at large, there will always be specific categories of consumers with particular sensitivities towards some of these. These consumers need to be provided with the necessary information either to avoid certain foods or to consume them safely.

For the purposes of international food law, a definition of food is given in the Codex Alimentarius [74] (Table 2.1).

Table 2.1. Codex Alimentarius definition of food [74]

Any substance, whether processed, semi-processed, or raw, which is intended for human consumption (including drink, chewing gum, and any substance that has been used in the manufacture, preparation, or treatment of food but excluding cosmetics, tobacco or substances only used as drugs)

Clearly, there is no attempt to include in this definition any functional, or quality, aspects of what constitutes a food. Rather, the focus is entirely on consumer behaviour, i.e., that which is eaten, or, for human consumption. There is a European Commission proposal for a basic framework of principles and definitions for future European food law. A definition of food forms part of this [75]. At the time of writing, the proposed definition of food, as amended, reads as follows: "food (or foodstuff) means any substance or product, whether processed, partially processed or unprocessed, intended to be, or reasonably expected to be ingested by humans". It includes drink, chewing gum and any substance intentionally incorporated into the food as well as water intended for human consumption. However, this does not address the concept of food grade, which is an important issue in the suitability of chemicals to be added to food, in particular, flavouring substances. The assessment of the correct level of purity of such substances therefore remains largely an issue of due diligence and risk assessment on the part of the supplier. A food, which is fit for human consumption in principle, may subsequently become unfit. As most foods are perishable, becoming unfit is a common feature of most foods at some point in their lives. What this means is that the food is unacceptable for human consumption according to its intended use, for reasons of contamination, whether by extraneous matter or otherwise, or through putrefaction, deterioration or decay [75].

Food represents an increasingly innovative and technological industrial sector [75]. When a food is placed on the market for the first time, one possible scenario is that a product is simply made more accessible to a consumer already familiar with it. A typical example would be the importation of a food which consumers may have previously encountered abroad. Alternatively, consumers will be confronted with an unfamiliar product, i.e, one to which they have not had any prior exposure. In this case, the shopper, and potential consumer, may be well aware of the nature of the product as a good and proper food in some other market. Whether they will wish to explore the novelty on offer to them will depend on personal, and perhaps contextual, factors. There is an inbuilt mechanism for people to be cautious in the adoption as foods of unfamiliar materials. This is an important issue in the context of food-related consumer value. The relevant concepts, i.e., neophilia and neophobia, are introduced in Chap. 3. They are further explored in Part 3 of the book, in particular, in the context of variety-seeking.

It will be found that authors writing on these issues, North Americans in particular, tend to refer to unfamiliar foods as novel foods [76]. However, this terminology is inappropriate in the EU context. Here, novel foods are foods without a significant history of consumption within the community. Typical examples are genetically modified foods, foods with modified primary structures and foods based on microbial biomass. European food law requires such foods to be assessed for safety before being placed on the market [69]. Novel foods first attracted official attention in the early 1970s because of the widening use of textured vegetable proteins and proteins from microorganisms [77]. The ACNFP was set up around the time of the passing of the Food Safety Act 1990. Its first annual report featured genetically modified bakers' and brewers' yeasts, irradiated wheat, a novel fat replacer, chymosine enzyme and a fructose syrup containing dextran, as well as ethical considerations of food products derived from transgenic animals [77]. Among the new technologies scrutinised by the ACNFP to date, genetically modified foods are perhaps the ones that have proved the most controversial [78]. By contrast, there is considerable demand for fungal protein products in the form of the vegetarian meat substitute Quorn [79]. One of the products investigated by the ACNFP is the fat substitute Olestra, i.e., Derek Cooper's so-called unfood [68]. This was not approved for the European market despite the fact that approval was granted in the USA for its use in specified snack foods [80].

An interesting perspective on the essential nature of what constitutes food may be gained by studying how different food related professions interpret the concept. The legal view has been shown to focus on the safe ingestion of materials (other than medicines), for whatever purpose. Bender and Bender may be regarded as representing the nutritional view [74] (Table 2.2). Bender and Bender identify two types of essential food function, i.e., the fuelling of metabolic processes – such as breathing, circulation, digestion and work – and the supply of building blocks for body matter. One might add to this that good food in nutritional terms would also be highly digestible. However, as with food safety, ease of digestion may be related to the individual consumer and their established food habits as much as on the food itself. For example, diets rich in olive oil, i.e., so-called Mediterranean diets, are currently considered to be particularly beneficial for good health. However, travellers encountering this type of diet for the first time are advised that, when consuming large amounts of olive oil for the first time, one may experience mild intestinal problems [81]. The nutritionist understanding of the functionality of food, as expressed in [74], clearly is only a partial view. As it negates eating motivations, e.g., the psychosocial functions of food, this can lead to failures in programmes attempting to promote dietary change for health reasons [82]. Chefs typically focus on the hedonic, rather than the nutritional, attributes of food. Many of the so-called TV chefs would be hard-pressed to "go easy" on cream and similar rich ingredients, arguing instead that viewers would be right in spoiling themselves every now and again. Chef and restaurateur Albert Roux aims to entertain his customers and rejects the notion that he ought to offer nutritional value [83]. In fact, in planning his own food pleasures, Roux demonstrates how he himself con-

Table 2.2. Definition of food from a nutritionist perspective [74]

Any solid or liquid material consumed by a living organism to supply energy, build and replace tissue, or participate in such reactions

sciously weighs up any likely benefits and disbenefits deriving from specific foods and meals. He accepts that enjoyment may come at a price. One of Roux's favourite dishes is cassoulet, but he knows that when he eats it, he "will enjoy it, but I also know that I am going to be bloated" [83].

The past decade has seen the entry into the global food market of so-called functional foods. Self-evidently, all foods are functional in that they perform a variety of functions. However, the term functional food, a marketing concept, refers to a range of foods with certain common features. These foods are designed to provide not only essential nutrients and palatability but, in addition, specific physiological functionality with various roles in the preservation of human health and in enhancing performance [84]. Many of these have in common that they aim to prevent some of the degenerative diseases common in industrialised societies, e.g., heart disease and certain cancers. The concept of functional foods was developed in Japan, where amongst the first products that were approved were those from which unfavourable components had been removed (hypoallergenic rice, low phosphorus milk) and those to which protective factors had been added (dietary fibre, calcium, oligosaccharides, lactic acid bacteria). Functional foods are discussed in some detail in Chapter 6. However, the selection of foods for specific health purposes is not a modern idea. Specific food items have been traditionally used for medicinal purposes, and in some cultures there is no clear dividing line between food and medicine. Whilst in the West this connection has been largely severed, it still survives elsewhere. For example, in China a detailed and widespread knowledge of the healing properties of different dietary items has been maintained to date [85]. Here, many foods have a dual use, with medicinal foods including ginger, cinnamon, garlic, vinegar, eggs, sesame, mung beans and rice. Foods with beneficial health effects are selected depending on whether the body needs "heating" or "cooling", to remedy lack of appetite or energy, and to alleviate various illnesses. Western governmental health policy does recognise the role of good nutrition in good population health, and dietary targets have been set for certain types of food [86].

There is wide-ranging public debate about the nature of food, i.e, whether food is something people want to eat or whether it is something that provides nourishment; whether modern or scientific methods of production and processing are inferior or superior to traditional ones; and whether foods should be consumed in a particular manner and context, and in a particular frame of mind. There is an extensive debate about the so-called goodness and badness, and the naturalness and artificiality of modern foods. This debate is pursued in Part 3 of the book. Fresh, perishable foods may be perceived as possessing greater vitality or wholesomeness than foods that have been stabilised by refining (white flour vs wholemeal flour) or by otherwise processing them [87]. At one end of the perceived "good-bad food" scale tend to be located foods referred to as health foods, wholefoods or similar. Junk food is a frequently used, derogatory term for certain types of food. Perversely, and to illustrate a point, these so-called junk foods tend to enjoy high levels of popularity; they are good and bad at the same time. A dictionary definition of junk food [67] is given in Table 2.3.

As indicated above, individuals in each society select their diet from a much larger range of potential foods. Culture determines what materials are accept-

Table 2.3. Definition of the term junk food [67]

Food that has low nutritional value, typically produced in the form of packaged snacks needing little or no preparation

able as foods as well as the appropriate contexts for their use. Few food preferences are innate (see Chapter 7) and almost all are learned within the context of one's social reference groups, in particular, the family. Through their food choices and by what they accept and do not accept as food, people make statements about their group affiliations and, within a slightly narrower frame, express their self-identity. Rules about what is and is not food may be flexible, as in more secular and individualistic societies, or strict, as in some religious dietary codes. As the use of animals for food usually involves killing, religious codes for meat consumption tend to be particularly explicit Table 2.4 lists some food prohibitions or taboos based on religious dietary rules [73].

Whilst Jewish dietary law forbids consumption of eggs with blood spots, the rule may be waived in times of starvation. Food taboos may have their origins in food safety and hygienic considerations, but the value derived from them may evolve with time. For examples, where the original circumstances that created a particular taboo cease to exist, the taboo may transmute into a social tool, reinforcing group membership and hence, assisting the survival of cultures. The term foodways refers not only to food and cooking but to all food-related activities, concepts, and beliefs shared by a particular group of people [88]. Adherence to specific food taboos commonly marks an ethnic group as different from its neighbours [88]. A food culture that has had recent in-depth attention from researchers is that of the Cajuns of south Louisiana. In studying the foodways of this group, Gutierrez explored the realm between their food habits and their ethnic identity, discovering in the process that whilst Cajun identity is strongly defined by what they do eat, their food taboos too are identity markers [88]. Cajuns are aware that some outsiders categorise certain Cajun foodstuffs as inedible or repulsive, e.g., boudin, hogshead cheese, turtle, frog, alligator, raccoon, crawfish and squirrel. Several factors appear to motivate some outsiders (especially middle-class Americans) to avoid certain Cajun foods and food events. Some view certain parts of edible animals as inedible, e.g., pig or beef hearts, spleen, tripe, brains, stomach, or tongue. Others are offended by the vis-

Table 2.4. Some religious dietary prohibition in relation to meat [73]

Buddhism	Meat generally – strictly observed only by monks and devout laymen
Christianity	No prohibitions
Hinduism	Cattle
Islam	Blood; pork; etc.
Jainism	Prohibition against the taking of all life, even in its smallest form
Judaism	Blood; animals that do not chew the cud and do not possess cloven hooves; certain birds and fish; creeping things; etc.

ible killing and dissection of animals at boucheries, Mardi Gras, and seafood boils. These observations suggest that some people have a tendency to avoid as inedible or repulsive foods that are clearly derived from living animals, that are clearly "of nature" [88]. These include foods made from wild animals that are rarely or never brought into the sphere of culture by being bred and cared for by humans, parts of farm animals that look like anatomically functioning organs, and foods made from animals that look like animals even after they are cooked. From the Cajun point of view, this type of outsider appears squeamish or naïve. Unlike more urbanised people, many Cajuns commonly derive some of their food directly from the environment and are therefore familiar with those preliminary food-processing steps, such as slaughtering, skinning, cleaning, and butchering, which are obscured from the view of people who always buy their more fully processed foods at grocery stores. At the same time, Cajun food habits have undergone a reversal in value, signalling environmental competence, oneness with nature, fun and culinary skills and sociability. Many cultures feature esteemed foods that outsiders would consider to be repulsive. The term Balut describes boiled fertilised duck egg, an oriental delicacy esteemed for the range of textures that are revealed during eating [73]. Any attempt to cross-culturally market such foods would be likely to encounter strong barriers.

Naturalness in foods is a complex concept, but despite this, a focus on naturalness is exploited widely in food advertising. There have been official attempts to regulate the term natural, and terms substantially the same. However, "nature" is commonly regarded as the most complex word in the English language, closely followed by "culture" [89]. An original meaning of culture is the tending of natural growth. Culture implies the opposing concepts of artificial, or synthetic, and natural, i.e., what is done to the world and what the world does to people. A food being referred to as unnatural usually incorporates a value judgement.

2.3
Food Quality Attributes

Positive quality attributes are those that are present in a food as positive features or selling points, whilst negative attributes, i.e., deficiencies or defects, are those that are absent. For example, negative features in fruit and vegetables include disease, mis-shapes, blemishes, over-maturity and the presence of foreign matter. Another, quite different, example of a negative quality attribute would be an overly complicated product formulation, apparent to consumers via long lists of ingredients and additives as manifested on food labels [56]. It is true that sometimes less is more! Quality attributes assume the value of conformity or nonconformity, within tolerance or out of tolerance, complete or incomplete, or a similar dichotomy [90]. Any interaction between a consumer and a food involves the consumer considering and evaluating a range of quality attributes in the food. These attributes will contribute, in differing proportions, to the overall level of satisfaction derived from purchasing or consuming the product. The current section is strongly product-focused as it aims to define and discuss some key categories of quality attributes relevant to foods and beverages. It also serves as a bridge to Part 2 of the book, which is entirely focused on food attributes. In specifying product attributes, the analytical methodology to be used to verify compliance must be specified at the same time. The measurement and verification of these attributes often employs physical, chemical or microbiological analytical techniques. However, human assessors are required for the sensory evaluation of foods. Fortunately, any food can be characterised by its appearance/texture/flavour profile in a value-neutral manner, i.e., without any acceptance or liking being implied. In addition to the exploration of inherent food product characteristics, the current section also introduces signalling criteria [10], which are often referred to as quality cues or quality indicators, and their role in communicating product quality to consumers. Such cues form an essential part of commercial product specifications. Obviously, their effectiveness is measured using consumer research methodology rather than product analysis.

Qualities (plural) are the distinctive attributes or characteristics possessed by a product [67], with the term product property carrying similar meaning. Quality management system standard BS EN ISO 9000:2000 employs the term characteristic, which it further defines as a distinguishing feature [8]. Porter variously refers to product features, attributes and characteristics, as well as to buyer purchase criteria [10]. These buyer purchase criteria are defined as attributes that create actual or perceived value for the buyer. Porter divides purchase criteria into two types, i.e., use criteria, defined as specific measures of what creates buyer value, and signalling criteria. Use criteria include product features, conformity to product specification, delivery time and product support; signalling criteria include the outward appearance of the product, packaging and labels, price, advertising of product characteristics and brand reputation [10]. Other examples of signalling criteria, especially in the food market, include product endorsements and quality marks [26]. In consumer goods in particular, use criteria are also likely to include intangibles such as style, perceived status and brand connotation [10]. Branding can thereby be seen to play a dual role, i.e., as a direct source of consumer value and as a quality signal or cue. Signalling criteria are all the more important in circumstances where it is difficult for buyers to assess and monitor product quality attributes for themselves. This situation is typical of foods, in particular, pre-packed, processed foods. Signalling criteria are also important where foods are bought on behalf of others, e.g., a bottle of wine taken by a guest as a contribution to a dinner party. Contextual information providing associations, and so-called scripts detailing why and how a product might be used, both are signalling criteria. Products, which do not signal their value effectively, are unlikely to be chosen, and it follows that foods that are not consumed do not confer nutritional or any other benefits upon people. In contemporary food retailing, very little pre-purchase sampling is possible, nor can personnel normally be relied on to be able to answer quality related questions competently. Signalling attributes draw attention to products, raise awareness about them, and initiate communications about them. Buyer purchase criteria, once identified, should be ranked and should include the price [10].

Quality attributes are variously categorised as inherent or assigned [8], intrinsic or extrinsic, and tangible or intangible. Inherent means existing in something, especially as a permanent characteristic. In contrast, characteristics assigned to a product (e.g., the price or owner of the product) do not count as quality characteristics of that product. Similarly, the term intrinsic refers to the physical attributes of the product, whereas extrinsic attributes are typified by brand name, price and sales outlet [91]. Consumers find out about the quality of the foods they buy in different ways, depending on whether they are evaluating search, experience or credence products or attributes. Search attributes allow quality to be evaluated prior to purchase or consumption, whilst experience attributes do not [91]. Except in situations where sampling is permitted, the flavour of a food is an experience attribute. On the other hand, the amount and distribution of visible fat or gristle in a piece of raw meat represent search attributes, but only if the consumer knows how to interpret them in terms of eating performance. Presence of a certain amount of intra-muscular fat is required if meat is to remain juicy and tender after cooking. This is a fact that is well appreciated by professional cooks and chefs. However, visible intra-muscular fat actually reduces consumer expectations of meat quality [92]. What serves as a search attribute for one individual does not necessarily do so for another person. It all depends on individual food knowledge and skills. Credence characteristics are those attributes that consumers cannot verify for themselves. In this case, whether a purchasing or consumption decision is made will depend on the credence given to the signalling criteria. In the case of food, this refers especially to food labelling information. Advertising commonly relies on credence. An example is the current TV advertising campaign involving a fat spread called Olivio. This consists of a series of mini dramas featuring groups of rather jolly and highly amorous, yet elderly, Italians. Although no explicit claims for the spread are made, the implication is that Olivio offers optimum nutrition and counteracts the ordinary frailties of old age.

The example of cue utilisation in respect of the visible fat and gristle in raw meat [92] illustrates the fact that consumers infer certain quality attributes from other attributes. In this way, it turns out that what one might think of as a use characteristic, i.e, fat content, takes on the mantle of a signalling characteristic, i.e, it cues textural attributes. French, German, Spanish and UK consumers use a number of similar cues in choosing beef. These cues relate to meat cut, colour, odour, use-by date, display hygiene, packaging, price, weight, visible bone, fat and blood vessels, the colour and consistency of the fat, and whether the meat is fresh or frozen [91]. Schröder and Horsburgh investigated the expected eating quality of pre-packed pork sausages in terms of cues consumers received from packaging [56]. Brand name was a key factor in subjects' expectations of eating quality, as was the quality of the packaging, i.e., in terms of product presentation, artwork, and tamper resistance. Ingredients lists constituted a further significant cue regarding eating quality. The presence of (large numbers of) so-called E Numbers (food additives) and of ingredients that were perceived as alien in the context of the product (whey protein), led to reduced expectations of eating quality. The packaging for the sausages was found to communicate more than specific product characteristics. It also delivered cues about manufacturer and/or retailer competence and integrity. Careless, as well as inappropriately sophisticated, pack design elicited negative attitudes about a product. However, subjects were highly receptive to quality marks and claims, although the signif-

Table 2.5. Key quality attributes applicable to meat [56, 92, 93]

Selling points	Defects	Quality cues
Tenderness	Pale soft exudative (PSE)	Country of origin
Juiciness	Dark firm dry (DFD)	Place of purchase (butcher,
Clean, typical flavour	Visible fat and gristle	supermarket)
Colour (cue flavour)	Boar taint (pork)	Quality assurance schemes
Leanness (cue flavour)	Blood splashes	and associated marks
Freshness (cue safety)		

icance of these was rarely understood. The visual evaluation of a food, even at point of purchase, is more than simply a cool headed quality screening mechanism. Rather, it should be seen as the first stage in the food consumption process itself in that it has a major impact on eating related mental states such as appetite (see Chapter 7). Table 2.5 lists some key quality attributes applicable to meat, based on three published papers [56, 92, 93].

Grunert takes the origin of meat and the type of animal production system used, to mean extrinsic quality [91]. However, this point of view is arguable because, according to the BS EN ISO 9000:2000 definition of quality [8], both are permanent characteristics of the product. The convention adopted in this book, in line with the international standard, is for "inherent" to mean all attributes relating both to the product and the process of production. Clearly, this includes the provenance of a product.

The last of the common, quality-related dichotomies to be explored here is that of tangible versus intangible attributes. This will be helpful in illum – inating Grunert's view of provenance as an extrinsic quality cue. As tangibility is generally understood to mean having a physical presence [67], literally, being capable of being touched, this might lead one to focus on a physical product. However, increasingly process as well as product matters to food consumers, as exemplified by the current debate about meat production and the role of animal welfare and feeding regimes [30]. Process specifications are just as real, and just as verifiable, as product specifications. The processes of food production are important in quality because palatability is determined not only by sensory properties but also by perceptions about what constitutes proper food production, e.g., proper feed, and the fulfilment of basic animal needs (freedoms). These processes are therefore also to be regarded as tangible. A typical example of an intangible quality attribute of a food product would be its brand image. Therefore, the concept of intangibility relates to cueing and signalling and is equivalent to attributes that may also be described as assigned or extrinsic.

The dividing line between tangibility and intangibility can be somewhat fluid. The quality attribute "availability" serves to illustrate this point. On the one hand, availability may refer to a specified stock level for some product. This type of availability is tangible in the sense that it can be controlled through process management. On the other hand, the availability concept may pressed into service as an indicator of the apparent popularity of some product. For example, the massed display of a new product can be exploited to suggest strong demand for it, which then becomes a self-fulfilling prophecy. The tactic was employed, with considerable success, during the recent market introduction of the juice drink Sunny Delight [94]. In a new product that is immediately highly visible, widely available and appearing to be popular, risk perceptions normally associated with first-time purchases are reduced. The massed display of a product therefore signals to the shopper endorsement by other consumers and may be equally, if not more, effective than celebrity endorsement or the use of quality marks. Holbrook does not distinguish between tangible and intangible products or product attributes [12]. By the time quality attributes have been translated into consumer value, they will all have acquired various dimensions of intangibility, as by then they will all be bound up with feelings and emotions.

Core quality attributes, in particular, absence of toxins and microbiological safety, apply to all foods; equally, all foods can be defined in terms of their composition, nutrients present, sensory performance and other aspects of functionality [95]. Added-value attributes might include authenticity of ingredients and recipes, convenience, novelty and rarity. A further dimension of added value, which is concerned with process more than with product, is ethical production. Whilst added value in this context stands primarily for added value in the consumption of such products, these products will tend to be more expensive than their standard versions [95]. Where added-value quality attributes are present in a particular food, and where these are accurately communicated to consumers, the conditions for product differentiation exist. In positioning a product in a particular market, its relationships to other products in the same category or market are defined. In focused products, all quality attributes are expressly targeted towards a particular market segment. Use of flavourings, rather than the

Table 2.6. Food quality attributes: product-focus

Composition	Dietary fuels	Fats, carbohydrates and proteins
	Dietary fibre	Fibre generally, prebiotics
	Nutrients	Minerals, vitamins, essential fatty acids and indispensible amino acids
	Bioactive compounds	Probiotics and flavonoids
	Chemical additions	Additives and other chemical additions, e.g., nutrient supplementation
Contaminants	Limits or exclusion	Pests, microorganisms, chemicals and foreign matter
Performance	Sensory	Appearance, texture, aroma and taste
	Shelf life	Safety, freshness and peak condition
	Performance in use	Effectiveness, e.g., spreadability of butter, or, melting, stretching and coverage of (mozzarella) cheese topping for pizza
	Weights and measures	Amount (weight or volume), size (e.g., eggs, portions), counted number

Table 2.7. Food quality attributes: process-focus

Service	Predictability	Batch-to-batch conformity
	Versatility	Substitutability and complementarity
	Product support	Customer service, recipe suggestions, instructions for use
	Stock control	Mixed packs of "eat now/keep" tropical fruits, e.g., mangoes, papayas and bananas; Seven single-shot portions of probiotic beverage (Yakult)
	Availability/access	
Provenance	Place of origin	Country, region
	Animal breed	Aberdeen Angus (meat breed) beef as opposed to beef from dairy cattle
	Production system	Conventional or organic, ethical production systems
	Place of purchase	Independent retailer, supermarket

equivalent basic ingredients, in products positioned as premium foods constitutes a possible source of disappointment. This is especially true if foods are positioned to emulate restaurant quality. Examples encountered by the author include a retailer apparently using onion powder in a very expensive beefburger,
butter flavour biscuits with a whiff of rancidity (presumably due to the use of
flavourings) about them, and the use of garlic oil in pickles conferring a vaguely
dirty flavour impression. Where individual products in a food category differ
from each other in one or several quality attributes, but these differences are not
communicated, differentiation cannot truly be said to exist. An example of this
has been discovered in the UK market for apple juice where, although each of the
major retailers stocks many different brands and varieties, substantial differences between them, especially in terms of sensory attributes, are not signalled
effectively to prospective consumers via the packaging [24].

Specific quality attributes apply to individual products and reflect both the
food and the consumer requirements of that food. Table 2.6 provides a general,
categorical overview of such attributes, focused on the product. The organisation of this table broadly reflects the structure, and anticipates the subject
matter, of Part 2 of the book. A selection of process-focused food quality
attributes is shown in Table 2.7.

3 Theoretical Perspectives on Consumer Behaviour and Food Choice

3.1
Introduction

The field of Consumer Behaviour is fundamentally concerned with the study of the nature of market exchanges between consumers and their immediate suppliers. It may be thought of as a toolbox more than as an academic discipline in its own right. This is because it draws for its theoretical underpinning on the major established social sciences, in particular, economics, psychology and sociology. To gain a proper understanding of consumer behaviour one has not only to study behaviour itself but also the motivation that lies behind it. As mental states are not amenable to direct measurement, ultimately, most consumer research begins with the observation of actions. In economics, this is where things may be left. Here, the observed action is unambiguously measurable as the demand for a product, and simple rules then link this with supply factors and with the resources the consumer has available. Other observable characteristics relevant to consumer behaviour include social class and demographic group. In psychology, the observed action is merely the first impulse for a much deeper analysis of the causes of the action. Behaviours thus become vehicles used to infer mental states, e.g., motivational states and associated cognitive factors, e.g., attitudes. Unlike economics, the psychological approach focuses not only on observed behaviour, but is also particularly interested in preferences and behavioural intentions, e.g., purchase intent. The economist's view is that preferences are revealed through behaviours. Clearly, actual behaviours can reveal preferences only within the context of opportunity sets. Economics therefore does not address the issue of behaviour barriers to consumption for specific items, e.g., a healthy diet. Such barriers may lie in ingrained individual consumption habits or in household routines rather than expressing true preferences for a product. Strictly speaking, psychology is concerned with personal factors that act upon the person so as to cause them to behave in specific, largely predetermined ways. However, individuals' personal psychology is influenced by the features and norms of the society within which they were raised, whilst later in life, peer group and other social pressures continue to influence consumption choices. Sociology and social psychology focus on the social aspect of consumer behaviour motivations. The theoretical approaches of these disciplines may be drawn on to answer questions about how society acts through consumers. In summary, the main objectives of Consumer Behaviour as a discipline are the observation and interpretation of consumption related behaviours and the prediction and influencing of consumer choices.

Consumers are, by definition, the ultimate users of goods and services and constitute the final link within supply chains. Food consumers eat food, drink beverages, chew gum and cook with food ingredients. Consumption may also occur on behalf of other, e.g., when a consumer obtains food which is to be eaten by somebody else. Food related consumption value, or disbenefit, does not derive solely from the actual eating of a food, but may lie in aspects of acquisition and disposal. Acquisition value refers to shopping and food preparation and to food provisioning from household stock (larder) foods. Food is also routinely acquired through eating out and it can be grown in domestic gardens. Food items may also be received, or given, as gifts, and they may form a focus for celebrations. Disposal value may exist in relation to nuisance or waste materials such as packaging, bones and other inedible parts of foods. In many households, one member will take on the role as the main food shopper and meal provider. This role may eventually grow into that of a gatekeeper. In that case, rather than shopping according to detailed instructions or lists, the person doing the shopping will tend to make assumptions about what will be suitable for the actual consumer of the food. Food choices, i.e., what to consume and when, and in what amounts, may respond to nutritional requirements or to some other kind of desire. Consumption may be consciously deferred, perhaps to increase the anticipation of enjoyment. On the other hand, prospective future health benefit, i.e., the avoidance of obesity and degenerative diseases, may be chosen over present gratification. It is very common for consumers to act against their better judgement when making food choices. Fieldhouse acknowledges that, in his work as a nutritionist, he often failed to effect the desired dietary changes in his patients [82]. He attributed this failure to a prevailing lack of appreciation of the non-biological meanings of food at the time. This is clearly not the sort of issue that the natural sciences can solve in isolation. On the other hand, the social sciences are well placed to offer a fuller understanding of the factors underlying food choice.

In the 1990s, The UK Economic and Social Research Council funded "The Nation's Diet: the social science of food choice", a research programme that was designed to foster greater coherence in the social scientific approaches to food choice. Projects covered a diverse range of topics, including microeconomic influences on food choice, cultural perspectives on food consumption, and the formulation of individual beliefs and attitudes and their impact on food choice. The disciplines represented in the programme included economics, human geography, psychology, social administration, social anthropology, sociology, as well as educational studies, marketing (as a branch of business studies) and media studies [96]. As previously indicated, the three social sciences that are most commonly discussed as contributing to the discipline of Consumer Behaviour are economics, psychology and sociology [97]. This convention will be followed in the present chapter, with some discussion of the anthropological aspects of food choice included in the section on sociology. Specific issues with relevance to food choice will be highlighted wherever possible. There will thus be no evaluation of food choice in the light of generic Consumer Behaviour theory. In fact, Fine and Leopold criticise Consumer Behaviour for its lack of success at integration and generation of independent grand theories and for being "parasitic upon" its contributory discipline [98]. However, there certainly is major overlap between psychology and sociology in terms of the causes of human behaviour. As Lewin points out, sociological influences deeply affect the psychology underlying economic behaviour [99]. In addition, rationality, one of the key concepts within economics, is seen as a psychological interpretation placed on observed behaviour. As will be demonstrated, consumption may be understood both as a form of social structuring and as an identity constructing activity [100].

The market may be understood as an enabling mechanism, with social interactions, intangibles and symbolic aspects being involved as well as the actual purchase of the product [98]. Markets are not static. Food is an example of a routine consumption item, where, if value is not perceived on consumption, repeat purchasing will be unlikely. Ever-increasing competitiveness in markets for foods has meant that consumers are afforded both greater opportunity for, and greater complexity of, choice. Product categories may come and go with changing perceptions about products. At the present time, the UK market for low-fat yoghurts continues to decline, matching equally pronounced growth trends in the so-called virtually fat-free yoghurt sector [101]. Low-fat yoghurts are becoming redundant because they do not appear to offer any significant advantages over virtually fat-free ones. New food categories, e.g., functional foods, are introduced to sell a concept, in this case, health benefits above and beyond those conferred by what is conventionally understand as nutrients. Within product categories, there is proliferation of brands and varieties. Providing the right kinds of choices for consumers is a major challenge for food suppliers, but making satisfying consumer choices also requires skills. In the end, as in any market, the supply and demand sides work together. But the interdependence of both goes beyond this. For example, the manner in which products are actually used by consumers, if communicated to the supplier, can provide value input into product development activities. In fact, it is highly advantageous for food suppliers to actively canvas consumers about such feedback. Consumer orientation in markets leads to segmentation, where each segment addresses increasingly focused consumer requirements. Consumer markets can be segmented by population profile (demographic, socio-economic and geographic) or by psychographic (lifestyle, personality) or behavioural criteria (benefits sought, purchase occasion, purchase behaviour, usage, perceptions and beliefs). All markets that are segmented initially by product category, e.g., raw chilled chicken, orange juice or ready meals, can be subdivided according to the type of consumer value. The consumer value focus within the food market finds expression in more or less self-explanatory marketing terms and prefixes, e.g., fast-, convenience- and health-. Others deliberately generate uncertainty and mystery in order to gain public attention, e.g., functional foods.

The chapter starts with the economics perspective on consumer behaviour. Economics aims to uncover consumption rules mainly in relation to resource allocation, with little emphasis on specific motivational factors. Theorising about human nature is based on the assumption of so-called rational man, who acts to maximise consumer value (referred to here as utility) to himself across a range of consumption choices. Out of this comes individual choice which, in turn, is aggregated over all the consumers of the product to give the market demand for that product. Economic man must accomplish the maximisation of utility within the bounds of resource limitations such as income, time, information and skills. Whilst this chapter will concentrate on mainstream, or neoclassical, theory, some attention is given to certain alternative approaches, in so far as these attempt to integrate aspects of economics, psychology and/or sociology. The so-called rational consumer of neoclassical microeconomics is seen to weigh up the various costs and benefits implied in all choices. Whether a cost-benefit analysis is done explicitly or implicitly, i.e., relying on well-used heuristics, depends, in part, on the search costs attached to acquiring information. Most importantly, rationality requires that the consumer must correctly identify the most valuable alternative that is sacrificed in pursuit of any given activity.

Psychological factors underlying consumer behaviour are examined next. Since important factors such as tastes, preferences and motivations cannot be observed directly, they have to be inferred. This is done either from actual behaviours or, where there are no observable actions, from statements obtained from subjects. One of the most important aspects of psychological analysis is the attribution of the causes of behaviour, as these are part of the motivational framework. Thus, dispositional (internal) causes tend to describe more settled, stable behaviours than do environmental, or situational, causes. Human nature in psychology is seen as the result of an interplay between biological and genetic inheritance and nurture [96]. Individual human development as the result of nurture points both towards psychological and sociological mechanisms. This can be illustrated with learning theory, which has both associative and cognitive (psychological), and social components. Psychology-oriented consumer studies currently tend to be dominated by cognitive approaches [100]. However, the full range of motivational factors and constraints in consumption choices may be investigated using psychological theories and methodologies. Among the social sciences, psychology is alone in the adoption of experimental research designs on any scale [96]. Consumer psychology is concerned with individual factors influencing consumption. The role of the human senses in perceiving both the internal (bodily) and external environments is crucial here. This particular aspect of psychology, together with the related physiological aspects of food consumption, is discussed in depth in Chapter 7.

The final section of the current chapter examines the role of reference groups and cultural filters in food consumption behaviours and includes both sociology and social anthropology. Foods can act as markers of cultural identity and as indicators of cultural differences [102]. Sociology-oriented consumer studies have traditionally conceptualised consumption as due to cultural and social structures, e.g., family, social groups, group processes, social class, culture and subculture, status, and lifestyles. More recently, there has been a shift towards the role of consumption in actively creating such structures [100]. Anthropology is concerned with the meaning of consumption as ritual, sociology with culturally constructed use value in relation to status or social position [103]. The anthropological perspective is bi- or multi-lateral as opposed to global, and marginal instead of central. The anthropologist may be said to take up a position as a marginal native [103].

As approaches to analysing consumer behaviour differ somewhat between academic disciplines, so does terminology. Thus the psychologist's choice, once aggregated over a population, becomes the economist's demand. However, there are many parallel concepts between individual disciplines. The discipline of economics, for example, argues that consumption bestows utility; in sociology, it may well be status or social position; in psychology, it is a conditioned response to gain a level of wellbeing; and in anthropology, it has been interpreted in terms of its symbolic role in rituals [98]. The economist's (marginal) utility becomes the marketer's (and the philosopher's) consumer value. Marginal utility, and marginal rates of substitution, relate to the concept of variety-seeking in psychology; in the context of food consumption, this may mean sensory-specific satiety. The economist's bounded rationality translates directly into the psychologist's cognitive limitations. In terms of research methodology, economists tend to analyse someone else's primary data, whereas both psychologists and sociologist are likely to collect their own, the former referring to their sources as subjects and the latter as informants or respondents [96]. Figure 3.1 presents schematically the contributions of food science, economics, psychology and sociology to the study of Consumer Behaviour in relation to food choice.

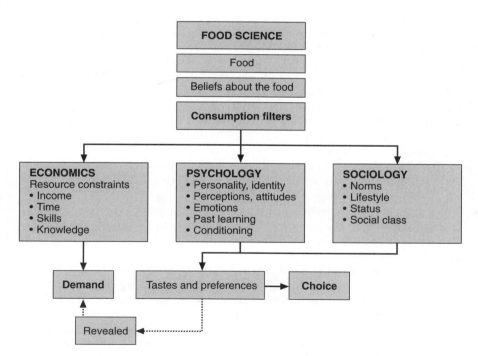

Fig. 3.1. Contributions of food science, economics, psychology and sociology to Consumer Behaviour in relation to food choice

3.2
Microeconomics of Consumer Behaviour and Food Choice

Economics is concerned with the study of what constitutes rational resource allocation among alternative choices in order to satisfy given wants. The contemporary mainstream perspective on the discipline is known as neoclassical, and
the field of economics that is particularly relevant to this book is microeconomics. The subject matter of microeconomics is the economic behaviour of individual economic units, including consumers and households [104]. In choosing
among alternatives, consumers supply the economist with tangible data based
on observable actions. For example, a good has been bought. What is not of direct interest to economic analysis in this context are the causes of, and motivations for, such behaviours; nor are consumers' future intentions in respect of the
purchase of a particular good investigated. It is not feasible to incorporate all
motives underlying behaviour into economic theory [99]. One of the key determinants of a consumer's behaviour is his or her tastes or preferences. Factors affecting these include age, gender, education, advertising and demonstration effect (observing the choices of fellow consumers) [104]. Preferences will normally be well established for goods if these have been frequently consumed.
Values and tastes are developed incrementally and the preferences that emerge
are influenced, through reinforcement, by the process of choosing. This learning
mechanism is typical of the emergence of specific food preferences. For the purposes of economic analysis, values, tastes and preferences relevant to a particular choice are taken as givens. These reveal themselves through economic
choices, e.g., purchase and consumption (Fig. 3.1). The theory of revealed preferences also avoids the problems associated with questioning consumers directly about their preferences [104]. Preferences may be established by varying
the prices of market baskets, e.g., the dishes on a restaurant menu, and noting
the effect on consumer choices. Ultimately, individual consumer choices translate into effective demand in the market place.

Three key assumptions are made about the nature of preferences in economics [104]. Faced with two or more so-called bundles of goods, consumers decide
which they prefer or whether they are indifferent between them. This is a familiar scenario in a supermarket or restaurant. Here, the shopper or diner might be
faced with having to allocate some monetary budget to best effect. The example
also serves to demonstrate that money does not constitute the only type of budget relevant to resource allocation issues. For example, if the shopper or diner
happened to be following some low-calorie diet, they might be allocating a
(maximum) calorie budget through judicious picking among alternative foods.
The second assumption about the nature of preferences is transitivity. To illustrate, a consumer who prefers Coca Cola to Pepsi Cola, and Pepsi Cola to some
retailer's brand of cola, must also prefer Coca Cola to that retailer's cola. Whilst
the assumption of transitivity may convince in this example, this is not necessarily always the case with foods. One of the reasons for this is that the present
desirability of a food is influenced by when it was last consumed. Factors such
as deprivation effects and sensory-specific satiety are likely to modify preference rankings. For example, a meat eater may nominate rib roast of beef as his

most desired meat, followed by beefburgers, pork chops and, finally, roast chicken. At the time of stating these preferences, this individual may not have eaten meat for some time, or they may have eaten a disproportionate number of roast chicken in the recent past. If rib of beef were going to being served to this individuals at several consecutive meals, it would be likely to lose its top ranked position either temporarily or permanently. The context of consumption will also influence which of the meats would be preferred, including other foods available for eating with the meat. For example, rib of beef might not be perceived as particularly enjoyable if the only accompaniments available were a salad and a bun; on the other hand, a beef burger probably would. The third assumption is that more is always preferred to less. In the context of food consumption, this too is controversial. Thus, a desirable food loses its appeal if an unmanageable portion of it is served. UK visitors to continental European countries may be overwhelmed when presented with a whole chicken (Brathendl in Austria), or a whole leg of pork (Eisbein in Germany), as a single portion. Again, sensory-specific satiety is an issue. Clearly, with food, more is not always better than less, especially when the actual consumption cannot be deferred. However, the third assumption would normally apply to food purchasing, especially when dealing with items with a reasonable shelf life. One question that would need to be asked in this context, is exactly how much more would be better. This would need to be answered on a case-by-case basis.

The usual definition, in economics, of a good is something consumers are willing to pay for, and a bad is something consumers pay to have removed or must be compensated to accept [105]. Attached to goods is utility, and to bads, disutility. Markets consist of the buyers and sellers of a good. Over the years, economists have increasingly recognised that even subtle product differences matter, and the trend in analysis has been toward ever narrower definitions of goods and markets. Two otherwise identical products are often classified as separate if they differ only with the respect to the times or places they are available, e.g., a drink of water in the desert [106]. An alternative expression to goods in economics is commodities. Goods, or commodities, are commonly defined by two primary attributes, i.e., their exchange value and their use value [98]. Neoclassical economics sets up a relationship between these two types of value, i.e., the ultimate (marginal) utility provided by the commodity is the explanation for its price [98]. Intuitively, economists have long understood preferences as being defined over general categories of needs, rather than over particular individual commodities [99]. According to Lancaster, it is not the commodity as such that brings satisfaction; instead, utility is derived from the configuration of individual attributes of the commodity [107]. Therefore, consumers base their decisions not on a product itself but on product attributes. This position mirrors that outlined by Holbrook [12]. The Lancaster approach has been used to explain purchasing behaviour for various products. In the case of breakfast cereals, it has been possible to predict the amount that consumers would be willing to spend based on the composition [97]. This focus on product characteristics is thought to be particularly appropriate to the marketing context [108]. Economists have emphasised that, by transforming market goods into the desired consumption commodities, the consumer produces his or her own commodity set. Though market goods may already exist as inputs in consumption, still the output (or outcome) is not inscribed in them once for all, but is the result of multiple, as yet unknown, combinatory consumption possibilities [109]. Stigler and Becker argue that the basic units that enter the utility function are non-market goods (Z goods), e.g., nutrition, that are produced by the household, rather than individual commodities, e.g., meat [110]. So-called household production functions describe the process by which commodities, combined with time, physical capital and human capital (defined as human skills and knowledge), are converted into these Z goods. Consumers maximise utility subject to the household production functions and a time constraint as well as incomes and prices. The model brings to the fore the importance of factors that are ignored the standard approach, e.g., demand for products such as convenience foods.

Preference relations may be described by means of a utility function, which expresses the preference ordering of bundles of goods – or of attributes in a type of good (Lancaster). A utility function $u(x)$ assigns a numerical value to each element X of a preference bundle, ranking the elements of X in accordance with the individual's preferences. There is a difference between total utility and marginal utility [104]. Marginal utility is the more important concept, as it expresses the additional satisfaction derived from consuming an additional unit of something. Marginal utility is therefore regarded as a key element in determining the level of consumption [99]. According to the law of diminishing marginal utility, the additional satisfaction declines with continuing consumption [104]. Presented with a set of alternatives, consumers have the option of a number of choices, all of which may be satisfactory. An indifference curve is a set of bundles of goods or product attributes that are preferred equally [106]. These curves are constructed from the observation of empirical data without knowing the consumers' utility. Figure 3.2 gives a theoretical example of an indifference curve for a consumer who has to choose between expenditure on indulgence foods for home consumption and restaurant meals.

An important property of a consumer's preferences is the rate at which they are willing to exchange one good for another [106]. This is represented at any point on any indifference curve by its slope at that point, and is called the marginal rate of substitution. This diminishes in the sense that for any pair of goods, the less a consumer has of one good, the more they must be given of the other good in order to give up a unit of the first good. Preferences in which goods are perfect substitutes, e.g., two similar varieties of bread, exhibit a constant rate of substitution.

The standard model of rational behaviour dominates demand analysis in economics. The model is normative, as opposed to positive or descriptive, in that it describes what consumers should choose if they were being truly rational. In reality, behaviour often deviates from the prescriptions of the rational choice model, as will be demonstrated later. However, the rational choice model can play an important role in guiding people towards better decision-making [106]. The overall motivation underlying consumer behaviour in economic theory is the desire to choose to best advantage, i.e., rationally. The concept of utility stands as a proxy for the degree to which an individual has reached whatever

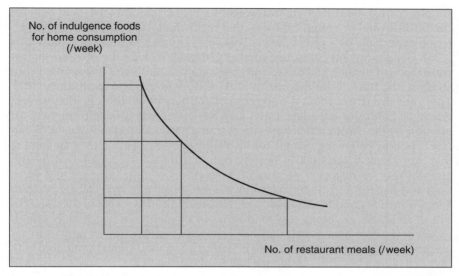

Fig. 3.2. Indifference curve for home consumption of indulgence foods and restaurant visits

goals he or she seeks [99]. Rational choice in economics implies the maximisation of utility within given constraints. The priority assigned to individual utility, both as an explanatory factor and as a desirable outcome, also gives rise to the idea of consumer sovereignty [98]. Choices are often made without advance knowledge of the consequences, e.g., when purchasing a particular food for the first time. Such choices are said to be risky choices. Since people do not always enjoy later (eating a food) what they prefer now (buying or ordering the food), there are two notions of utility in modern economics, i.e., experienced and decision utility [111]. The experienced utility of an outcome is a measure of the hedonic experience of that outcome, whilst its decision utility is the weight assigned to the outcome in a decision. As the model of rational consumer choice relies on the assumption that individuals have a desire to choose wisely, some form of cost-benefit calculation regarding alternative choices is implied by it. There are two important refinements of the definition of rationality, i.e., the self-interest standard and the present-aim standard [106]. The self-interest standard of rationality suggests that people are selfish. Accordingly, rational persons assign significant weight to only those costs and benefits that accrue directly to themselves. The present-aim standards of rationality requires only that persons act efficiently in the pursuit of whatever aims, objectives, motives or emotions they happen to have at the moment of action. This standard allows for emotional and altruistic behaviours to be classed as rational.

There is a simple decision rule for carrying out cost-benefit analyses. If $C(x)$ denotes the costs of doing x and $B(x)$ the benefits, it is: If $B(x) > C(x)$, do x; otherwise don't [106]. The relationship between costs and benefits is reciprocal, i.e., avoiding a cost equates to receiving a benefit. Whilst the decision rule itself is simple, correct identification and weighting of the relevant costs and benefits can be difficult. For example the cost that really matters is the most valuable alternative forgone by a given choice. This cost is known as the opportunity cost [106]. Constraints on consumer choices generally include the obvious budget constraints of income and prices, but also cognitive, skills and information constraints, and time, technical, institutional and socio-cultural constraints [107]. There are also issues about the object(s) of any benefit or cost, e.g, whether self, reference groups or outsiders. Costs may be deliberately hidden from view, for example by the way in which advertisers frame product messages, e.g, "85% fat-free" biscuits. Following careful consideration, a biscuit with a 15% fat content may not seem a particularly healthy option after all. Ordinary cost-benefit analysis may be distorted because of mismatches of time, saliency and scale. Cost and benefit may be separated by a substantial time interval, e.g., dietary habits aimed at good health in old age. This example also serves to illustrate the role of vividness, i.e., being easy to imagine, in a benefit or a cost. Finally, one element of the cost-benefit pair may be perceived as having impact only in an aggregate sense, e.g., the cost to health of eating a bad food (once) compared with that of having a bad diet (eating the bad food habitually).

Consumers generally do not allocate their resources efficiently. When confronted with complex choices, they tend to search in a haphazard way for potentially relevant information, and usually quit once understanding reaches a certain threshold. March and Simon consequently describe consumers as satisficers, rather than as maximisers, of utility [112]. The perceived importance that the right decision is made will determine the desired level of rationality and the search costs that consumers will be willing to expend. Earl notes the following limitations of rationality: a limited capability for mental processing; a limited number of alternatives that can be considered concurrently; limits to information processing capability; ignorance; and uncertainty [108]. Rationality is therefore thought of as bounded, rather than global, and choices likely to be suboptimal. A prime example of people systematically violating the prescriptions of the rational choice model is that of sunk costs. Because they are not recoverable at the time of choosing, the rational consumer would be expected to ignore such costs. Frank illustrates this phenomenon as follows [106]. An all-you-can-eat lunch at a pizza parlour cost $3. In a particular experiment, some customers had their money refunded before ordering, whilst others had not. It turned out that those who had received refunds ate substantially less than those who had not. Clearly, the two groups had different motivations for their behaviours. Both might initially have chosen the lunch with the intention to fill themselves up with food at a good price. However, with the price factor removed, individuals might have felt free to relax, ignore the all-you-can-eat expectation, and eat only until they felt satisfied. However, those who had paid for the food, would have felt rational in eating all they could, thinking they must get value for money as originally planned. People may not always choose rationally, but they do like to perceive themselves as rational, smart individuals.

Supply and demand analysis is the economist's analytical tool for explaining the prices and quantities of goods traded in markets. The demand curve for a good is a summary of the various cost-benefit calculations that buyers make with respect to that good. Preferences translate into demand, i.e, the quantity of a good or service which purchasers will be prepared to buy at a given price and in a given time interval. Individual purchase decisions add up to so-called demand curves. Normally, the quantity demanded of a good increases as the price falls, and decreases as the price increases. However, if current rarity, or exclusivity, of a good constitutes a significant part of its consumer value, a drop in price may reduce demand from the current customer base, despite the overall increase in demand. This situation is typical of so-called fashion goods, where a high price may form part of a consumer's preference bundle. People may also prefer expensive goods because they find it convenient to use price as a heuristic for quality. Figure 3.3 shows a theoretical demand curve for rib of beef. The curve is downward sloping, i.e., as the price of rib of beef falls, people buy more of it. Whilst demand implies preference, it does not necessarily mean satisfaction, because all possibly options might be unsatisfactory in an absolute sense.

The income-consumption curve summarises purchase responses to variations in income [106]. Figure 3.4 shows a theoretical income-consumption curve for low- and high-grade minced meat. It can be seen in this figure that, as income increases, the budget constraint moves outward. The income-consumption curve traces out how these changes in income affect consumption by joining up all the points at which budget constraints are tangential to indifference curves. In Fig. 3.4, as income increases, the amount of low-grade mince consumed also increases. However, with further increases in income, it is increasingly substituted for high-grade mince. The low-grade mince initially behaves as a so-called normal good, but turns into an inferior one when its consumption starts to decrease in response to increased income. The prototypical inferior good is one for which there are several, strongly preferred, but more expensive substitutes [106].

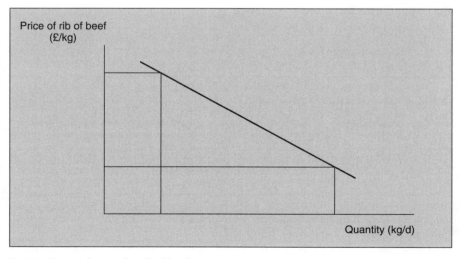

Fig. 3.3. Demand curve for rib of beef

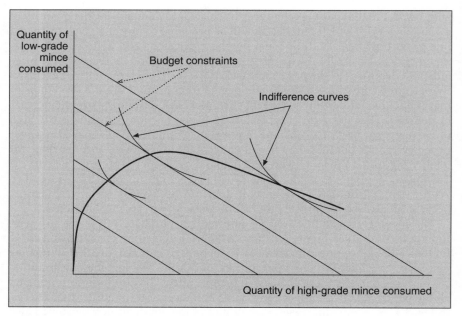

Fig. 3.4. Income-consumption curve for low- and high-grade minced meat

The economic analysis of food choice has a long history, dating back at least to the work of Ernst Engel on the analysis of household budgets in the mid-nineteenth century [113]. The so-called Engel law, which states that the poorer the family, the greater the proportion of its total expenditure that must be devoted to the provision of food, has proved remarkably robust [113]. In Engel curves, the expenditure for a food is plotted against income. Ritson and Hutchins show Engel curves for beef, pork, broiler/chicken and other poultry [114]. These are derived from National Food Survey data. To a surprising extent, products become inferior goods at high incomes. Figure 3.5 is an approximation of the Engel curves for chickens and for poultry other than chickens.

There are several possible explanations linking rising income with decreasing expenditure on meat as discovered in the National Food Survey. One of these derives from the fact that the survey only investigates household food consumption. The first explanation therefore is that high income, and high-income jobs, are associated with less home cooking and more eating out. Secondly, income level may be related to tastes and preferences, both in terms of the actual food and of the amount of food consumed. Individuals occupying the wealthier part of the income scale may exhibit greater sophistication and/or variety in their dietary habits. They may eat more seafood and other alternatives to meat, e.g., pasta. They may be eating more vegetables and cutting down on meat, perhaps in pursuit of a trimmer body shape or better health. Whilst economic analysis does not answer these question, it certainly does raise them, something which in itself is useful. A behavioral policy in regard to some action is a rule or

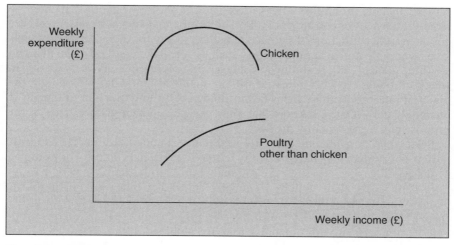

Fig. 3.5. Examples of Engel curves

principle if it overrides cost-benefit calculation [115]. In routine consumption particularly, consumers will eventually internalise the rule of the game without necessarily being aware of the economic theories underlying it. Whilst customs, rather than deliberate cost-benefit analyses, are often the underlying causes of economic behaviour, these customs owe their very existence to their efficiency [99]. Cultural norms and parental control of their children's sweets consumption are imposed; whereas we may construct rules for our own alcohol consumption or calorie intake [115].

Extensions to the basic economic model take into account risk and uncertainty in decision making. Information is an important input into decision making. However, no matter how much time and energy is spent gathering information, most choices must be made without complete knowledge about the relevant alternatives, i.e., decision making is often a gamble [106]. Some product information cannot be obtained in advance, e.g., whether one will like a food not previously tasted. Before 1975, UK consumers did not drink fat-reduced milk, the reason being that it was not being supplied. Since its introduction, demand has increased steadily, year on year, as consumers substituted the product for traditional wholemilk. Where information can be obtained prior to consumption, but is costly to gather, it may not be rational to make fully informed choices. There is a formal economic theory of choice between uncertain alternatives, and this has as its central premise that people choose the alternative that has the highest expected utility or satisfaction [106]. Risk refers to a situation where the outcome is not certain, but where the probability of each alternative possible outcome is known or can be estimated [104]. As a result of risk aversion in uncertain choices, existing habits may assume the character of preferences. Signals of quality, e.g., brand names may be adopted to guide consumer choices so that preference is for the brand or agent, rather than the product per se. Becker and

Murphy include (noninformative) advertising as one of the goods that enter the fixed preferences of consumers [105]. They argue that such advertisements are goods in utility functions if people are willing to pay for them. Advertisements give favourable notice to other goods, such as cornflakes, and raise the demand for these goods [105]. Branding and advertising represent sunk costs for a firm which, if it goes out of business, it will be unable to liquidate. Accordingly, the material interests of these firms favour doing everything they can to remain in business [106]. If buyers know that, they can place much greater trust in the promise of a higher-quality product.

Even when possible outcomes are well defined, and choice problems transparent, consumers may have difficulty in making rational choices. For example, as discussed earlier, although the rational choice model requires people to ignore sunk costs, their actions frequently contradict this rule. Furthermore, choices among risky prospects exhibit several effects inconsistent with utility theory [116]. In particular, people underweight outcomes that are merely probable in comparison with outcomes that are obtained with certainty, i.e., that maintain the status quo. In addition, people generally discard components that are shared by all prospects, leading to inconsistent preferences when the same choice is presented in a different form. Kahneman and Tversky developed an alternative theory of choice to utility theory, i.e., prospect theory [116]. In prospect theory, value is assigned to gains and losses rather than to final assets, and probabilities are replaced by decision weights. The value function is normally concave for gains, convex for losses, and steeper for losses than for gains. According to the value function, people tend to attach considerably less weight to a gain than to a loss, and the impact of incremental gains or losses diminishes as the gains or losses become larger. Figure 3.6 shows a typical value function similar to that in [116].

Prospect theory distinguishes two phases in the choice process: a phase of framing and editing, followed by a phase of evaluation. Many of the errors consumers make in choosing between alternatives are systematic, and lead to systematic bias in the estimates made. Kahneman and Tversky identified three particularly simple heuristics that people use to make judgements and inferences about the environment [106]. They are availability (from memory), representativeness (of an object or person of a category) and anchoring and adjustment. In anchoring and adjustment, people first choose a preliminary estimate – an anchor – and then adjust it in accordance with whatever additional information they have that appears relevant.

Behavioural theorists see choice as an ongoing process of problem-solving, during which consumers' views of the world and of their wants may undergo considerable evolution. In the orthodox approach, the bounds of the analysis are narrowly defined, determinants of consumer choice being limited to prices and income, and with a strong focus on purchasing decisions [113]. The behavioural perspective on consumer theory in economics, as opposed to the neoclassical, mixes economics with other disciplines, particularly psychology and management science [108]. This approach is inductive in that it relies on first gathering facts and then constructing theories around these. It encompasses the whole decision process and tries to incorporate factors such as learning, memory and at-

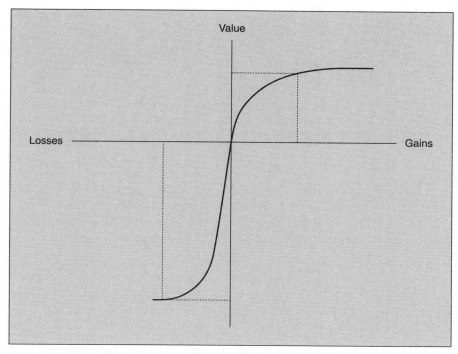

Fig. 3.6. A typical value function according to Kahneman and Tversky [106, 116]

titudes, as well as social factors [113]. Behavioural economists tend to think about choice in terms of movable aspirational targets and rules of thumb, downplaying the significance of substitution and emphasising the roles of search processes, personal principles and habits [108]. However, despite the inadequacies of the rational choice model as a way of analysing consumer behaviour, there is no single, well-articulated, alternative theory for the analysis of the consumer [117].

3.3
Consumer Psychology and Food Choice

People's behaviours generally reflect a desire to experience pleasant sensations and cognitions, and to avoid unpleasant ones. Jill Dupleix's so-called new gourmet illustrates the point [118]. He or she has developed not just a discerning palate, but in addition, an equally discerning ethical and aesthetic awareness about the sources and nature of the food they are willing to consume. Consequently, caviar has lost its past palatability to them. This is because of a perception that Beluga sturgeon is close to extinction and that the Caspian Sea is highly polluted. Similarly, luxury in foods has ceased to mean simply a combination of product features, and has become linked to the process of production, e.g., the skills of the artisanal food producer. People strive to interpret their environment in a coherent manner, and where this coherence is lacking, they are likely to feel psychological discomfort. People also strive for congruence between the values they hold. Dissonance is experienced, in particular, where individuals act against their beliefs or attitudes, or against what they perceive to be in their own best interest. Examples include diverging from a healthy diet and the consumption of meat from production systems perceived as cruel to animals because the meat is cheap. People may use commitment devices, or consumption rules, in order to implement decisions perceived as correct but that may be difficult to adhere to.

The hedonic experience associated with a particular stimulus is susceptible to large changes over time and over varying circumstances; and hedonic predictions for food, and other types of, consumption may be poor [111]. For example, people tend to buy too much food when shopping while hungry [119]; or they may select a greater variety of items than the restauration of their preferences at the time of consumption would warrant [111]; or their attitudes towards an unfamiliar, or less preferred, food improve with the number of exposures to that food [111]. To some considerable extent, experienced pleasure depends on the time and the context. This is very much the case with food consumption, where the phenomenon of sensory-specific satiety (SSS), a negative change in the hedonic response to a food, is well established. In experimental terms, SSS is defined as the difference between the change in the pleasantness of the food that has just been eaten and the change in the pleasantness of foods that were tasted but were not eaten [120]. The desire for variety in meals occurs across species, i.e., individuals, when presented exclusively with one palatable food for a long time, will avoid that food when additional, normally less preferred food is also provided [120]. SSS is an important aspect of the motivational system underlying food consumption.

Memories of past hedonic events, whether positive or negative, influence later evaluations through, on the one hand, an endowment effect, and the other, a contrast effect [119]. In other words, a positive experience gives satisfaction but, at the same time, renders future similar experiences less positive; and a negative experience puts subsequent experiences that are less bad in a better light. The hedonic impact of any event reflects a balance of both of these effects [119]. For example, the occasional purchase of a piece of fruit at the peak of ripeness and with perfect flavour will result in a highly satisfactory consumption experience, but will, at the same time raise the evaluation standard for future purchases. Expectations about the liking for a food profoundly affect actual liking, and such expectation effects are generally interpreted as contrast [119].

Consumer Psychology is the systematic inquiry into the mind of the consumer. It depends on the discovery of the principles that underlie how people construct meaning about both the material and the social worlds around them. Individuals' perceptions, both of themselves and of others, influence their feelings and actions; and objects and issues, as they too are endowed with meaning, become part of this perceptual world. Although foods are relatively mundane items of consumption, what kinds of foods people eat, especially special-occasion foods, forms part of their self-perception as well as influencing how others perceive them. For example, observing a shopper at the supermarket check-out till allows one to draw inferences about that person's general tastes and attitudes. The nature and causes of human social behaviour are the field of study of Social Psychology. This includes the activities of individuals in the presence of others, social interaction between two or more persons and the relationship between individuals and their reference groups [102]. It is well understood that groups shape their members' values through group norms or rules. In the same context of socialisation, groups regulate their members' learning.

So-called middle range theories in Social Psychology consist of narrow frameworks used to explain specific social behaviour, e.g., the role of persuasion in attitude change. By contrast, grand theories, or theoretical perspectives, are rooted in explicit assumptions about human nature. These offer more general explanations of behaviour, as is the case with role theory, reinforcement theory, cognitive theory and symbolic interaction theory [102]. Whilst reinforcement is a key aspect of learning, cognitive theory underpins our understanding of information processing. Symbolic interaction theory posits that in their interactions with others, individuals continually reaffirm established, and negotiate new, meanings. As well as interacting with others, people act towards themselves through mechanisms such as self-perception, self-evaluation and self-control. According to role theory, where there are role expectations for some position, the behaviour of anyone occupying that position can be largely predicted. This may be seen as applicable to certain food habits, e.g., of those typical of high status individuals in a society compared to those typical of low status individuals, and food habits typical of certain occupational groups, e.g., athletes and fashion models. Both role consistent behaviour and stable self-perceptions can act as a barrier towards behavioural change, e.g., towards adopting a healthier diet. In one case, it may be part of someone's role in business to frequently attend lavish business lunches and business-related dinners, that constitutes a barrier towards change. In another case, an obese person may find it difficult to think of themselves as a (future) slim person, which would involve doing all the things that slim persons are meant to do. In some sense, this example too relates to role consistent behaviour. In order to achieve permanent success, an obese person embarking on a weight-loss diet will therefore also need to engage in other activities to promote a more radical change of their self-image.

The human mind represents a complex system of information processing. The study of mental states such as beliefs, attitudes, preferences and intentions is particularly relevant to Consumer Psychology [121]. Such states, and the processes that generate them, intervene between stimuli, such as food on a plate, and behavioural responses, such as eating or rejecting the food. Relationships between sensory stimuli and perceptions, i.e., the interpretation of environmental cues via sensations, are the subject matter of psychophysics [122]. The standard route in the stimulus-response model starts with an environmental stimulus, which is registered by the relevant sense organ. Perceptions are generated from sensations via processes of stimulus recognition and interpretation, and some behavioural response then occurs. If sensory inputs are incomplete, the brain has to fill in details from memory. However, the direction of the events in the stimulus-response model may be reversed, at least in part. For example,

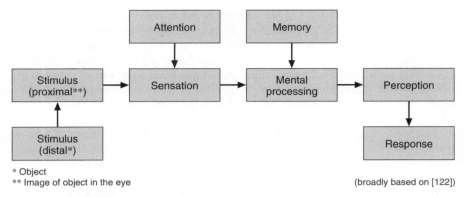

* Object
** Image of object in the eye   (broadly based on [122])

Fig. 3.7. Stimulus-response model for vision

when someone is very hungry, or if they happen to be thinking about some favourite food, memory, rather than any actual odorous molecules in the environment, may generate the perception of a food smell. A number of psychological factors and states influence perception, e.g., memory, personality, mood, emotions, prior knowledge and expectations. Figure 3.7 illustrates the key stages that link stimuli with behavioural responses. However, an in-depth discussion of these stages, in particular, sensation and perception is provided in Chapter 7.

Information processing involves the decoding, transmitting, encoding and storing of representations of the world. Cognitions include knowing, perceiving and judging [123], and thinking, understanding and reflecting [124]. Product perception is based on recognition, categorisation and evaluation; evaluation asks whether a product is right for a specific user and use context. There is a natural tendency to perceive stimuli as members of groups, where they are categorised through comparisons with prototypes. Categorical perception relies on abstraction, generalisation, closure and continuation. In assigning meaning to stimuli, mental models, also known as cognitive structures, are typically utilised, in particular, schemas and scripts. How a person, an object, an issue, or an event, are categorised through language plays an important part in how each is perceived. Many UK parents meet with resistance from their children when extolling to them the virtues of vegetables or, possibly worse, so-called greens. Categorising all vegetables as greens, and perhaps associating them all with vegetables such as over-boiled cabbage, may cause the rejection of other vegetables that many children do like, e.g., sweetcorn and carrots. A concept is defined as an awareness of a quality formed by abstracting the quality from the various other qualities with which it is associated. Vegetable is a concept, and so is crunchiness in vegetables.

Schemas are well-organised cognitive structures, or models, describing some social entity [102]. Like categories, schemas apply to persons, objects, issues and events. Self-schemas apply to the self, i.e., one's self-image, and include both role identities and personal qualities. Scripts are event schemas, and they typically detail the activities that constitute the event itself, as well as the objects and players participating in it [102]. Anyone who takes their friends along to a first date should know that they are violating the accepted script for a first date. Schemas apply both to foods and to food consumers, e.g., organic foods and so-called green consumers. Product knowledge structures are schemas that describe how a product is used and what benefits are derived from using it. Meal scripts are familiar, especially for celebratory meals such as Christmas lunch. Any violation of food-related script expectations, e.g., serving the courses of a meal in reverse order of the expected, will cause confusion, and possibly offence [125].

Lowe and his research group studied the causal factors that lead children to refuse vegetables and applied various aspects of general psychological theory to effect changes in both the attitudes and the behaviours of a group of five- to seven-year-old children [126]. Their working assumption was that, since food choice is a learned behaviour, almost any child can learn to eat almost any food. Children's food preferences should therefore be subject to verbal and other cognitive and learning factors. The group's experimental approaches included peer modelling using video tapes and rewards to reinforce desired behaviour, i.e., eating of previously rejected vegetables. The observational learning element took a playful, fun approach, in that the tapes featured both Food Dudes (children eating vegetables) and an evil Junk Food Junta, against whom the Dudes fought. However, despite being (externally) influenced to change their vegetable consumption, the children did not, strictly, respond to environmental pressures, e.g., a parent expecting, or even forcing, them to eat up the food on their plate; rather, they had internalised the new behaviours. This distinction between dispositional and environmental causes of behaviours is highlighted in attribution theory [102]. This is an important aspect of consumer psychology closely linked to cognitive structures, e.g., people's self-schemas, and to motivational structures. There is a crucial difference in the nature of actions taken in response to someone else's wishes and of doing something because one really wants to do it. In making an attribution of a behaviour, it appears that people examine their beliefs and attitudes towards it. Dispositional (environmental) attributions reflect stable personality characteristics, or traits, and the associated behaviours are therefore more stable than behaviours that are merely expected in a given situation. Attribution theory also explains the mechanism underlying the so-called foot-in-the-door persuasion tool or strategy. This aims to guide the consumer from acting on someone else's wishes, e.g., accepting a food sample in a supermarket, towards developing a stable preference.

For sensory information to be registered by the brain, minimum threshold levels of stimulation must be achieved, e.g., the lightness level or the concentration of an odorous substance in the air; this is the absolute threshold. Beyond this, psychophysical perception is largely based on stimulus differentials rather than absolutes [122]. It is often important to understand whether a change in a product attribute or in a marketing stimulus is of sufficient magnitude to be noticed by consumers. The difference threshold, or just-noticeable-difference (JND), is the smallest change in stimulus intensity that will be noticed [127]. The JND is dependent on the absolute intensity of the two stimuli (Weber's law). For example, with increasing intensity of the initial stimulus, a greater amount of change is necessary to produce a JND [127]. JND is an important concept related to consumer satisfaction. In foods, it is as relevant to attributes such as flavour and quantity as it is to price. JND in food product reformulation raises the question of what level of difference, if any, would be acceptable. Sensory stimulation is sometimes perceived subliminally, as in the case where a car is being driven along a familiar route without much thought, until something unusual, e.g., a road sign or an obstruction catches the driver's attention. Clearly, before a stimulus can be registered and processed, it needs to be attended to. Merely presenting potential consumers with a product is not enough to cause them to evaluate it; their attention must be gained first. In the case of food, smells, as well as visual information, may be employed to achieve this. The smell of bread baking and chickens roasting in supermarkets are familiar examples of this. A pleasant aroma permeates into the consumer's consciousness sooner of later, sooner if he or she feels hungry, reflecting a motivation to eat. Attention is both selective and dynamic. It is captured by stimulus attributes such as saliency, vividness, size, colour, intensity, contrast, change, novelty and variety. Information that is salient and/or vivid attracts and holds attention because it is direct and emotionally interesting. Personal determinants of attention include need, motivation, attitudes, adaptation, and attention span. Automatic activities do not benefit from conscious attention, e.g., swallowing. The greater the attention paid to a stimulus the more likely is its impact on comprehension and memory.

Perception is the process through which individuals are exposed to information, attend to it and comprehend it. The understanding of perceptual processes is strongly influenced by the ideas of the gestalt movement [102]. Central to gestalt is the principle that people respond to configurations of stimuli rather than to single, discrete ones. In other words, people comprehend the meaning of a stimulus by viewing it in context, with comprehension being helped by focusing on more or less complete patterns. Framing structures provide the context within which a stimulus is presented and understood. Framing and anchoring effects also explain some of the misperceptions that commonly lead to distorted evaluation of alternative choices. There are examples of framing approaches and their effects both in the field of commercial product positioning and in public information campaigns. At the time of writing, use of gain frames to present negative product attributes to the consumer, e.g., a ready meal being "80% fat-free" instead of it containing "20% fat", is very popular among companies marketing food in the UK. This is justified from their point of view because consumers have been shown to respond positively, i.e., they buy more of the product. Clearly, there is no overt deception, yet consumers are apparently allowing themselves to be misled. This strategy can backfire however if it introduces dissonance and confusion into the mind of the consumer. As an example, milk from a Scottish dairy breed, the Ayrshire, has been marketed simultaneously as a premium brand and with a "96% fat-free" claim [128]. In the public policy arena, it has been found that the way a health message is framed can enhance the impact of that message [129]. For example, focus on the positive consequences of performing a certain health enhancing behaviour may be more effective in primary health education aimed at disease prevention. The difference in perceived value of gains and losses is expressed by the value function of Kahneman and Tversky (Fig. 3.6).

Learning is the mechanism that enables the members of a group to acquire more or less shared meanings and perceptions about each other and about their environment generally. Learning means a relatively permanent change in someone's memory and/or behaviour. Learning theory tends to be approached from several different angles, with associative and cognitive learning the two main theories [130]. These will provide the focus for discussions here, although social learning theory will also be illustrated. Associative learning is learning about the relationship between two events that occur together, whilst cognitive learning describes learning involving complex thought processes [130]. However, there are certain overlaps between these two mechanisms. For example, memory and attention, which are basically cognitive in nature, also play a role in associative learning. In associative learning, associative bonds are strengthened by repetition. However, this might also be said of cognitive learning, as thinking solutions to problems are often forgotten again.

There are two important varieties of associative learning, i.e, classical conditioning and instrumental, or operant, conditioning. In classical conditioning, a conditioned stimulus becomes associated with an unconditioned stimulus and its unconditioned response. In the case of Pavlov's dog, the unconditioned stimulus was the dog's dinner and the unconditioned response, the dog's salivating [130]. The conditioned stimulus that, over a period of time, became associated with the salivating, was a bell that happened to be sounded when it was the dog's feeding time. The dog had learned to salivate on hearing the bell whether or not any food was present. In operant conditioning, the stimulus-response order is reversed, i.e., the behaviour occurs first, then a stimulus is given. The mechanism obeys Thorndike's law of effect, according to which a response will increase if it is followed by a satisfying outcome [130]. In operant conditioning, remembered hedonic experiences have an adaptive function, i.e., behaviours that have been found satisfactory in the past are reinforced and therefore likely to be repeated. Whether a stimulus will serve as a reinforcer depends on whether its current level is below the preferred level, i.e., whether one has been recently deprived of it [130]. In this way, an ice cream offered to a child as a reward for good behaviour may be effective, but only if the child has been recently deprived of ice cream. If the child is to be rewarded for a period of quiet behaviour, but is already satiated with ice cream, a more effective incentive for the child to repeat the behaviour would be to let them run around and make a lot of noise.

Shaping refers to a gradual, focused approach towards establishing and motivating certain desired behaviours in a subject. This is the approach required to train animals in performing certain routines, but is also appropriate for use with small children and with consumers generally. What shaping tries to achieve is to reinforce whatever aspect of a subject's behaviour is closest to the desired response, gradually building up to achieving this desired response. Up to one in twenty children under five in Britain are affected by so-called faltering growth, with refusing to eat the cause in the majority of cases [131]. The eating habits of one child suffering from faltering growth, Abi, were transformed within just a few months using an approach based on operant conditioning and shaping. She was given three meals a day interspersed with snacks. Tantrums and food-throwing were ignored and praise given whenever she ate anything. Meals were limited to 15 minutes per course, and leftovers taken away without fuss. Two years later Abi had become something of a gourmet, unlike most of her friends. Shaping of buying behaviours might involve the handing out of free samples, special offers, or the provision of seating areas or of pleasant toilets in a store. These are examples of tools for gradually shaping the ultimate, desired behaviour, i.e., for the visitor to the store to spend their money there. Consumers store information in memory in the form of associations [124]. For example, a brand name may be associated with a variety of product attributes as well as emotions. Emotions related to products or brands may be generated through advertising or in the individual consumer, the product may be linked to personal, emotive associations, e.g., associations with childhood memories.

Unlike associative learning, cognitive learning is deliberate, i.e., it results from active mental processing. Here, to learn can mean to discover or to invent something, to commit something to memory or to become efficient in carrying out some procedure. With the exception of rote learning, cognitive learning is usually insightful, involving generalised problem-solving approaches, but depends for effectiveness on feedback that is both immediate and correct. The last stage in cognitive information processing, or learning, is called elaboration [124]. This refers to the creation within memory of complex networks of ideas and feelings about people, products, issues and events. Here the new information may be extensively integrated with existing schemas. Schema-irrelevant information tends to be more difficult to memorise than information that is either schema-consistent or schema-inconsistent.

Social, or vicarious, learning contains aspects of both associative and cognitive learning [127]. In this type of learning, individuals imitate the behaviours of others. This type of learning underlies much of modern advertising [127]. Imitating others may be little more than a reflex, e.g., entering an unfamiliar, but evidently popular restaurant without really thinking about one's choice. On the other hand, choosing that restaurant may have been the outcome of some explicit deliberations. These might have included that searching for objective information about all possible alternative choices would be both costly and impractical. Furthermore, we may assume that other people have similar pay-offs from alternative choices, so that if they appear to be satisfied with a particular choice, we too are likely to be satisfied. Learning the rules for making rational choices is an aspect of the cognitive brain. One of the social outcomes of social learning is conformity, i.e., the convergence of individual behaviours. This is one of the mechanisms underlying the emergence of fashions and fads.

Whether a particular consumption item comes to mind when required depends in part on its availability from memory which, in turn, is affected by the vividness of the memory, and this is usually associated with its emotional content. Brand associations thus become practical tools for consumers, as they make the process of choosing, e.g., in a supermarket, more efficient. Memory is built up slowly however, so that the first entrant into a market, or the market leader, who is supported by heavy advertising, will occupy more space in consumers' memories. It is easy to imagine how consumers forget about products, so that constant investment in reinforcement of these memories by the producer is necessary. When a person is motivated, there can be a range of accompanying feelings, emotions and moods that shape behaviour; these influences, which may be positive or negative, are referred to as affect [127]. Thus, positive affect speeds up information processing. Motivations represent the flip side of attributions, only this time the causes of behaviour are examined from the actor's point of view, rather than from that of the observer. Internal motivational states include hunger, thirst and emotions. Motivations are the driving forces that propel individuals towards the satisfaction of their needs. These needs exist at different levels, e.g., in the case of Maslow's hierachy of needs, there are five levels [127]. In this hierarchy, physiological (food and sleep) needs occupy the lowest level. Once these needs are satisfied, motivation turns to higher-level needs. According to Maslow, these are, in ascending order, safety (security and shelter), belongingness, esteem and self-actualisation. The need for consistency among an individual's cognitions has already been mentioned, and this may be seen as being part of the need for self-actualisation. Another need with major relevance to consumption behaviours, which also fits into this particular needs category, is the need for stimulation, or novelty [109]. If motivation is conceptualised as a general state of arousal, then involvement may be viewed as a state of arousal aimed at a particular object or objective. Involvement may be cognitive and lead to raised attention and more extensive information processing, or it may be affective, with heightened expectations and excitement. The greater the involvement in a consumption decision, the greater the degree of disappointment if the product fails to live up to expectations. Although emotions can be strong motivators of behaviours, because they represent immediate responses to a stimulus, they fade quickly when arousal subsides, e.g., joy, anticipation, surprise, sadness, fear, disgust and anger. On the other hand, complex, socially significant, feelings or sentiments, e.g., grief, love and jealousy, are more enduring. As well as representing responses to already experienced stimuli, emotions and mood also modulate people's responses to further stimuli, including their readiness to be persuaded by arguments [132].

An individual's psychological needs may be an expression of certain personality traits. Traits, e.g., sociability and level of self-control, are defined as any distinguishable, relatively enduring way in which one individual differs from another [127]. Traits differ from attitudes in that they are not directed at anything in particular [123]. Traits vary in the strength in which they are present in different individuals, and they interact with other traits. For example, aggressiveness may be modified by sociability and co-operation [123]. An important trait that is especially relevant to consumption generally, and food consumption in particular, is the sensation-seeking trait [133]. According to Eysenck, differences between extroverts and introverts may be characterised in optimal levels of stimulation, with extroverts feeling best and functioning most efficiently at higher levels of stimulation or arousal, and introverts feeling and functioning best at lower levels [134]. Zuckerman suggests that the stimulus quality of novelty and complexity, and the need for change, variation and intensity are motivating qualities for sensation seekers [134].

Novelty describes that which has not been experienced before. Therefore by choosing it, one also chooses to learn new things. Novelty may be high, as when something is new relative to everything that has been experienced before, or low, as when a new combination of already experienced events is encountered [109]. Entirely novel products are challenging as they require extensive learning, including reassessment and modification of present household routines. By contrast, marginally new products will fit comfortably into established schemas, as well as household patterns, and a more gentle learning process therefore ensues. Novelty allows for discovery, which is a form of creativity and a highly pleasant and motivational state. In food consumption terms, the notion of extreme novelty might be attached to genetically modified foods at present. However, these do not offer the sort of experience discussed here. A more appropriate example would be someone being introduced to foods, or dishes, representing an as yet unfamiliar food culture which may be very different from one's own. The desire for variety has been highlighted already in the context of sensory-specific satiety. However, the concept is more widely applicable, as it relates to individuals alternating among all kinds of consumption activities, e.g., restaurants, music, or leisure activities [135]. All of these may be familiar already, i.e., there is a clear distinction between novelty seeking and variety seeking.

Attitude enters as an important factor into a range of models of consumer choice in psychology. Attitudes are learned predispositions to respond favourably or unfavourably to a particular person, idea or object [124]. Evaluations of some kind are therefore inherent in all attitudes. They differ in this from beliefs, which merely describe what someone holds to be true, whatever their evaluation of this truth. Someone stating that smoking is unhealthy may be defining both a belief and an attitude, because the word unhealthy may be seen as an evaluative term. However, because attitude systems are complex and interconnected, this individual may still arrive at an overall judgement that smoking is a good thing for them to engage in given present circumstances. People with positive attitudes towards themselves, who possess high self-esteem, may act differently from those with low self-esteem, especially if the context is a social one. Attitudes grow out of individual and group norms and values, i.e., the perception of each of what is the right thing to do in a given situation. In common with many other cognitions, attitudes are context dependent. Attitudes are of interest in Consumer Behaviour theory for two major reasons. The first is the question of how attitudes affect choices, and the second is to do with attitude change and its causes and mechanisms. In marketing a product, the positive attitudes of existing customers towards it need to be reinforced, whilst favourable attitudes need to be created among potential customers. An important consideration in both of these activities is the fact that attitudes have a schematic function. Therefore, persuading someone to become a customer for a product is likely to involve schematic change in how they perceive the product. Persuasion is also linked to attitudes in respect of the effectiveness of the source of persuasion. Source factors have been found to play a role in persuasion, especially if the recipient lacks the motivation to process the message carefully. Source likeability can therefore be conceptualised as an attitude; i.e., an affective evaluation (likeability) is linked to an object (source) [136].

Stable attitudes serve as decision heuristics in that they reduce the mental effort required in making rational choices. However, one of the fundamental problems in eliciting attitudes from subjects accurately is the fact that attitudes cannot be observed, but must be inferred either from behaviours or from answers provided in response to subjective questions. In fact, attitudes may not even exist in a coherent form, as measured attitudes are quite unstable over time. Part of the problem comes from the respondents' reluctance to admit lack of an attitude, part that people may be wrong about their attitudes [137]. Through reporting attitudes consistent with behaviour rather than actual attitudes, a subject may be, consciously or subconsciously, dissipating cognitive dissonance. It is therefore not surprising that espoused attitudes do not necessarily predict behaviour, because they are not necessarily stable evaluations that people hold. In common with all learned responses, attitudes will become increasingly refined and stable the more frequently and consistently they are reinforced. However, only a small proportion of attitudes are acquired by direct experience with the object, many being acquired through socialisation and involving sources of attitudes such as family, friends, role models and the media. Attitudes may predict behaviour only when they are cognitively accessible to the actor at the moment of action [138], that is, when the actor is performing a behaviour that has not become routinised.

The theory of reasoned action (TRA) was developed by Fishbein and Ajzen based on the assumption that volitional behaviour is rational [102]. According to this model, behaviour is determined directly by behavioural intention. Intention is the result of a weighing up of attitudes towards the behaviour and subjective norms, i.e., the individual's perception of others' beliefs about the appropriateness of a behaviour. Attitude derives from a person's beliefs about the likely consequences of the behaviour, subjective norm from their beliefs about the reactions of others to the behaviour [102]. Both kinds of belief also contain an evaluative element, i.e, whether the outcome would be regarded as either positive or negative, and the importance of complying with external expectations. The components of the TRA and the causal links between them are illustrated in Fig. 3.8.

When combined with quantitative measures of the components of attitudes, this model can predict a specific behaviour under specific circumstances [102]. However, even its authors acknowledge that the link between intention and behaviour can be weakened by a lack of temporal correspondence between the

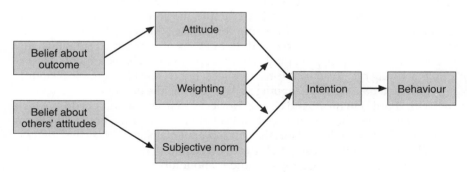

Fig. 3.8. Theory of Reasoned Action according to Ajzen and Fishbein [102]

two. Ajzen's theory of planned behaviour (TPB) is an extension of the TRA in that it contains an added component that further balances attitude and subjective norm, i.e., perceived control [139]. In this model, perceived control is determined by specific control beliefs in a parallel manner to attitudes being determined by behaviour beliefs and subjective norms by normative beliefs. Control factors derive from both internal causes (e.g., self-control) and from external ones (e.g., opportunity). Both issues are familiar in the area of food choice, e.g., in the difficulties encountered when transforming an understanding of the nature of a healthy diet into an actual healthy diet for oneself. On the one hand, one might find the new, approved foods less appealing to eat than the ones one has been used to; on the other hand, access to healthy food may be difficult, e.g., over the course of the working day.

The TPB therefore also applies to non-volitional behaviours, i.e., behaviours that are not entirely under the control of the individual. Other influences, e.g., demographic factors, are assumed to act through the model variables. Whilst the TRA and the TPB may be useful in explaining reasoned choice, choice very often is not rational. For example, sales encourage impulse buying, with the individual frequently regretting their purchases after they have been made. This may or may not be a control issue. Past behaviour is often more influential than behavioural intentions [140]. This is in part because much mundane behaviour, including eating, is mindless and automatic. Such routine behaviours may have started as a result of a reasoned assessment, or they may not; and even if they have, social and other circumstances may have changed such they have ceased to be rational. Deferred consumption is often recognised as beneficial, yet consumers often lack the self-control to defer consumption of food items such as chips and sweets. Recently, the TRA and the TPB have been used extensively in food choice research; for example, the role of perceived control as a factor predictive of behaviour has been confirmed in some studies of weight loss and dietary health behaviour [141]. In a study of people's attitudes towards two common sweet biscuits and wholemeal bread, biscuit consumption was affected by perceived control problems, but bread consumption was not [142]. The TRA provided a reasonable, but limited, degree of prediction of behavioural intention and (self-reported) behaviour for the consumption of low- and full-fat flavoured milks [143].

Attitudes vary not only according to the social context of a behaviour but also according to the physical context. For example, the acceptability of a particular food is greatly dependent on the eating context, i.e., the type of eating occasion and the other items to be served with it. In terms of sensory acceptability, different breads are appropriate for different toppings or accompaniments. For example, a strong rye bread can be neatly sliced, but its strong, aromatic flavour is not complemented well by a topping such as jam. On the other hand, it goes very well with sliced cold meats and cheese. However, a cheese such as cheddar, which is crumbly, cannot be readily sliced and is better eaten with a bread that is not thinly sliced. This surely explains the development of dishes such as the "Ploughman's Lunch", which consists of a hunk of bread and a chunk of cheese. It is easy to see that in eliciting consumers' attitudes about foods, careful attention needs to be paid to context. People's attitudes towards a particular food can be ambivalent, i.e, they may approve of some of the product attributes, whilst disapproving of others. Some beliefs about, and attitudes towards, foods result directly from interaction with foods, e.g., taste aversions. However, most are transmitted socially, e.g., beliefs about healthy foods and generally acceptable foods. Eating can either be a mundane or a highly involving experience. How a food is perceived depends on the individual, their current physiological and psychological states, the nature of the food itself and the context of consumption. General attitudes towards specific foods or eating patterns represent a summary of many feelings, under a variety of conditions. To really understand the underlying motivations of specific food choices, all of these must be carefully disentangled. Finally, product acceptance in experimental settings is insufficient to predict actual consumption because, in order to introduce a new product into current consumption habits, existing ones need to be displaced. Positive attitudes alone cannot predict behaviour since, with food in particular, one is limited in how much one can consume.

3.4
Social Influences on Consumer Behaviour and Food Choice

The social meanings of specific foods and of food-related behaviours can differ considerably between groups of people. In fact, the foodways of a culture are among its most important determinants, a useful tool of routine reinforcement of a shared way of life. Sharing foodways identifies group members and, by the same token, non-sharing or rejection marks outsiders. Celebrations are often centred around communal eating and frequently specify particular foods or dishes, e.g., the wedding cake and the Christmas turkey. The term foodways indicates that it is not just the foods themselves that matter in shared food culture, but that processes related to food consumption are equally important. Examples of such key processes include the Halal slaughter of animals and the requirements of the Kosher kitchen to maintain separate utensils for meat and dairy products [82]. Foodways can therefore include styles of eating, details of food preparation, the number of meals per day and the times of eating [82]. More generally, foodways refers to the shared beliefs about foods and eating within a particular culture [88]. Being tied in with hospitality, food may be given, or shared, to define relationships between individuals, e.g., love or friendship. Prestigious foods may be consumed to signal aspirations or social status, and may also be given to, or shared with, others for the same purpose. Prestigious foods are likely both to reflect current eating fashions and to be exclusive, and hence, expensive. An example of a food that is now considered prestigious due to its rarity is wild salmon. Yet, this was once a very common food.

Cross-cultural communications using food as a medium can be problematic however, as illustrated by travel writer Colin Thubron when he recalls en-route meals shared with strangers as a symbol of friendship. On one occasion, during the course of some travels through Turkmenistan, Thubron was invited by some local men to join them in a picnic to be held in a local desert [144]. On offer was mutton shashlik, which turned out to be both tough and "dribbling blood and fat". Whilst Thubrons's eating companions "celebrated every mouthful with a carnivorous burp", he himself recoiled even from the relatively tender pieces of meat which they had reserved for him, so that he was compelled to remove these from his mouth, hiding them wherever possible. At the end of the meal, Thubron passed around some cheese biscuits among his new friends, presumably as a treat, only to find one of them busy secreting his in the sand. In a different sort of cross-cultural communication problem, a US multinational tried to sell baby foods to West African villagers [145]. The labels of these foods showed pictures of smiling babies. Accustomed to seeing the contents of packaged foods depicted on product labels, the prospective customers for the baby food were, not surprisingly, horrified.

The two academic disciplines with particular relevance to the examination of the social influences on consumer behaviour and food choice are Sociology and Social Anthropology. The sociological perspective raises the question about how individuals interact to create, sustain and change social patterns [16], whilst Social Anthropology concerns itself with the comparative study of human societies and cultures and their development [67]. The term culture describes the values, beliefs, behaviour and material objects that constitute a people's way of life, where values represent the standards by which people assess desirability, goodness and beauty [16]. Values are internalised to form part of people's personality, influence general perceptions and serve as broad guidelines for social living. Personal values reflect people's judgements of what is important in life, and what would be the correct thing to do in a given situation. Material culture constitutes the tangible objects created by members of a society [16]; this includes foods and recipes.

Culture is the end result of myriads of individual choices [146]. The theory of culture emphasises the constitution-making capacity of individuals, where any choice constitutes an act of allegiance [146]. At the same time, individual choices may also be seen as a protest against undesired models of society. In Douglas' view, all consumption behaviour is inspired by cultural hostility, and protest is therefore the aspect of consumption which reveals the consumer as a coherent, rational being [146]. Often, people do not know what they want, but they always know what they do not want, and this includes ideas and cultures they do not want to be associated with. Cultural theory assumes four distinctive lifestyles (individualist, hierarchical, enclave and isolate), each in conflict with the others, and their mutual hostility and incompatibility accounting for the stability of each [146]. Whilst the fast-lane individualist culture is motivated to expand individual alliances, the hierarchical, traditional household culture cannot readily accommodate this. Both are at conflict with each other, "each seizing a victory where it can" [146]. The egalitarian lifestyle, preferring simplicity, frankness, intimate friendship and spiritual value, is rejected by both individualists and hierarchists. Whilst the isolate may well be benign in his attitudes to the cultures he does not want to adopt, by standing aloof he or she manages to give offence to all. However, membership of a group involves the individual in accepting a degree of conformity, whilst absolute compliance is rare [124]. Lifestyle analysis has become an important aspect of consumer product differentiation and market segmentation. From examinations of lifestyle, labels typically emanate to be attached to different groups of consumers. Examples of such labels are self-starters, materialists, nesters (pre-nesters, full-nesters and empty-nesters), survivors and sustainers, inner- (I-am-me, experiential and socially conscious) and outer-directed (belongers, emulators and achievers), young sophisticates, cabbages, traditional working class, middle-aged sophisticates, Coronation Street housewives, self-confident, homely, penny pinchers, and so on [123]. These labels are used to help to target marketing focus. They are based partly on social factors and partly on individual personality traits, e.g., the need for approval. Preference for spiritual values might cause an individual to adopt a lifestyle which would incorporate a wide variety of schema-consistent consumption goods and behaviours, e.g., "all-natural" clothes and domestic products, organic foods, yoga for contemplation and (intelligent) exercise, complementary medicine, aromatherapy, and so on.

A culture gives meaning to the actions of its members and to the material things that they use, and meaning therefore arises from cultural applications. Conversely, these things and actions can be utilised by an observer to decode the general principles underlying cultures. The structuralist approach to the sociology of food and eating is represented by Lévi-Strauss, Douglas and (partly) Bourdieu and Fischler [147]. This approach recognises that taste is culturally shaped and socially controlled. Food in Anthropology is seen as a means of expressing group identity, of the relationship with other groups and with the gods [147]. In Anthropology, the researcher traditionally takes up long-term residence in a community. The anthropologist, as a "marginal native", thereby takes up a position on the border of a culture, looking both ways [145]. Social anthropological studies consistently highlight the symbolic meaning of food and food-related events. At the most basic level, cross-cultural definitions of what is considered edible vary enormously [148]. The treatment of food as analogous to language was particularly influenced by the work of Claude Lévi-Strauss, who considered that foods were "goods to think with" [149]. Accordingly, foods and eating events, such as meals, embody symbolic meaning, which can be decoded to reflect on cultural values. In the work of Mary Douglas, food and eating are symbolic of a particular social order [148].

A child's family is the most important factor in facilitating its social learning, with schooling and peer groups building on this foundation [16]. For example, family meals play an important role in teaching children appropriate food-related behaviour patterns. This includes what foods to eat, in what contexts to eat them, as well as food- and meal related roles and manners. The considerable influence that children now have on family consumption and expenditure is attributable, at least in part, to the peer pressures to which they are exposed outside the home, and to food advertising which reinforces such pressures. People's values derive largely from norms transmitted through the family and other social institutions. However, norms can be either descriptive of injunctive in nature; i.e., the term norm is used to describe both common behaviours and behaviours that are socially sanctioned [150]. Descriptive norms work on the basis of the demonstration effect discussed previously. The logic is that if an observer imitates what large numbers of people are already doing, that particular course of action is likely to prove satisfactory. Once a particular behaviour has been repeated many times, what was originally a case of social mimicking is likely to be transformed into a personal preference. This is the case with the development of food preferences, as mirrored in the term acquired taste. In contrast, injunctive norms determine what ought to be done, and the motivation in adhering to these is the prospect of social rewards and punishments, including a feeling of virtuousness. The predictive value of social norms remains controversial. However, persons who are, either dispositionally or temporarily, focused on normative considerations, are more likely to act in a norm-consistent manner than persons that are not [150].

Coca Cola is promoted as a universal product but, at the same time, it is closely identified with the culture of the USA [145]. Globalisation is emerging as a central concern in Sociology, with global brands, e.g., McDonald's fast-food restaurants and Coca Cola, prime examples of the ever expanding geographical coverage achieved by consumer goods. Global homogenisation describes the cultural effect of the migration of goods within the world market system in which cultural differences are increasingly being eroded [145]. However, whilst McDonald routinely adapts itself to local markets, e.g., selling fish in Norway, for their part, the locals adapt these restaurants to their cultures. This process of re-contextualising, whereby foreign goods are assigned new meanings and uses, is known as creolisation [145]. Coca Cola has been indigenised in various parts of the world by mixing it with drinks, e.g., with rum in the Caribbean, whilst in Russia it is recognised for its ability to smoothe wrinkles [145]. So-called fusion cooking, in which "East meets West" is a currently popular example of the creolisation of cultures. The cultural meaning of individual foods can also vary within the same person, depending on the context and general perspective. For example, someone raised and living in the UK will perceive baked beans as an everyday, cheap and filling snack. However, the same individual, after permanently relocating abroad may find that, having become difficult to obtain and expensive, baked beans now represent a rare treat clothed in nostalgia value. Foods like that exist in many nations, and they are the ones that expatriates are most likely to miss, and that have a tendency of springing to mind when generally feeling homesick. However, the still familiar pasta-mangiares-, krauts- and frogs-related European national stereotypes are becoming distinctly dated. As for the practical value of global goods, e.g., McDonald's burgers, these offer reassurance through familiarity and predictability, which is especially valued when one finds oneself in unfamiliar surroundings where one may already feel generally disorientated.

In a recent study of food choice criteria among consumers in upstate New York, five key types of value were identified, i.e., taste, health, cost, time and social relations [151]. In each choice, several of these factors are commonly balanced against each other. The managing-social-relationships value describes the properties of a food choice behaviour that influence how well it is accepted by other people with whom one shares eating. People weigh this up because the preservation of household harmony is important to them. Considerations of this sort are not restricted to what everybody is actually eating, however, but also relate to what might offend someone simply watching, or smelling, a fellow diner eat. On the Planet Ketchup website, users are encouraged to communicate their ketchup related cravings [152]. Denise from New York likes to use ketchup as a topping on vanilla ice cream and as a dipping sauce for chocolate chip cookies, but finds that, if she does this in front of other people, "they get upset for some reason". Yet, at about the same time, a top Parisian chef braises tomatoes for dessert by stuffing them with a dessert-type farce [153].

Eating episodes do not necessarily evoke social feelings or associations. The ability to satisfy one's appetite for ice cream with tomato ketchup exemplifies the advantages of eating solitarily. Similarly, the eating of an apple or an orange, or the mindless nibbling of a packet of peanuts, are mostly mundane activities. The reason for this may be that the foods in question are more or less natural, i.e., culturally unaltered, whilst social meanings tend to be associated with dishes, meals and, in particular, festive or ritual dishes or meals. All of these latter imply the presence of culturally determined structures. For example, Scotch pies are the essential snack to be eaten when watching a Scottish football match; a salad simply would not do. One of the most highly ritualised meals is the Burns Supper, on or about 25th January, which celebrates the birth of Robert Burns in 1759. This has rules not only regarding the food itself, i.e., the haggis and its traditional accompaniments, but about the whole elaborate ceremony surrounding the serving of the food, including the grace and loyal toast, the speeches and songs and recitations of Burns' work [154]. Many food-related customs may not register with the participants in the conscious way that taking part in a Burns Supper does. There are, nevertheless, many examples of such customs. In the patriarchal society of the Auvergne, it was traditional for the head of the household to serve soup, and the practice survives in some families today [155]. In Crete, a host inevitably offers his guest, or customer, a glass of Raki in order to conclude a meal, part of the ritual being that the recipient would sip this slowly [156].

Traditional meals are food events and social events at the same times. Sociability in the meal context has three dimensions, i.e., facilitation, commensality and interaction [157]. Social facilitation is concerned with how people's performance is influenced by others. One of the effects of eating in company is that people eat more than they otherwise would. This is thought to be due to social modelling, i.e., each diner copying the others, possibly setting up a spiral leading to over-consumption. Whether this occurs depends in part on exactly how a meal is served, in particular, how portions are controlled. Better control over intake is exerted in meals in which courses are served in succession, compared with meals where many dishes are placed on the table simultaneously, to be shared by the diners. Social facilitation of eating varies according to who is present at a meal, factors including gender and familiarity [157] as well as status. Commensality refers to the selection of eating partners, defining desirable, acceptable, avoided and prohibited meal partners. This selection may be based on personal preferences or on cultural, or religious, inclusivity or exclusivity ("our foods" vs. "their foods"). Knowledge about what, when, where and with whom a person ate constitutes potentially risky biographical information [157]. Preparing and cooking for others is risky too, especially for strangers, because the outcome is relatively uncertain. There is always a chance that guests will not like some of what is on offer, or how it is being offered. In fact, inviting people for a meal in the UK is referred to as entertaining guests, highlighting the expectations that surround such events. Interaction is the dimension of meals that refers to the social interchanges that occur with others during the course of a meal. Meals are subject to highly ritualised social interactions, generally summarised as table manners, and they also hold participants captive. These generate pressures that may be perceived as either positive or negative. Family meals in particular can provide the stage for interpersonal conflicts being highlighted and fought out. Although meals are still seen as social events, modern living also increases the value individuals place on their private space. A new category of meal eater, largely relying on the ready meals provided by food retailers has thus emerged recently. Categorised by food marketers as "dapys" – dine alone and please yourself – these are consumers who are "adventurous and like to pamper themselves with foods from around the world" [158]. The stigma that was once attached to eating alone [157] may no longer hold for many individuals. Individuals may also feel more relaxed towards dining out alone. Restaurants in the vein of sushi bars, where the chef also assumes the role of the host, offer a template of relaxed, non-committal interaction the solitary diner can engage in if he or she so wishes.

Apart from social organisation, there are two other important aspects of meal taking, i.e., meal format and eating pattern. Meal format refers both to the composition of the main course and the sequence of the whole meal, i.e., the starter, the main course and the dessert [159]. Meals can vary from simple to elaborate, ranging from single-course meals to meals with a large number of courses. There are many different types of meals, e.g., the "main meal of the day", family meals (implying the comfort of the familiar), celebratory meals, and restaurant meals (implying surprise and possibly, novelty). The contemporary custom of eating meals in precise helpings, with courses appearing in a predetermined order dates back to the late 19th century, when it was introduced as "service à la Russe" [159]. Meal patterns are made up of three basic components, i.e., the rhythm of eating events, their number, and the pattern of alternations of hot and cold meals and snacks [159]. Meal patterns vary with social class, between occupational groups and over time. The "full English breakfast" is no longer the traditional English breakfast, and most people who still take it occasionally either consider it as a treat, or they may use it to lay a foundation for a day's energetic leisure pursuits, e.g., (hill) walking. In fact, many UK restaurant chains offer the dish as a simple meal available throughout the day.

Tastes in food depend in part on the ideas people have about the body, and about the effects of food on the body, e.g., strength, health and beauty [160]. These perceptions may vary between classes, ages and genders. Bourdieu puts forward the example of fish as a food regarded by (working class) men as unsuitable for themselves [160]. Fish is perceived as light rather than filling, and it is also fiddly to eat. Above all, fish has to be eaten with restraint, in small mouthfuls, chewed gently, with the front of the mouth, on the tips of the teeth (because of the bones), and all of this contradicts the image of masculinity. Similarly, in the UK, in the 1970s it was jocularly said that "real men do not eat quiche". Apparently, they still don't, at least when attending, or watching, football matches in Scotland! Dishes featuring a large proportion of meat, e.g., steak, still tend to be perceived as masculine. However, whilst household provisioning and cooking largely remain part of the female domain, it is normally expected that the man of the house will take charge of the barbecue [153], as well as of the carving of the Sunday roast.

A very modern issue related to food, gender and body consciousness concerns the current slimness ideal which, particularly in girls and young women, has been connected to eating disorders, such as anorexia nervosa, which are becoming increasingly common. Today, both slimness and obesity are strongly invested with social meaning, with contributory factors such as beauty, health and being in control of oneself. Eating disorders may be made worse by the social expectations placed on the individual, e.g., the expectation to be seen to tuck heartily into the family meal and to enjoy the foods served during such meals. Traditional food patterns and their social correlates often erect barriers towards dietary change, change that could lead to more sensible eating patterns and better health. This example contradicts Lewin's contention that overall, customary behavioural patterns are likely to be rational and efficient simply because they have survived [99]. Times change, and in the food context, the high-energy foods appropriate to people performing hard physical work [155] are no longer relevant in the modern context, at least not in the portions and with the frequency they used to be served. Women's roles in society have also changed. However, even high-achieving women, such as the politician Hilary Clinton and the popular singer Cher, still feel compelled to make public statements about whether "staying at home to bake more cookies" might deliver greater happiness to a woman's family than her pursuing a successful career.

Migration and travel modify traditional foodways. Ethnic groups are defined by origin of birth or descent and have common cultural traits. By eating, or not eating, certain foods, individuals can express their ethnic affiliation. Even where ethnic groups are fully integrated into the national culture, distinctive cuisines still play an important role in celebrating one's origins and reaffirming one's identity. This has recently been described for the descendants of the German Lutherans who settled the Barossa valley in Australia [161] and for the Cajuns in Louisiana in the USA [88]. The Barossa is the only place in Australia where a distinctive regional cuisine was established in the historic past, which still exists as a vital component of its present-day community and cultural heritage [161]. The foods include pickled dill cucumbers, sausages, pickled pork and breads and Streusel cakes. Gradually, over a time span of some 150 years, traditional and new ingredients and preparation techniques blended to create a new kind of cuisine. A similar pattern evolved in Cajun cooking, although it has remained as more separate when viewed from the dominant culture, partly because it has remained closely tied to activities such as hunting and the public slaughter of animals. Cajun foods include boudin, hogshead cheese, turtle, frogs, alligator, raccoon, crawfish and squirrel, which may be considered as repulsive by outsiders. Similarly, outsiders may view certain parts of edible animals as inedible – hearts, spleen, tripe, brains, stomach, or tongue. On the other hand, Cajun foodways have undergone a reversal in evaluation, having become fashionable with many outsiders. To these admirers, Cajon foodways represent environmental competence, oneness with nature, fun, culinary skills and sociability and enjoyable lifestyle.

Consumer ethnocentrism accounts, at least in part, for the country-of-origin effect on product evaluation, being positively related to the purchase of domestic products [162]. This is particularly true of commodities, e.g., beef, and assuming that there is no significant price differential. Where differentiated goods are concerned, consumers often deliberately opt for "something different". In particular, the intermingling of cultures and cuisines in modern countries allows for an interesting mixture of personal food choices. In this way, it has become easy to either dip or fully delve into another food culture. By changing one's perception of what constitutes proper food and proper eating from the points of view of others, one is helped to develop a diet that suits one's individual requirements. Recently, Britain's foreign secretary made use of a popular dish called chicken tikka masala as a metaphor for British multiculturalism. In the lively debate that followed his remark, the dish itself was unveiled as an "innocuous curry", a dish with no genuine provenance [163]. In this context, by referring to the British appetite for the easiest form of exoticism, Meades accuses the British food consumer of both "magpie indiscriminacy" and "amended insularity" [164].

Conclusion

Quality in a food defines those attributes that influence a consumer's satisfaction with the product, or differently put, that represent consumption value to him or her. Consumption value is therefore closely related to the motive(s) for buying a particular food, or for eating it. Quality has turned into a key issue in modern food markets in recent decades, largely because these markets have become increasingly competitive. This has meant that consumers are now offered extensive choice in their food purchasing. Both the analysis and the delivery of food quality and consumer value contain subjective elements. The quality of a product derives from particular constellations of certain product (and process) attributes. Subjective choices are involved in identifying significant product attributes. Weighting them is an aspect of consumer value. Consumer requirements and tastes in relation to food change all the time, due to political, economic, social and technological factors. This is why today a purely natural-scientific approach to food quality management is no longer appropriate. In fact, as this first part of the book has shown, all the major social sciences, in particular, economics, psychology and sociology, offer important insights into consumer behaviour. In economics, the individual buying/consumption motive is summarised as maximising utility. The discipline is particularly successful at modelling the behaviour of buying/consuming populations, albeit within rather narrow constraints. Here only preference orderings can be observed, rather than the actual magnitude of any individual preference. Unlike marketing approaches, which are largely based on psychological principles, economics does not ask people about their future buying, or consumption, intentions. It is, nevertheless, good at predicting future behaviour, reflecting the fact that one of the most reliable predictors of future behaviour is past behaviour. The psychological analysis of consumer behaviour addresses both individual and social motivations, although social factors are also the topic of sociology. Social factors enter psychological analysis, because beliefs, personality, and other individual determinants of behaviour are, to a large extent, shaped by group dynamics. In food marketing, psychological approaches are particularly appropriate. This is because, in order to be able to position a particular product correctly, a marketer requires a detailed understanding of how target consumers perceive that product.

Competitive markets are characterised by segmentation, and market orientation reflects customer focus. Consumer oriented markets predominantly produce products that are desired, rather than selling products that are easy to produce. The latter approach is characteristic of product orientation in markets. Segmented markets target either groups of consumers (e.g., infants, young adults) or purchase occasions (e.g., Christmas goods). Trying to buy a nutcracker in the UK in April is difficult, because nutcrackers are still associated with Christmas, even though the nuts themselves have been repositioned as a healthy snack, and are available year-round. The value of a food to a consumer is context-depended. A food may be appropriate in one use situation, but totally inappropriate in another. Consumer marketing is characterised by the need to satisfy emotional requirements as well as more tangible individual preferences. This is picked up on in advertising, e.g., in adverts with sensual and/or seasonal associations. Individual food choices reflect individual tastes. These in turn are influenced by personal psychological factors and by social factors, such as the culture someone has been raised in. The social sciences are concerned with preferences, but recognise that preferences cannot be observed. Instead they must be inferred from observed behaviours or elicited directly by interviewing the consumer. The economist's observed demand cannot be equated with consumer satisfaction, because in any given choice situation, all the alternatives may be unsatisfactory. Nevertheless, this will not stop the behaviour from being repeated, at least until a consumer identifies a more satisfactory alternative. Preferences can be inferred from a particular choice, from the frequency with which a choice is made, and from the duration of a behaviour, i.e., eating a particular food. The liking for a food decreases during the act of consumption. Both economics and psychology recognise this phenomenon, and have their own terminology to describe it, i.e., marginal rates of substitution and sensory-specific satiety (SSS), respectively.

The perceived value of any food is strongly influenced by the category in which it is placed, e.g., functional foods (strong attention to good health). The framing of product information therefore affects product evaluation. All information is framed, and all choices take place in the context of framed information. The framing of food benefits is likely to be provided externally, through the supplier's marketing function. If the information is only weakly framed, or if the consumer or purchaser of a food actively rejects the way in which it is framed, they may re-frame it for themselves before evaluating a product. Food habits are learned from an early age. They become tied in with emotions and personal and cultural values, which makes them all the more difficult to unlearn in later life. However, perceptions, and therefore behaviours, can be changed, and this is part of the role of food marketing and advertising, as well as social marketing. In persuading someone to change an established behaviour, schematic changes are normally required, i.e., the new behaviour, if initially schema-inconsistent, has to be integrated into existing belief and attitude systems, and aligned with self-schemas. Beliefs are states of information about (more or less uncertain) events. In the ideal market, producers and consumers have equal access to all relevant information. Such modern markets, with competent and confident consumers, reflect the expressed aspirations of the current UK government. In such markets, the sovereign consumer will deal with unfavourable asymmetries of information by availing themselves of the assistance of consumer agents, e.g., the Consumers Association, the Soil Association, food retailers, etc. Consumption itself leads to learning about how to consume more rationally, i.e., through reinforcement. The consumer as producer of his or her own value from products is actively engaged in discovery. For them, novelty therefore has both a hedonic and a productive value [109]. Concern is sometimes expressed that the large variety of food products on offer, whilst not representing true choice, is likely to mislead people into believing in the reality of free and meaningful choice. Certainly, to opt for salmon rather than trout from a menu, or to go to Pizza Hut rather than Pizzaland conveys no social meaning. On the other hand, one might as well argue that competition among companies disadvantages consumers. Some segments may just be too small, too unprofitable or not sufficiently committed for the major retailers to be able to satisfy them. E-commerce may provide the answer to this as it offers a means of establishing direct communication between a primary producer and a consumer, even though they may be located in different parts of the country (or the world).

References

1. Milmo C. McDonald's bans use of meat from GM-fed animals. *The Independent*, 20th November 2000
2. Hayes D. Pig help on way, promises minister. *The Scotsman*, 22nd October 1999
3. Tester M. Clarifying the GM Debate. *Food Science and Technology Today* 2000; 14(1): 7
4. HMSO. *Food Safety Act 1990*. London: HMSO, 1990
5. Adams R. 1993 and the Changing Scene. *Sheep Dairy News* 1991; 8(3):91–92
6. Grunert KG. Food Quality: A Means-End Perspective. *Food Quality and Preference* 1995; 6: 171–176
7. Cardello AV. Food Quality: Relativity, Context and Consumer Expectations. *Food Quality and Preference* 1995; 6: 163–170
8. ISO. *BS EN ISO 9000:2000. Quality management systems – Fundamentals and vocabulary.* Geneva: International Standards Organization, 2000
9. BQF. *Links to the business excellence model.* London: British Quality Foundation, 1998
10. Porter ME. *Competitive Advantage.* New York: The Free Press, 1985
11. Gorsuch T. The Genesis of a New Food Product Idea. *Food Science and Technology Today* 1988; 2 (1): 52–53
12. Holbrook MB. Introduction to consumer value. In: Holbrook MD (ed). *Consumer Value. A framework for analysis and research.* London: Routledge, pp 1–28, 1999
13. Stiglitz JE. *Economics.* 2nd edition. New York: Norton, 1997
14. Solomon MR. The value of status and the status of value. In: Holbrook MD (ed). *Consumer Value. A framework for analysis and research.* London: Routledge, pp 63–84, 1999
15. Fletcher M, Elliott V, Webster P. French meat at risk over filthy farming. *The Times*, 23rd October 1999
16. Macionis JJ, Plummer K. *Sociology.* 6th edition. London: Prentice Hall, 1998

17. McEachern MG, Schröder MJA. The role of livestock production ethics in consumer values towards meat. *Journal of Agricultural and Environmental Ethics* 2002; 15: 221–237
18. Vasu SC (tr). *Gheranda Samhita*, Ayar, Madras, 1933. Cited in: Hewitt J. *The Complete Yoga Book*. London: Ryder, 1991
19. Yoneda S, Hoshino K. *Zen Vegetarian Cooking*. Yokyo: Kodansha International, 1998
20. Arthur C. Animal feed ban to halt spread of BSE. *The Independent*, 18th December 1997
21. House of Commons. Select Committee on Agriculture: Second Report. *Organic Farming*. London: Agriculture Committee Publications, 2001
22. Health Which. *Why Buy Organic?* London: Consumers' Association, April 1997
23. Krebs J. Country File. *BBC Television*, 3rd September 2000
24. Schröder MJA, Jack FR. Adding vale to commodities through product differentiation: the example of apple juice. *Appetite [Abstracts]* 2000; 35: 211
25. Schröder MJA. Food Standards Agency in Scotland: the consumer's viewpoint. *Food Science and Technology Today* 2001; 15 (1): 54–58
26. Which. *Marks of Approval*. London: Consumers' Association, April 2001
27. Jobber D. *Principles and Practice of Marketing*, 2nd edition. London: McGraw Hill, 1998
28. Sloan AE. Why New Products Fail. *Food Technology* 1994; January: 36–37
29. MAFF *Working Together for the Food Chain: Views from the Food Chain Group*. London: Ministry of Agriculture, Fisheries and Food, 1999
30. Schröder MJA, McEachern MG. ISO 9001 as an Audit Frame for Integrated Quality Management in Meat Supply Chains: The Example of Scottish Beef. *Managerial Auditing Journal* 2002; 17(1/2): 79–85
31. Shaw SA. New attitudes in the food industry. *Food Science and Technology Today* 1998; 12(1): 37–40
32. Britton A. Economic Prosperity and the Quality of Production. *National Institute Economic Review* 1993; August: 6–10
33. Murdoch J. Networks – a new paradigm of rural development? *Journal of Rural Studies* 2000; 16: 407–419
34. Kahn BE, McAlister L. *Grocery Revolution*. Reading, MA: Addison-Wesley, 1997
35. IGD Consumer Unit. *How do Consumers Feel About Grocery Shopping?* 1999. Watford, UK: Institute of Grocery Distribution
36. Griffiths J (ed). *Key Note. UK Food Market. 1999 Market Review*. 11th edition. Hampton, Middlesex: Key Note, 1999
37. *Council Regulation (EEC)* No 2081/92 ("Scotch Beef")
38. Davidson A. *The Importance of Origin as a Quality Attribute for Beef*. 2001. Honours Project. Queen Margaret University College, Edinburgh
39. Anon. *Social Trends. No 31*. London: The Stationery Office, 2001
40. Summers J. Cryogenics and tunnel vision. *Food Technology International* 1998; 73–75
41. Blythman J. *The Food We Eat*. London: Michael Joseph, 1996
42. Http://www.food.gov.uk
43. van Schothorst M, Jongeneel S. HACCP, product liability and due diligence. *Food Control* 1992; 3(3): 122–124
44. *The Food Safety (General Food Hygiene) (Butchers' Shops) Amendment regulations 2000* (S.I. 2000 No. 930) London: The Stationery Office
45. Murphy R. Enforcement and good manufacturing process – solution or smokescreen? *Food Science and Technology Today* 1999; 13 (1): 41–43
46. ISO. *BS EN ISO 9001:2000. Quality management systems – Requirements*. Geneva: International Standards Organization, 2000
47. Seddon J. Ten arguments against ISO 9000. *Managing Service Quality* 1997; 7(4): 162–168
48. Silcock S. Acceptance and application of ISO 9002 in Europe and throughout the world. *Food Control* 1992; 3(2): 76–79
49. Tague NR. Using ISO 9000 to Drive Total Quality. *Managing Service Quality* 1994; 4(1): 24–27
50. ISO. *BS EN ISO 9002:1994. Quality systems – Model for quality assurance in production, installation and servicing*. Geneva: International Standards Organization, 1994

51. Grigg NP, McAlinden, C. Bridging the gap: ISO 9000 in the food and drinks industry. *Proceedings of the 6th International Conference on ISO 9000 and TQM (6-ICIT)*, Scotland: Paisley Business School, 2001; 536–542
52. Price R. Real need for marketing switch. *The Scotsman*, 19th November 1999
53. Dale B, Oakland J. *Quality Improvement Through Standards*. Cheltenham: Stanley Thornes, 1991
54. Hofmeister KR. Quality Function Deployment: Market Success through Customer-Driven Products. In: Graf E, Saguy IS (ed). *Food Product Development. From Concept to the Marketplace*. New York: Van Nostrand Reinhold, pp 189–210, 1991
55. Grigson J, *English Food*. London: Penguin Books, 1977
56. Schröder MJA, Horsburgh K. Communicating food quality to consumers. *Journal of Consumer Studies and Home Economics* 1997; 21: 131–139
57. Hooley GJ, Hussey MK. *Quantitative Methods in Marketing*, 2nd edition. London: Thomson, 1999
58. Buncombe A. There's beef in your French fries, says McDonald's. *The Independent*, 25th May 2001
59. No. 8 *Code of Practice on food standards inspections* under [4]
60. Van Hecke A. The Harmonization Process in the European Community. *World Food Regulation Review. Special Supplements*. June 1991
61. Shapton N. Implementing a food safety programme. *Food Manufacture* 1989; August: 47–50
62. Codex Alimentarius (*Codex Alimentarius Commission website*)
63. Regulation No. 10/65/EEC of the Council of 26th January 1965 laying down common quality standards for garlic *Official Journal of the European Communities* 246/65
64. Everitt M. Sensory science in quality control: a standardised approach. *Food Science and Technology Today* 1997; 11(1): 39–40
65. BSI. *BS 7373:1991. Guide to the preparation of specifications*. Milton Keynes: British Standards Institution, 1991
66. Hobhouse H. *Seeds of Change: Six Plants That Transformed Mankind*. 4th edition. London: Papermac, 1999
67. Pearsall J (ed). *The New Oxford Dictionary of English*. Oxford: Oxford University Press, 1998
68. Cooper D. Unfood for the Unfit. In: Cooper D. *Snail Eggs & Samphire*. London: Macmillan, pp 198–200, 2000
69. *The Novel Foods and Novel Food Ingredients Regulations 1997* (S.I. 1997 No. 1335). London: The Stationery Office
70. Philbrick N. The hunger that consumed a crew of cannibals. *The Sunday Times*, 7th May 2000
71. Kavasch B. *Native Harvests*. New York: Vintage Books 1979
72. Mabey R. *Food for Free*. UK: Fontana/Collins, 1975
73. Davidson A. *The Oxford Companion to Food*. Oxford: Oxford University Press 1999
74. Bender AE, Bender DA. *A Dictionary of Food and Nutrition*. Oxford: Oxford University Press, 1995
75. European Commission. *Proposal for a regulation of the European Parliament and Council laying down the general principles and requirements of food law, establishing the European Food Authority, and laying down procedures in matters of food*. 2000/0286 (COD). Commission of the European Communities, 8th November 2000
76. Pliner P, Hobden K. Development of a Scale to Measure the Trait of Food Neophobia in Humans. *Appetite* 1992; 19: 105–120
77. Turner A. Novel foods, novel regulations. *Food Manufacture* 1997; May: 25–26
78. Branson L. Baby boomers begin backlash against GM foods. *The Scotsman*, 5th January 2000
79. Squires S. Sounds Like Corn. Tastes Like Chicken. *The Washington Post Online*, 12th March 2002
80. IFST: Current Hot Topics. Olestra. June 1999. Retrieved on 21st November 2001 from http://www.ifst.org/hottop29.htm

81. Oliver J. *Crete*. London: Lonely Planet Publications, 2000
82. Fieldhouse P. *Food and Nutrition. Customs and Culture*. London: Chapman & Hall, 1986
83. Schafheitle JM. Meal Design: A Dialogue with Four Acclaimed Chefs. In: Meiselman HL (ed). *Dimensions of the Meal. The Science, Culture, Business and Art of Eating*. Gaithersburg: Aspen Publishers, pp 270–310, 2000
84. Chesson A, James WPT. Physiological functionality in foods: a public health perspective. *Food Science and Technology Today* 1998; 12 (1): 34–37
85. Hom K. *The Taste of China*. London: Pavilion, 1996
86. *The Scottish Diet*. The Scottish Office Home and Health Department, 1993
87. Walker C, Cannon G. *The Food Scandal*. London: Century Publishing, 1985
88. Gutierrez CP. *Cajun Foodways*. University Press of Mississippi, 1992
89. Eagleton T. *The Idea of Culture*. Oxford: Blackwell, 2000
90. Evans JR, Lindsay WM. *The Management and Control of Quality*, 3rd edition. St Paul: West Publishing Company, 1996
91. Grunert KG. What's in a Steak? A Cross-Cultural Study on the Quality Perception of Beef. *Food Quality and Preference* 1997; 8(3): 157–174
92. Bredahl L, Grunert KG, Fertin C. Relating Consumer Perceptions of Pork Quality to Physical Product Characteristics. *Food Quality and Preference* 1998; 9(4): 273–281
93. Cowan C, Mannion M, Langan J, Keane JB. *Consumer Perceptions of Meat Quality*. Project FAIR-CT 95 – 0046, "Consumer behaviour and quality policy – meat", October 1999
94. Anon. Winners and Losers. Sunny D is a Hit. *Food Innovation Bulletin*, Leatherhead Food Research Association, September 1999
95. Foster A, Macrae S. Food Quality. In: National Consumer Council (ed). *Your Food: Whose Choice?* London: HMSO, pp 116–134, 1992
96. Murcott A. Food choice, the social sciences and 'The Nation's Diet' Research Programme. In: Murcott A (ed). *The nation's diet: the social science of food choice*. Harlow: Addison Wesley Longman, pp 1–21, 1998
97. Dubois B. *Understanding the Consumer. A European Perspective*. Harlow: Prentice Hall, 2000
98. Fine B, Leopold E. *The World of Consumption*. London: Routledge, 1993
99. Lewin SB. Economics and Psychology: Lessons For Our Own Day From the Early Twentieth Century. *Journal of Economic Literature* 1996; Vol. XXXIV: 1293–1323
100. Uusitalo L. Consumption in Postmodernity. Social structuration and the construction of the self. In: Bianchi M (ed). *The Active Consumer. Novelty and Surprise in Consumer Choice*. London: Routledge, pp 215–235, 1998
101. Anon. Market Survey 2. Yoghurt. *Consumer Goods UK* 2000; No.512, October
102. Michener HA, DeLamater JD, *Social Psychology*. 3rd edition. London: Harcourt Brace College Publishers, 1994
103. James A. Cooking the Books. Global or local identities in contemporary British food cultures? In: Howes D. (ed). *Cross-Cultural Consumption. Global Markets Local Realities*. London: Routledge, pp 77–92, 1996
104. Mansfield E. *Microeconomics*. 8th edition. New York: W.W. Norton, 1994
105. Becker GS, Murphy KM. A Simple Theory of Advertising as a Good or Bad. *The Quarterly Journal of Economics* 1993; CVIII(4): 941–964
106. Frank RH. *Microeconomics and Behavior*, 4th edition. Boston, Irwin McGraw-Hill, 1999
107. Lancaster KL. A new approach to consumer theory. *Journal of Political Economy* 1966: 74(3): 132–157
108. Earl PE. *Microeconomics for Business and Marketing*, Aldershot: Edward Elgar, 1995
109. Bianchi M. Introduction. In Bianchi M (ed). *The Active Consumer. Novelty and Surprise in Consumer Choice*. London: Routledge, pp 1–18, 1998
110. Stigler GJ, Becker GS. De Gustibus Non Est Disputandum. *The American Economic Review*. 1977; 67(2): 76–90.
111. Kahneman D. New Challenges to the Rationality Assumption. *Journal of Institutional and Theoretical Economics*. 1994; 150 (1): 18–36
112. March JG, Simon HA. *Organizations*. New York: John Wiley & Sons, 1958

113. Young T, Burton M, Dorsett R. Consumer theory and food choice in economics, with an example. In: Murcott A (ed). *The nation's diet: the social science of food choice.* Harlow: Addison Wesley Longman, pp 81–94, 1998
114. Ritson C, Hutchins R. Food choice and the demand for food. In: Marshall D (ed). *Food Choice and the Consumer.* London: Blackie Academic & Professional, pp 43–76, 1995
115. Prelec D, Herrnstein RJ. Preferences or Principles: Alternative Guidelines for Choice. In: Zeckenhauser RJ (ed.). *Strategy and Choice.* Cambridge, Massachusetts: The MIT Press, pp 319–340, 1991
116. Kahneman D, Tversky A. Prospect theory: an analysis of decision under risk. *Econometrica* 1979; 47(2): 263–291
117. Gualerzi D. Economic Change, Choice and Innovation in Consumption. In: Bianchi M (ed). *The Active Consumer. Novelty and Surprise in Consumer Choice.* London: Routledge, pp 46–63, 1998
118. Dupleix J. Follow me, I'm a New Gourmet. *The Times,* 7th April 2001
119. Tversky A, Griffin D. Endowment and Contrast in Judgments of Well-Being. In: Zeckenhauser RJ (ed.). *Strategy and Choice.* Cambridge, Massachusetts: The MIT Press, pp 297–318, 1991
120. Rolls BJ. Sensory-Specific Satiety and Variety in the Meal. In: Meiselman HL (ed). *Dimensions of the Meal. The Science, Culture, Business, and the Art of Eating.* Gaithersburg, Maryland: An Aspen Publications, pp 107–116, 2000
121. Conner M, Povey R, Sparks P, James R, Shepherd R. Understanding dietary choice and dietary change: contributions from social psychology. In: Murcott A (ed). *The nation's diet: the social science of food choice.* Harlow: Addison Wesley Longman, pp 43–56, 1998
122. Goldstein EB. *Sensation and Perception.* 5th edition. Pacific Grove: Brooks/Cole Publishing, 1999
123. Chisnall PM. *Consumer Behaviour.* 3rd edition. London: McGraw-Hill, 1994
124. Foxall GR, Goldsmith RE and Brown S. *Consumer Pychology for Marketing,* 2nd edition. London: International Thomson Business Press, 1998
125. Cline S. *Just desserts: Women and food.* London: Andre Deutsch, 1990 (cited in 159)
126. Lowe CF, Dowey A, Horne P. Changing what children eat. In: Murcott A (ed). *The nation's diet: the social science of food choice.* Harlow: Addison Wesley Longman, pp 57–80, 1998
127. Engel JF, Blackwell RD, Miniard PW. *Consumer Behavior. International Edition.* 8th edition. Forth Worth: The Dryden Press, 1995
128. Morrison I. Milking the Ayrshires. *The Scotsman,* 16th March 1998
129. Brug J, Ruiter R, Martens M, van Assema P, Kools M. Message framing in nutrition education. Abstracts. The 9th annual multidisciplinary Conference on Food Choice. *Appetite* 2000; 35: 195
130. Lieberman DA. *Learning. Behavior and Cognition.* 2nd edition. Pacific Grove, California: Brooks/Cole Publishing Company, 1993
131. Stuart J. "Meal times were a battle field". *The Guardian,* 6th July 2000
132. Schwarz N, Bless H, Bohner G. Mood and Persuasion: Affective States Influence the Processing of Persuasive Communications. *Advances in Experimental Social Psychology* 1991; 24: 161–199
133. Zuckerman M, Buchsbaum MS, Murphy DL. Sensation Seeking and Its Biological Correlates. *Psychological Bulletin* 1980; 88(1): 187–214
134. Zuckerman M. The psychobiological basis of personality. In: Nyborg H (ed). *The Scientific Study of Human Nature: Tribute to Hans J. Eysenck at Eighty.* Oxford: Elsevier Science Ltd, pp 3–16, 1997
135. Ratner RK, Kahn BE, Kahneman D. Choosing Less-Preferred Experiences for the Sake of Variety. *Journal of Consumer Research* 1999; 26: 1–15
136. Roskos-Ewoldsen DR, Fazio RH. The Accessibility of Source Likability as Determinant of Persuasion. *Personality and Social Psychology Bulletin* 1992; 18(1): 19–25
137. Bertrand M, Mullainathan S. Do People Mean What They Say? Implications for Subjective Survey Data. *The American Economic Review* 2001; 91(2): 67–72

138. Griffin DW, Ross L. Subjective Construal, Social Inference, and Human Misunderstanding. *Advances in Experimental Social Psychology* 1991; 24: 319–359
139. Ajzen I. *Attitudes, personality and behaviour.* Milton Keynes: Open University Press, 1988
140. Gärling T. The importance of routines for the performance of everyday activities. *Scandinavian Journal of Psychology* 1992; 33: 170–177
141. Shepherd R. Psychological aspects of food choice. *Food Science and technology Today* 1995; 9(3): 178–182
142. Sparks P, Hedderley D, Shepherd R. An investigation into the relationship between perceived control, attitude variability and the consumption of two common foods. *European Journal of Social Psychology* 1992; 22: 55–71
143. Shepherd R, Sparks P, Bellier S, Raats MM. Attitudes and Choice of Flavoured Milks: Extensions of Fishbein and Ajzen's Theory of Reasoned Action. *Food Quality and Preference* 1993; 3: 157–164
144. Thubron C. *The Lost Heart of Asia.* London: Penguin Books, 1994
145. Howes D. Introduction: Commodities and Cultural Borders. In Howes D (ed). *Cross-cultural consumption. Global markets, local realities.* London: Routledge, pp 1–16, 1996
146. Douglas M. *Thought Styles. Critical Essays on Good Taste.* London: SAGE Publications, 1996
147. Mennell S, Murcott A, van Otterloo AH. *The Sociology of Food, Eating, Diet and Culture.* London: Sage Publications, 1992
148. Caplan P, Keane A, Willetts A, Williams J. Studying food choice in its social and cultural contexts: approaches from a social anthropological perspective. In: Murcott A (ed). *The nation's diet: the social science of food choice.* Harlow: Addison Wesley Longman, pp 168–182, 1998
149. Leach E. *Claude Lévi-Strauss.* New York: The Viking Press, 1970
150. Cialdini RB, Kallgren CA, Reno RR. A Focus Theory of Normative Conduct: A Theoretical Refinement and Reevaluation of the Role of Norms in Human Behavior. *Advances in Experimental Social Psychology* 1991; 24: 201–232
151. Connors M, Bisogni CA, Sobal J, Devine CM. Managing values in personal food systems. *Appetite* 2001; 36: 189–200
152. Mortishead C. US food giants thrown in the mixer. *The Times,* 18th September 1999
153. Gopnik A. *Paris to the Moon. A Family in France.* London: Vintage, 2001
154. Marshall N. *Chambers Companion to the Burns Supper.* Edinburgh: Chambers, 1992
155. Graham P. *Mourjou. The Life and Food of an Auvergne Village.* London: Penguin Books, 1999
156. Balistier T. *Kretischer Raki – Raki-Kultur.* Mähringen: Verlag Dr Thomas Balistier, 1999
157. Sobal J. Socialbility and Meals: Facilitation, Commensality, and Interaction. In: Meiselman HL (ed). *Dimensions of the Meal. The Science, Culture, Business, and the Art of Eating.* Gaithersburg, Maryland: An Aspen Publications, pp 119–133, 2000
158. Murphy E. Britons are dining in style, but now they just want to be alone. *The Scotsman,* 19th October 1999
159. Mäkelä J. Cultural Definitions of the Meal. In: Meiselman HL (ed). *Dimensions of the Meal. The Science, Culture, Business, and the Art of Eating.* Gaithersburg, Maryland: An Aspen Publications, pp 7–18, 2000
160. Bourdieu P. *Distinction: A Social Critique of the Judgement of Taste.* London: Routledge, 1992
161. Ioannou N. *Barossa Journeys. Into a valley of tradition.* Kent Town, SA: Paringa Press, 1997
162. Verlegh PWJ, Steenkamp J-BEM. Country-of Origin Effects: Review and Meta-Analysis. In: AIR-CAT. *Workshop: Consumer Preferences for Products of the Own Region/Country and Consequences for the Food Marketing.* CEC, DGXII, 1998
163. Wahhab I. The truth about tikka masala. *The Independent,* 24th April 2001
164. Meades J. Goodness gracious! Chicken tikka smacks of our capability for self-delusion. *The Times,* 21nd April 2001

Part 2

Food Quality Attributes

Part 2 provides the "objective" background information on food, that is needed to underpin the arguments about food quality exercised in this book. It presents an overview of important aspects of food science and technology, which are relevant to the quality of foods and beverages. However, as demonstrated in Part 1, in the context of quality, even objectivity is not to be understood as an absolute. Objective definitions and measurements typically follow on from some sort of subjective categorisation and prioritisation of product and process attributes. It is important that significant product attributes, in terms of their importance to consumer satisfaction, are identified with care. In practical quality management terms, product quality needs to be quantifiable and measurable so that specifications can be both set and verified. This middle section is split into three chapters, each adopting a specific, focused perspective on food product quality, but with thematic linkages within each chapter and between different ones. Chapter 4 focuses on food components, beginning with structural ones, and then moving on first to nutrients and secondly, to so-called bioactive compounds. This particular structure has been adopted to emphasise the focus on quality. For example, carbohydrates are first discussed in their role as structural food components and then again in that of dietary fuels. This approach differs from standard food science textbooks, where one might find a section, say, on carbohydrates or one on nutrients. Chapter 5 sets out the origins and nature of important performance attributes, in particular, attributes related to the sensory perception of foods. It also provides illustrations of fitness for purpose, a familiar quality-related concept, linked here to the objectives of food processing. Factors affecting the keeping quality of foods are outlined, as shelf life is a key quality attribute of many modern food products. Finally, Chapter 6 deals with food additives, functional ingredients and food contaminants.

The structure of most original foodstuffs, such as the edible parts of plants and animals, is basically due to the organisation of cells and cell communities. Cells provide a fixed framework, which further serves to contain aqueous or fatty systems. In man-made foods, such structures are often mimicked or approximated in order to achieve palatability, particularly acceptable texture. Structural features therefore tend to be complex both in original and in man-made foods. The compositional elements in foods that determine structure belong to three main chemical groups, i.e., carbohydrates, proteins and fats. Water too plays a key part, both as a solvent and dispersing agents for low molecular weight compounds, and because of its involvement in lipid/aqueous structures, such as membranes and emulsions. Wheat flour is an example of a refined food component, extracted from the original foodstuff. As such it lacks a suitable structure for direct consumption. A structure has to be created through appropriate processing, in this case perhaps using gluten-assisted dough leavening, followed by baking, to make bread. Molecules identical or similar to those that provide structure in original foodstuffs (carbohydrates, proteins, fat and water) – some native, some refined, some modified – figure in the structure of processed and fabricated food products. The complexity of the structure of most foods has major implications for their sensory attributes, in particular, appearance, texture, and flavour release during eating. It also affects the patterns of microbiological colonisation of the foods and therefore, keeping quality. Multiphase systems mean that different types of microflora may exist in different parts of the food. Structural factors also impact on suitability of a food for different types of processing. So-called structural additives may be used to underpin/stabilise the structure of many man-made foods.

Water holding capacity (WHC) is an example of a key structure-related food quality attribute. It is important in a number of different foods, in particular, meat and meat products, where it is directly related to perceived juiciness. The crunchiness of fruits and vegetables is largely related to the turgor within plant cells and depends on their physical integrity. In meats and meat products, structural additives such as polyphosphates may be used to confer suitable WHC where the WHC of the raw material is insufficient. As it is generally impossible to reconstitute dried foods to achieve the original structure and related functionality, it is interesting to ponder whether the same holds true, in vivo, for dietary fibre. For example, is a prune as good a source of fibre as the plum from which it was made? Does the prune achieve perfect reconstitution and functionality during its passage through the human digestive system?

Many of the carbohydrates, proteins and fats that provide structure to a food, are in addition sources of nutritionally important food components, e.g., macronutrients, metabolic fuels (energy sources) and fibre. As well as macronutrients, food also contains micronutrients, i.e., vitamins and minerals, in varying proportions [1]. It has been known for some time that nutrients, the classic essential dietary factors, are not the only food components exerting positive effects on human health. Currently, there is considerable interest in so-called bioactive food components, in particular, flavonoids and other secondary plant metabolites, and pre- and pro-biotics [2]. This interest extends both to discovering and revealing the intrinsic value of foods containing these compounds naturally and to so-called functional foods, new food concepts where bio-active compounds have been introduced into conventional foods. The fact that many original foodstuffs exhibit various kinds of biological activity that is beneficial to humans, as well as to the plants producing them, should not come as a surprise. The close affinity between foods and medicines has been understood by many cultures around the world. For the purposes of this book, the term bioactive will be applied both to compounds with beneficial and to those with undesirable effects. Included are food allergens, natural toxins of plant origin, and dietary estrogens and anti-estrogens. Often, whether ingestion of a particular substance produces a desired or an undesired effect, depends on the dosage and on subjective factors, such as the health status of the person consuming them and the context of consumption. This is evident with common stimulants and toxins such as the caffeine contained in coffee and tea. Also discussed here are intrinsic toxins that must be removed through processing in order to transform a potential food into an actual one, a familiar example being heat inactivation of enzyme inhibitors in red kidney beans. It is interesting – perhaps revealing – to note the official view of mould-derived toxins in foods as naturally occurring toxins [3]. In this book, mycotoxins will be discussed as extrinsic toxins, i.e., as biological environmental contamination.

Although the evaluation of the sensory character of foods and beverages ultimately requires the involvement of the human senses, these attributes are regarded as objectively determinable; they relate mainly to the appearance, texture and flavour of foods. Whilst standard sensory terminology is available for some common foods and beverages, it needs to be remembered that there will always be a subjective component in sensory product profiling, which relates to attribute selection and weighting. For most foods, sensory descriptive terminology needs to be newly developed, making use, as far as possible, of generally accepted sensory terms. Hedonic considerations of foods are excluded from Part 2 of the book, as these are clearly dependent on individual preferences and therefore part of the subject matter of Part 3.

Food additives and food contaminants include chemicals that are of concern to sections of the current UK consumer market for food. Whilst the intentions in applying each category of chemicals differ, any negative effects on human health may nevertheless be similar. Additives perform specific technological functions in foods, are added deliberately for that purpose and remain in the food [4]. Additives are not the only food ingredients with specific technological function. Both in domestic and restaurant cooking and in industrial food processing, refined food ingredients are commonly used to create dishes, or at least parts of dishes, e.g., sauces. Complex, man-made foods such as extruded cereal products rely on these to a large degree, if not entirely. The larger the scale of production the more beneficial are chemical purity and specific functionality of ingredients in order to ensure predictable outcomes of processing.

Unlike additives, food contaminants do not perform a technological function in the finished food, and their presence there is not desired. Chemical food contaminants include residues of chemicals used as pesticides, veterinary medicines and processing aids, environmental contaminants including those introduced through malicious tampering, and additives carried over from ingredients, but which have no technological function in the finished foods. Veterinary medicines include antibiotics for the treatment of sick animals and for prophylactic purposes, i.e, to promote animal growth through the prevention of disease [5]. It is the live animal that receives treatment, whilst any contamination of foods made from the animal is incidental. A related situation exists in the feeding of pigments to laying hens and to farmed salmon in order to enhance rich yolk and flesh colour, respectively. It may be argued that if pigments such as canthaxanthin (E161 g) are transmitted to foods [6] by such indirect (and undeclared) means, they should be classified as contaminants, not additives. In reality – officially – they are neither, the Food Standards Agency's farm-to-fork connection not being made in this instance. At worst, this is an example of an officially sanctioned system of the misleading of consumers. Also considered under chemical contaminants are food additives, where these are used illegally in a food, whether as such or in terms of permitted levels being exceeded. Salty foods can be toxic to babies, with fatal consequences, so that salt would have to be understood as a dangerous contaminant of weaning and infant foods. Finally, contamination of a food can be desirable as is the case with oak-aged wines and whiskies. Biological contaminants include microorganisms, algae, nematodes, worms and protozoa and the toxins that some of these produce, e.g., bacterial and algal toxins and mycotoxins.

Part 2 therefore discusses food and beverages from the perspective of objectively definable and measurable attributes, but seen through the filtering device of the quality concept. It provides the scientific, and largely value-neutral, background for the analysis of specific types of consumer value derived from product attributes, which is the main focus of Part 3. Consumer value judgements cannot be made on the basis of scientific facts about foods. For example, oily fish are desirable dietary components because of their richness in beneficial fatty acids (benefits); at the same time, that same fat has become a sink of toxic environmental pollutants such as dioxin (costs). What action is to be taken by consumers and by those advising consumers? This debate (cost-benefit; risk assessment), among many others, is taken up in Part 3.

4 Food Composition

4.1
Introduction

Food composition is commonly discussed from a nutritional point of view, focusing on the three main chemical groups present in foods, i.e., carbohydrates, proteins and lipids, and on their role as sources of macronutrients and of food energy. The approach adopted in this book differs somewhat from that convention. The first three topics discussed in the present chapter adopt a focus on those food components that represent, or contribute significantly to, food structure. Structure after all is the primary identifying feature of any object, including foods and beverages. Structure expresses the relationships between the individual components that make up complex systems. It plays an important role in various aspects of food quality, in particular, sensory quality attributes, such as appearance and texture, and the microbiological quality of foods. Texture relates to the ease with which a food may be manipulated in the mouth, e.g., chewability and swallowability, important in terms of foods targeted at young children. Man-made foods are often designed specifically to create textures that consumers find interesting or attractive, and these are not always modelled on familiar natural foods. For example, extruded snack products tend to be valued especially for their texture which, in the case of starch based products, is typically light and crispy. Gels, such as tofu, also provide man-made structures with textures attractive to consumers. Food structure affects the order and rate of release of flavour-active components from the food matrix during eating. Integrity of structure plays an important role in the access to foods by spoilage microorganims, and different areas and phases (aqueous, lipid-based) within complex foods allow different types of microflora to become attached to, and develop in them. Structure also affects the suitability of a food for different types of processing. Different cuts of meat demand different culinary treatment or processing, and different varieties of a fruit or vegetables have different suitability for processing, e.g., drying, freezing or heat preservation. Food processing often affects structure in a way that alters appearance. For example, meat and fish turn from a translucent to an opaque state as a result of heat- or freezing-induced protein denaturation.

The structure of plant and animal tissues is due primarily to their biological design, in particular, cell organisation and assembly. There are two levels of structure here, i.e., the supporting framework, consisting of large macromolecules or biopolymers, and a complex system of smaller structures associated with the cells and their physiological function of dissolution, transport and exchange of metabolic products. Turgor of plant cells is directly related to texture, e.g., the crunchiness of apples. Destabilisation of the aqueous phase, which is one of the results of food being frozen, promotes molecular mobility. This enables migration and concentration of solutes, and usually leads to permanent cell damage. An example of this is the so-called drip of frozen products, which may be witnessed during thawing. The molecular basis of biological food structures is confined largely to four chemical categories, i.e., carbohydrates, proteins, lipids and water. In plant cells, the predominant structural elements are carbohydrates (polysaccharides); in muscle and other fleshy foods, proteins are the main structural components. Nevertheless, both plants and animals contain members of all four categories, in varying proportions, but performing a range of different functions. Structure may be of a more flexible, fibrillar, "spun" nature as is the case with muscle fibres; or it may mean a more rigid network providing strength as is typical of plant tissues [7]. Although plant exudates and animal secretions, such as gums and milk, appear structurally simple to the naked eye, on close inspection they too contain complex elements. The milk fat globules and casein micelles of milk, which allow fat and protein to be emulsified and dispersed within an aqueous phase, are a case in point [8]. Plants can be categorised according to their contents of the above basic food component groups, although this mainly reflects an interest in nutrient and energy stores rather than structure [9]. High-protein, vegetarian foods are represented by pulses and their sprouts, nuts and seeds; high-starch foods by grains and their sprouts, potatoes and carrots; high-fat foods by avocados, olives, sesame seeds and flaxseeds; and non-starchy vegetables by leafy greens, celery, asparagus, cucumber, radish, onion and green beans.

Although the macromolecules are important in various aspects of food structure, aqueous and lipid systems, and the interactions between them, have been a recent focus of food structure research. In man-made foods particularly, structural elements may result from physical and/or chemical interactions between one or more classes of food components [10]. In food dispersions, interactions may take place either between the continuous and dispersed phases or between components within the dispersed phase. This affects stability as well as texture, e.g., in terms of spreadability and viscosity (pourability, clinging to surfaces). Structure formation in dispersed systems, e.g., foams and emulsions, is driven by the presence of interfaces, hydrophilic-hydrophobic balances, net charge on surfaces, and so on [7]. Foods may be liquid or solid, or they may contain both a solid and a liquid phase. Food solids are divided into lipid-soluble and water-soluble substances. Carbohydrates and proteins may be present either in amorphous, or in ordered, crystalline states, or they may be dissolved. Native starch is composed of a mixture of amorphous amylose and partially crystalline amylopectin and as such, it is an example of a partially crystalline material [11]. Fats, oils and frozen foods are typical of solids being present in the crystalline state within the food matrix [11]. The mechanical and rheological properties of fats are important to their behaviour during processing, packaging and storage, and to their spreadability and oral melting properties. Ice crystals in frozen foods can damage the cell structure of the food, leading to drip during thawing. The mechanical properties of food powders, cereal foods, snack foods and boiled sweets, among others, are important in defining various quality parameters, including free-flowing properties of powders, stickiness, and perceived texture. Food powders are often produced by rapid dehydration, and they then contain amorphous carbohydrates and proteins, which become plasticised by water. Hygroscopic food powders become sticky, decrease in volume and lose their free-flowing properties above a critical water content or water activity [11].

Structural considerations take up the bulk of Chapter 4. The role of nutrients as food components is discussed next. In some instances, this means reviewing previously encountered food compounds, albeit from an alternative point of view. For example, food carbohydrates, proteins and lipids make a second appearance here. These structural macromolecules are broken down during the maturation and senescence of plants, in certain types of food processing, during spoilage and ultimately, in the human digestive tract. The essential nutrients derived from these macronutrients are the indispensable amino acids (IAA) and the essential fatty acids (EFA). These, together with vitamins and minerals, are classed as micronutrients. Referring to carbohydrates, proteins and fats as macronutrients, as is common practice, is somewhat misleading, since strictly speaking, it is only the IAA and EFA which exhibit any nutrient functionality. The non-essential portions of the macromolecules do however represent dietary energy value. The chapter closes with a discussion of a selection of the so-called bioactive food compounds. Bioactivity in foods may or may not be physiologically beneficial, and in some cases, the harmful effects of the substances concerned are undisputed. On the other hand, many of these compounds are increasingly believed to play important positive roles in diet-related human health.

4.2
Structural Carbohydrates

Carbohydrates are, or derive from, simple sugars and the latter are chemically defined as $(CH_2O)_n$. Their basic building blocks are monosaccharides, e.g., glucose. Polysaccharides, referred to as complex carbohydrates by nutritionists, are macromolecular, polymeric carbohydrates composed of repeating units of simple molecules [1]. Homopolymers, such as starch, are made up of only one type of monomer, in this case, D-glucose. Glycoproteins and some algal polysaccharides are examples of copolymers, i.e., polymers composed of different types of monomer [7]. The polysaccharides that are important in food science are found mainly in plants, where they have three major roles, to give structure (e.g., cellulose), as an energy reserve during dormancy and germination (e.g., starch), and to stabilise aqueous systems by binding water (e.g., gum) [7]. In addition, simple sugars help regulate the osmotic pressure within biological cells [12]. The so-called structural polysaccharides, i.e., cellulose, hemicellulose and lignin, constitute, or are part of, rigid, mechanical plant structures [12]. There are however two levels of structural organisation, i.e., a supporting framework of large biopolymers, and a complex system of smaller structures associated with the cells and their physiological functioning. Carbohydrates such as starch which, from the point of view of the plant, are more important as energy reservoirs than as structural elements, do however assume important structural functionality in man-made foods. In fact, the structure of starches themselves is related to the functional properties which they demonstrate during processing and when present in different foods. Starches thicken when heated in the presence of water, and they are then capable of gelling aqueous food systems such as sauces and puddings [13]. Polysaccharides in man-made foods include starches (raw, pregelatinised and modified), cellulose and cellulose derivatives, seaweed extracts (alginates, carrageenans, agar, and furcellaran), plant exudates or gums (arabic, karaya, and tragacanth), seed gums (locust bean and guar), plant extracts (pectins) and microbial gums (xanthan) [12]. The corresponding simple carbohydrates are the hexoses, i.e., glucose, galactose and mannose, the hexuronic acids, and the pentoses, i.e., arabinose, rhamnose and xylose. Some of these polysaccharides are chemically modified for closer control of functionality; these are controlled as food additives (see Chapter 6).

Cellulose, an essential cell wall component of all higher plants, has an average of about 3000 glucose units per molecule, is insoluble in most solvents and exhibits high mechanical strength [7]. Together with hemicellulose, lignin, pectin and proteins, it provides structural integrity. Long cellulose chains may be held together in bundles forming fibres, such as in "stringy" celery [14]. The cellulose and hemicellulose contents of fruit, vegetables and cereals increase with maturity. Cellulose structures are often ruptured by the growth of ice crystals when vegetables are frozen [14]. Hemicelluloses in plants are closely associated in cell walls with cellulose. Three types are recognised: the xylans, the mannans and glucomannans, and the galactans and arabinogalactans [7]. Hemicelluloses include polymers of mannose, galactose, xylose and arabinose, and of uronic acids and methyl- or acetyl-substituted monoses [12]. Xylans are major components of seed coats and cereal grains. A considerable proportion of the lignins in plants appears to be linked to hemicelluloses and pectins [7]. They act as cementing components between cells, harden cell walls and shield against water, and they do not soften when food is cooked. In fruit and vegetables, only small amounts of lignin are present; however, large amounts occur in the bran of some cereals. In the context of the texture of beans, the term "stringless" refers to varieties with low lignin contents [15].

Pectic substances are present in the middle lamella of the cell walls of fruit and vegetables, their main constituent being galacturonic acid, with rhamnose units inserted into the main chain [7]. They serve as cementing agents of the cell and regulate its water content [12]. During the ripening of fruit, the cell pectins are solubilised enzymatically. They are therefore present in fruit juices where they form gels readily. Pectins can be divided into pectic substances, protopectins and pectin [12]: Pectic substances comprise all polygalacturonic acid-containing material; protopectins comprise all bound, water-insoluble materials which yield pectin upon hydrolysis; and pectin refers to partly esterified galacturonic acids. For practical purposes, pectins are classified as either low-methoxyl or high-methoxyl. In pectinic acids all carboxyl groups are in the free form, and they are water insoluble; the salts of pectinic acids on the other hand are water-soluble [12]. Pectic acid is the simplest of the pectic substances; it is soluble in water and contains an abundance of carboxyl groups, making it acidic and capable of forming water-soluble salts [15]. Pectinic acids are similar to pectic acid except that some of the carboxyl groups are esterified with methyl groups; they too are capable of reacting with metallic ions to form salts, indeed, many of the pectic substances exist in plants as calcium of magnesium salts [15]. The term pectin designates those pectinic acids that are capable of forming jelly with sugar and acids. Pectins are esterified to varying degrees, with the remainder of the carboxyl groups present uncombined or combined to form salts [15]. Pectin is present in unripe fruits mainly as its precursor protopectin; this is responsible for their hardness [16]. Fruit ripening is generally accompanied by softening of the tissues. In pears, the change in cell wall integrity involves the increase in the soluble pectic substances, with an overall loss of cell wall arabinose and uronic acids [17].

Cell walls cemented together impart a degree of rigidity, but the main means of maintaining shape in plant tissue is the turgor pressure within individual cells; this is sustained by osmotically active constituents, principally sugars and ions [18]. As with breakdown of the cell walls themselves, breakdown of pectin in the middle lamellae leads to cell separation [18]. Middle lamella breakdown occurs earlier in potato varieties giving a mealy texture than in those giving a firmer texture [18]. High-methoxyl pectins tend to be less water soluble and hence more resistant to this breakdown than low-methoxyl pectins; therefore, the ratio of high and low methoxyl pectins in the tissue influences the rate at which cell separation occurs [18]. In navy beans (*Phaseolus vulgaris*), the so-called "hard-to-cook" phenomenon has been attributed to the failure of the middle lamella to dissolve during cooking [18].

At the cellular level, the key determinants of texture comprise the size of cells, cell wall thickness and cell-cell adhesion; what is important at the molecular level is the chemical nature of the cell walls and the interrelations between component polymers [19]. The texture of cooked potatoes is influenced by several structural factors, in particular, cell wall composition, the nature of interactions between adjacent cells, and starch gelatinisation. Two extremes with respect to texture after cooking have been studied, i.e., a mealy and a firm potato variety [19]. The firm variety had significantly less cell wall material than the mealy one, with no significant differences in the overall chemical composition of cell wall material. The cell walls of the mealy variety were both thicker and denser. In cooked samples of the firm variety, large intercellular contacts were present, with almost no intercellular contacts in the mealy variety [19]. Application of heat will usually induce softening of plant tissues caused by the solubilisation of pectic substances of the middle lamella, which is followed by cell separation [20]. Unlike potatoes, Chinese water chestnuts maintain a crunchy texture after cooking, since processes such as canning fail to induce cell separation [20]. This is thought to be due to the involvement of ferulic acid dimers in the cross-linking of polysaccharides, and to the presence of arabinose-containing polysaccharides such as arabinoxylans [20].

Native starch is composed of two polymeric fractions of repeating glucose units, i.e., amylose, which is linear, and amylopectin, which has a multiple-branched structure. Plants lay down starch in the form of granules, normally 10–50 μm in diameter, with molecules being organised into a semi-crystalline unit [7]. The starch of unripe fruit is converted into glucose as the fruit ripens. It acts as the storage reservoir of glucose in plants, but does not occur in animal tissues; the animal equivalent of starch is glycogen. Starch occurs in most green-leafed plants, in seeds (cereal grains), roots and tubers (tapioca and potato), stem-pith (sago), and in fruit (banana) [12]. Granules of different starch species (e.g., corn, potato, rice, sago, tapioca, wheat) have characteristic size, shape, and markings [12]. The amylose:amylopectin ratio is generally fairly constant in any one species, although rice starch may vary more widely, especially between *Japonica* and *Indica* varieties [12]. So-called waxy varieties of cereals have starch composed entirely of amylopectin, for example corn, rice, barley, and sorghum. On the other hand, starches from wrinkled-seeded garden peas and from so-called high-amylose sweetcorn are predominantly linear [12].

Suspensions of starch granules in water begin to swell when heated to a certain critical temperature, which is dependent on the species of starch. Eventually the swollen granules become so large that they begin to crowd one another, producing a thick-bodied consistency [12]. This process, called gelatinisation, occurs over defined temperature ranges; potato starch gelatinises at 56–67 °C, cornstarch at 62–72 °C, and rice and sorghum starches at 68–78 °C [12]. There are wide variations in the gelatinisation temperatures of different types of rice starch [12]. If water is reduced or solutes are added, the gelatinisation temperature increases [7]. After cooling, dilute solutions of starch will precipitate, but concentrated dispersions may form a firm, visco-elastic gel with crystallites as junction zones [7]. All native starches contain semi-crystalline regions caused by the ordered packing of adjacent branches of amylopectin molecules. The structure of legume starches differs from that of cereal and potato starches in the way that these crystalline regions are constructed and packed within the starch granule [13]. The structure of starches is related to the functional properties which they demonstrate during processing and when present in different foods. Changes in the mechanical properties of starch gels, which may occur during storage, reflect changes in crystallinity, often caused by the slow crystallisation of amylopectin. Cooked starch pastes owe their character to the persistence of undissolved swollen granules as elastic gel particles [12]. Only by pressure cooking at 120–150 °C can swollen starch granules be truly dissolved. During cooling, starch molecules aggregate and crystallise out of solution. The linear amylose chains in particular associate closely, forming aggregates of low solubility, and in high starch concentrations, they form gels. This is the familiar phenomenon of retrogradation, and it primarily involves amylose. Retrogradation is one of the mechanisms involved in bread staling, where the texture of bread increases in both firmness and crumbliness, as WHC decreases [12]. In gelatinised starch slurries, previously bonded water may be displaced and appear as free liquid. This phenomenon is known as syneresis or, in more everday language, weeping [21]. Retrogradation takes place throughout the storage of food products such as cakes and biscuits, as the starch competes for water with other ingredients, such as protein and sugar. Because of this competitive mechanism, starch contributes to structure formation in a range of man-made food products [22]. In fried potato products, one of the functions of blanching is the gelatinisation of starch in the surface layers, reducing fat absorption [23]. The functional properties of starches are usually described in terms of their viscosity, and their swelling and solubility behaviour under heating and subsequent cooling and storage [13]. Chemically modified starches, which have more tailored functional properties than native starches, are discussed in Chapter 6.

The position occupied by pectins in higher plants is mirrored in algae and seaweeds by either alginates, agars or carrageenans [7]. These substances too occur in the cell walls and intercellular spaces, providing for both cell flexibility and strength. In the global context, many seaweeds are consumed as foods in their own right, but seaweeds also serve as raw materials for a wide variety of functional food ingredients and additives. Pectins and gums are often referred to as soluble fibre because they form gels in water. Gels may be conceptualised as "soft solids". They occur in many different types of high-moisture, man-made foods, e.g., jellies, jams, confectionery, yoghurt, frankfurters and surimi [7]. The term gum is a generic name for polysaccharides that show great affinity for water and high viscosity in solution without forming gels [7]. These compounds, e.g., gum tragacanth, are also termed hydrocolloids. Seed gums from guar and locust bean are galactomannans. Plant exudates include the gums arabic, karaya and tragacanth. Fructosan defines indigestible polysaccharides of fructose, such as inulin. Inulin is soluble and is found particularly in Jerusalem artichoke, and to a lesser extent, other root vegetables [1]. Glucans are soluble, but undigested complexes made up of glucose units, and are found especially in oats, barley and rye. Gums can be dispersed in water to form a viscous mucilaginous mass [1]. They are present in seeds, plant sap and exudates and seaweed. Several sources of microbial polysaccharides are also used in foods, e.g., dextran, gellan and xanthan gums. Lactic acid can be used to generate "ropy" textures in sour milk products, such as Finnish national staple, viili.

4.3
Structural Proteins

The structural functionality of proteins is expressed through their macroscopic properties, such as gelation and water holding capacity (WHC) [24]. Native proteins provide the structure of meat, whilst isolated proteins from a variety of sources can be re-structured into a range of man-made food systems based on, for example, spun fibres (e.g., texturised vegetable proteins, TVP), gels, bean curds and cheese curds, and foams. Polymers of a small number of amino acids are called peptides, with oligopeptides containing up to about 50–100 amino acids [1]. All living tissues contain proteins – or polypeptides – with a total of twenty amino acids commonly occurring [1]. There are four levels of organisation [25]. The amino acid sequence of a protein constitutes its primary molecular structure, whilst secondary structure means the spatial (steric) relationships between amino acids located close to each other within the sequence. The most familiar examples of secondary protein structure are the α-helix and the β-pleated sheet. Such configurations arise from the regular repetition of intramolecular hydrogen bonds. Tertiary structure describes spatial configurations between more distantly located amino acids, and quaternary structure refers to the aggregation (packing) of proteins with several polypeptide chains. Specific three-dimensional configurations adopted by proteins in solution derive from these interactions [24]. In biological systems, proteins mediate a range of functions additional to mechanical support, e.g., transport and storage, co-ordinated motion, immune protection, excitability, and the control of cell growth and differentiation [25]. Muscle contraction is made possible by the sliding motion of two proteins, i.e., actin and myosin, and the strength of skin and bone is derived from collagen. Enzymes, the catalysts of chemical reactions in biological systems, are proteins, and many small molecules, such as oxygen (haemoglobin, myoglobin) and iron (transferrin, ferritin), are transported or stored as complexes with specific proteins [25]. Some hormones are polypeptides, e.g., insulin, which regulates the metabolism of glucose, and β-endorphin, which acts as a natural pain reliever [26]. In terms of the protein functionality in foods, quaternary structure is particularly important. For example, whilst dissolved casein does not gel readily, when integrated into micelles, as in milk, it can be induced (dependent on pH) to form strong cheese curds [24].

The native structure of a protein is the result of both intra-molecular forces, as described above, and of interactions with the surrounding aqueous environment. Changes in the secondary, tertiary and quaternary structures, i.e., without the cleavage of primary peptide bonds, are summarised under the term denaturation. This unfolding of the native structure of a protein is associated with a loss of functionality, both in live biological systems and in foods. In the latter, a degree of denaturation may in fact be desirable, e.g., in terms of improved digestibility, biological availability and/or technological performance [7]. Heating causes many proteins to denature, usually irreversibly, whilst changes in solvent conditions can lead to protein precipitation [24]. Food processes dependent on protein structure include gelation, texturisation, dough formation, emulsification and foaming, all of which can lead to stable food structures [7]. These processes are complex in the context of foods because of the intentional or unintentional modification of proteins resulting from processing steps such as heating, which can lead to associations with other compounds, including carbohydrates and lipids, as well as an unfolding of protein structures [7]. One of any protein's most important technological characteristics is its thermal stability; this can be altered through chemical modification [7].

Structural proteins tend to be fibrous, with highly organised internal structures, and with the polypeptide chains being lined up approximately in parallel. They may consist of single, linear polypeptides or of a number of cross-linked ones, cross-linking providing added structural support [27]. Examples of fibrous proteins are protective tissues such as hair, skin, nails and claws (keratins), connective tissues such as tendon (collagen), and the contractile material of muscle [26]. Structural proteins are also present in cellular membranes. The most important conformation of fibrous proteins is the α-helix, in which the polypeptide chain coils about itself in a spiral manner [26].

Muscle cells are cylindrical fibres, 10–100 μm in diameter and 20 μm to several centimetres in length; they are arranged in parallel bundles to form whole muscle [7]. A membrane with an associated connective tissue compound sur-

rounds the muscle cell; this is the sarcolemma [7]. The two major muscle proteins are actin and myosin. Both these proteins, either singularly or combined as
actomyosin, are prized components of man-made foods, because of their functional properties, which include binding and emulsification ability and the ability to form heat-set gels [7]. Actin and myosin overlap as a result of contraction,
and the degree to which this overlapping occurs is reflected in sarcomere length
[7]. The more post-mortem muscle sarcomeres are contracted, the tougher meat
becomes. For this reason, in the case of species where toughness of meat can be
a problem (e.g., beef), muscles are left on the carcass at least until rigor mortis
is complete, thus eliminating unwanted contraction during butchering [7].
Pelvic suspension of carcasses during hanging further counteracts muscle
shortening and related toughness of meat [28]. One consequence of meat comminution (chopping) is the release and solubilisation of muscle proteins that are
capable, upon heating, of forming a stable matrix for a variety of meat products;
addition of salt further encourages protein extraction in such systems [7].

Connective tissue includes formed elements and an amorphous ground substance in which formed elements are embedded [29]. The formed elements consist of the fibres of collagen, which are straight, inextensible and non-branching;
and of those of elastin, which are elastic, branching and yellow in colour. Collagen almost entirely composes the connective tissue in tendons and muscles,
where it is needed for the transmission of tensile stresses; this strength is a major contributor to meat texture. Post-mortem events dictate the alignment of collagen fibres that contribute to the toughness associated with contracted muscle
[7]. Collagen forms the basis of the gelatine industry, as solubilised and leached
collagen causes formation of gelatinous mixtures [7]. Three left-handed helices
coil around each other in a compact triple helix, i.e., tropocollagen; five cross-
linked tropocollagen molecules make up a collagen microfibril; bundles of fibrils
then form the rod-shaped collagen fibres [7]. The cross-linkages with tropocollagen increase with the age of animals, resulting in increased toughness of meat,
but collagen structure is weakened upon heating [7]. Different meat cuts also represent different collagen-related toughness, at least in dry-heat cooking methods.
In fact, a high collagen content may be appropriate, depending on the recipe.
Moist cooking has a dramatic effect on collagen properties, since it sequentially
produces softening, shrinkage, and conversion to gelatine, which is composed of
much less structured molecules [7]. Gelatine swells in contact with cold or warm
water, and when heated to temperatures above its melting point, the swollen gelatine dissolves. Upon cooling, the molecules re-form into triple helices and give
rise to a transparent and elastic gel [7]. Shin of beef adds a smooth jellied texture
to stewed beef, unlike the more expensive stewing beef [30]. Muscle blocks in fish
are held together, and their contractile forces are transmitted to the vertebral column, by thin collagenous membranes known collectively as myocommata [31].
Surimi technology involves repeated washing of fish mince, followed by de-wa-
tering to increase the concentration or myofibrillar proteins. The gels derived
from surimi by the action of salt and heat have an elastic, cohesive texture, which
is valued in Japanese ethnic foods such as kamaboka. This property has been exploited for the production for Western markets of so-called seafood analogues,
e.g., mock crabs, scallops and prawns [32].

Globular proteins are designed by nature either to be soluble in aqueous body fluids or to perform certain functions in intercellular membrane structures [26]. Enzymes and transport proteins tend be globular, where the polypeptide chain is folded around itself in a such a way as to give the entire molecule a rounded shape. These proteins often carry a non-protein molecule (the prosthetic group) as part of their structure [26]. Conjugated proteins are complex compounds of globular proteins with non-proteinaceous materials, e.g., phospho-proteins (casein), glyco-proteins (in blood plasma and connective tissues), chromo-proteins (flavo-proteins), metallo-proteins (haemoglobin) and lipo-proteins (parts of membrane structures) [27]. Globular proteins are susceptible to denaturation (unfolding) through heat or a change of pH, resulting in decreased solubility and increased susceptibility to attack by proteolytic enzymes [27]. Fully denatured globular proteins resemble random coils [7]. The gelling of myosin is induced by heat ($>60\,°C$), as is the case generally for globular proteins, and also strongly depends on factors such as the pH and ionic strength [7].

Protein accounts for about 10–15% of the dry weight of mature cereal grains, and storage proteins for about half of the total; in legumes such as soybean, protein levels are even higher (up to 40%) [33]. Gluten, the protein portion of wheat doughs, consists of over 50 individual proteins, classified as either monomeric gliadins or as glutenins; the latter consist of subunits assembled into polymers stabilised by inter-chain disulphide bonds [33]. Gliadins are responsible mainly for gluten viscosity, and glutenins for elasticity [33]. These proteins are synthesised in the developing starchy endosperm of the wheat grain and deposited in discrete structures called protein bodies; these bodies are disrupted as the cells become extended with starch, and the protein forms a matrix in the dry mature grain [33]. Milling, wetting and kneading result in the formation of a continuous visco-elastic gluten network within the dough which traps carbon dioxide, leading to the essential light, porous crumb structure which is characteristic of leavened bread [33]. Similar proteins also affect the quality of barley for malting, brewing and distilling [33]. Hordein storage proteins comprise about half of the total nitrogen in the mature barley grain [34]. Hordein consists of four groups of polypeptides, called B-, C-, D- and γ-hordeins, which differ in their structures, properties and contribution to the total. High levels of grain protein are disadvantageous for malting, but hordeins may contribute positively to the brewing performance of barley, by supplying peptides involved in foam stability and cling [34].

Plant proteins, most notably from soybean, are a major source of refined proteins for food and feed in the form of flours. The potential of these amorphous proteins lies mainly in their ability to be structured into man-made foods through extrusion, spinning, gelation or baking [7]. The re-structuring of concentrated plant proteins by extrusion has been employed since the late 1960s as a commercial process for manufacturing plant-based products for use as meat substitutes and meat extenders [7]. Structurally, the soybean is typical of oilseed legumes in that the cotyledons store protein in the form of separate globoid protein bodies that average 5–10 μm in diameter and contain up to 90% protein: the protein bodies account for 60–70% of the total protein of the soybean [7]. It may have been the desire to destroy bioactive components such as trypsin inhibitors in soybeans that first led to the preparation of soy milk [7]. Soy milk can be consumed as such, i.e., as a beverage, or it can be used as the starting material for several structured foods. Tofu is made by precipitating the proteins in soy milk with a calcium salt to form a coagulum that is then drained, pressed and washed [7]. The texturisation of de-fatted soybean grits by thermal extrusion is caused by protein fibre formation, resulting from thermally induced inter-molecular cross-links; extruded soy meal is an aggregate of insoluble carbohydrates within a continuous protein matrix [7]. Protein fibres may be spun from alkaline extracts of soy isolate (>90% protein) to form meat analogues [7]. The mycoprotein Quorn is derived from the body mass of the mould *Fusarium graminearum* [35]. It is manufactured in the form of fine fibres which are arranged so as to produce a texture which is reminiscent of that of meat.

Milk contains two basic types of protein, i.e., caseins and whey proteins, and these are present with an approximate ratio of 4:1. Caseins have the unusual feature that the hydroxyl groups of serine are phosphorylated. The clustered phosphoserine residues are responsible for the hydrophilic areas of strong negative charge; the molecules also contain blocks of hydrophobic residues [36]. Caseins (α_s-, β- and κ-) generally lack large areas of regular structure; they are rather flexible molecules which exhibit surfactant properties [37]. Caseins interact with each other and with calcium phosphate to form highly hydrated spherical complexes known as micelles. Alpha$_s$-caseins are sensitive to calcium, precipitating at neutral pH if Ca^{2+} ions are around [36]. Whey proteins (α-lactalbumin, β-lactoglobulin, (bovine) serum albumin, immunoglobulins) are characterised by well-defined, three-dimensional structures held together by disulphide bridges; these proteins are much more rigid than casein [37]. Both of the major whey proteins, α-lactalbumin and β-lactoglobulin, adsorb to oil-water interfaces and are capable of giving stable emulsions [37]. Whey proteins do not aggregate strongly and, unlike caseins, are not ambiphilic [36]. Casein micelles are relatively stable at temperatures up to 140°C, whilst whey proteins undergo extensive denaturation at around 80°C. The destabilisation of casein micelles is desirable in some circumstances, as in the case of gel formation in fermented dairy products, including cheese. In other cases it is undesirable, e.g., aggregation during age thickening of concentrated milks [36]. When oil is homogenised in a protein solution, the protein molecules are adsorbed at the surface of the oil droplets, thereby stabilising the emulsion [38]. In some cases, a protein is denatured at the surface during emulsification, whilst in others, biochemical function is retained. Various types of lipid-protein interaction take place at biological structures such as cell membranes [38]. Water-soluble proteins have an excellent capacity for stabilising foams and emulsions [39]. Protein-stabilised oil-in-water emulsions, in which protein molecules are adsorbed at the interface, form a steric stabilising layer and constitute the most important class of food colloids [40]. The stabilising layer inhibits the aggregation and subsequent coalescence of individual droplets. Due to its excellent emulsifying properties, sodium caseinate is freqently used in the preparation of emulsions [40]. Gelation of a globular protein, such as β-lactoglobulin, is the result of an aggregation process, which is generally triggered by a conformational change of the protein, induced by a modification of solvent conditions, usually a temperature rise (where it is an example of a heat-set gel) [41]. One of the key functions of egg albumen in cakes is the formation of cake structure by gelation. At high temperatures, in the final stages of baking, the albumen binds ingredients such as flour proteins, starch, sugar and water. This leads to the formation of a matrix which supports air cells and ensures the light texture required in cake [42]. Because of their heterogeneous structure and interaction with other food components, proteins exhibit a broad spectrum of functional properties [43]. The molecular basis for the functionality of proteins is related to their composition and structure.

4.4
Aqueous and Lipid Structures

Most interactions between food molecules are water-mediated [44]. Carbohydrate- and protein-based systems are formed, and interact, in aqueous media [7]. Foods obtained from biological tissues with minimal processing are generally high in moisture, as water is necessary for proper cell functioning. Some fruits, e.g., melons, contain more than 90% water in the edible portion; most tree and vine fruits contain 83–87% water; the water content of meat and fish depends primarily on the fat content and ranges from 50 to 70%; fluid dairy products contain 87–91% water [12]. Man-made foods on the other hand are often characterised by lower water levels. Dried fruits contain up to 25% water, bread 35–40% and breakfast cereals less than 4%; in intermediate-moisture foods (IMF), water content is lowered to a level that prevents microbiological damage, but that essentially preserves the texture of the raw material [12]. In IMF, most of the so-called bound water is retained. Jams and cheeses are traditional IMF, but the technology is now widespread in producing ready-to-eat foods, and has expanded to include pet foods. Biological tissues which are in contact with the ambient atmosphere continually adjust their moisture contents by either (ad)sorbing or desorbing water [45]. This process depends on the temperature, the actual amount of water involved and the nature and concentration of water-soluble substances present; the amount and rate of loss and gain also depend on the concentration of water vapour present in the atmosphere. Carbohydrates and proteins greatly influence the sorption properties of foods [11]. Humectants are used to adjust the water activity of a food, and include glycerol, sodium chloride, propylene glycol and various sugars, such as fructose, sucrose and corn syrups; these must be compatible with the food as well as being effective at low concentrations [45].

The abundance of water in foods plays a major role in their deterioration, and early food preservation technologies tended to focus on the removal of water, e.g., through drying, binding of free water by salt and sugar, and certain transformations, such as that of milk into butter. Another motivation for water removal is the reduction of weight, for example in fruit juice concentrate manufacture. Water is expensive to transport and water content is part of the specification of many manufactured foods, it being the cheapest ingredient of a food barring air. Many man-made foods are formulated for minimum water content, and may be lipid-based, e.g., chocolate, margarine and ice cream. Water therefore affects many food attributes, e.g., surface tension, viscosity, behaviour during processing, microbial, chemical and physical stability, palatability, and phase transitions [11]. Among the most important characteristics of water are its ability to act as a solvent (for sugars, ions etc.), dispersant or plasticiser, and the fact that water exists in the solid, liquid and gaseous phase at common food processing and storage temperatures [11].

A proportion of the water in food is strongly bound to specific sites, for example the hydroxyl groups of polysaccharides and the carbonyl and amino groups of proteins. Only mechanically trapped, so-called free water is truly mobile in the sense that it is readily removed during drying, or immobilised during freezing of a food; this is also the portion of the water that is available for microbial and enzymic activity. The available water in a food is expressed as its water activity (α_W), which is defined as the ratio of the water vapour pressure of a food to the saturated vapour pressure of water at the same temperature [46]. Water activity is widely regarded as representing the available water in hydrated systems [44]. The presence of various components in solution or colloidal suspension may result in decreased water pressure, surface tension and freezing point, and in increased viscosity or boiling point [12]. Packaging and storage are often designed to control or prevent moisture exchanges between a food and its environment in order to prevent deterioration, e.g., wilting (lettuce) and becoming soggy (biscuits). Many confectionery products contain layers of different sugary materials such as nougat, caramel, marshmallow, fudge, fondant cream, chocolate, wafer and biscuit. In order to avoid migration of water between layers, each of these layers should ideally have the same water activity [45]. However, as solute-water interactions are increasingly being studied at the molecular level, the concept of bound water is progressively falling into disuse; these studies have shown that even in low-moisture products, the mobility of water molecules is high [44]. Nevertheless, it may not be possible to separate this water from the other components of a food, and this is due to the plasticising role of water. Plasticisation means a general increase in molecular mobility, and water may affect the structure of food components through its plasticising effect. The lower the water content, the lower the free volume available for motions, mobility of solutes and water, and the degree of mobility of all the components of the system. Physical, chemical and biochemical changes in frozen foods are strongly affected by temperature, and this is associated with the increasing concentration and viscosity of the liquid phase, resulting from the separation of ice as temperature is lowered [44]. The slowing down of many diffusion-controlled processes has been attributed to this freeze concentration. The thermal conductivity of foods is influenced by cell structure, intercellular air and moisture content, and any reduction in moisture content causes a substantial reduction in thermal conductivity, with important implications for processes involving the removal of water, such as drying, as well as for canning [46].

Water is a permanent dipole with a polar O-H bond, i.e., its electrons are not symmetrically distributed about the molecule. In ice, water molecules tend to order themselves into a three-dimensional network, and whilst in liquid water the tendency to retain that organisation remains, the structure of liquid water is more labile as solute molecules perturb the local ordering of its molecules [44].

Transitions in foods due to processing and storage include both phase- and state-transitions. Phase transition refers to changes in the physical state (solid, liquid, gas) of materials, e.g., crystallisation and melting, whilst examples of state transitions include protein denaturation, starch gelatinisation, and glass transitions [11]. A glass is a solid, brittle material that has a structure similar to that of a liquid, with only short-range order, and in which structural defects are fixed [44]. Many food technologies, e.g., drying, freeze-drying, freezing and cooking, result in products that are mostly amorphous (as opposed to crystalline) and hence, may be relatively unstable. If these products have, or develop, enough mobility through hydration or temperature increase, structural reorganisations like crystallisation may take place with deleterious consequences for food quality [45]. Starch retrogradation is a temperature- and time-dependent phenomenon, which involves at least partial recrystallisation of starch components, and which plays an important part in the staling of bread [11].

It is well known that melting of crystalline polymers results in the formation of an amorphous melt, which can be supercooled to yield either a visco-elastic, rubbery state or a solid, glassy state; at low temperatures, amorphous materials are glassy, hard and brittle [11]. The glass transition offers a basis for understanding the role played by water in food-processing operations such as drying, freezing, extrusion and baking, and during food storage; most polymeric materials, but also low-molecular weight compounds, e.g., sugars, may be cooled below their melting point without crystallising [44]. Molecular mobility of amorphous compounds becomes apparent at the glass transition temperature T_g. If the temperature is raised above T_g, molecular mobility increases, and this is expressed in decreased viscosity and increased flow [11]. Temperature increase here has a plasticising effect. Plasticisers often have lower molecular weight than the bulk material and may decrease T_g, causing plasticisation [11]. Food solids often become soft if water content is increased, so that water may be considered a plasticiser of food materials; water has also been found to act as a plasticiser in natural polymers such as cellulose and collagen [11].

Loss of crispness and crumb toughening of cereal-based products are controlled mainly through plasticisation by water [44]. Fresh bread, biscuits, cakes, extruded flat breads and snack foods are mostly amorphous. Even if the most hydrated among them show recrystallisation with time, they remain mostly amorphous. Bread exhibits visco-elastic behaviour similar to that of synthetic polymers. At low temperature or low moisture contents, bread is glassy and brittle; as moisture content increases, it becomes more leathery. Bread crumb that has been progressively dehydrated until it vitrifies remains stable (glassy) during storage; in contrast, extruded cereal products, in which the macromolecules have been oriented, exhibit noticeable changes in texture and mechanical behaviour after only a few days in the glassy state [44]. Glass formation from melt is typical of boiled sweets, which are formed by rapid cooling of concentrated syrups [11]. In such materials, crystallisation during cooling can be avoided by using sugar combinations that retard the process of crystallisation, and that allow clear, transparent glasses to form. Food powders are often produced by rapid dehydration and contain amorphous carbohydrates and proteins, which become plasticised by water. Amorphous food solids may be used for encapsulation purposes [11]. Lactose for example may be incorporated into carriers in dry flavourings production. It was recognised early that changes in the physical state of lactose in milk powders were responsible for loss of quality [11]. This applies equally to lactose powders. Most common fats are partially crystalline materials. Many of the most important characteristics of edible fats and oils are related to their phase behaviour, e.g., solid-liquid and liquid-solid transitions during melting and solidification. Food fats are mixtures of triglycerides with melting and crystallisation properties that are governed by composition [11].

Lipid is a general term, which encompasses fats and oils, waxes, phospholipids, steroids and terpenes; the common properties of all of these are insolubility in water and solubility in hydrocarbons, chloroform and alcohols [1]. Lipds are smooth substances of greasy consistency. Nearly all lipids are less dense as liquids than water so that, when not emulsified, they float above the aqueous phase. Fat and oil are the terms used for lipids that are either solid or liquid at room temperature, respectively. In most natural oils and fats, the species present in by far the greatest amounts are triglycerides; these are esters formed by the combination of glycerol and fatty carboxylic acids [47]. When melted fat cools, it solidifies into a crystalline material, usually in the form of needles or platelets, which may associate into networks [7]. Unlike structural carbohydrates and proteins, lipids do not represent polymers of repeating molecular units. Molecules can often pack in a crystal lattice in a number of different ways; this ability to exist in a number of different crystal forms is known as polymorphism. In fats, the most familiar example is that of cocoa butter, which can exist in six different polymorphic forms, i.e., six different forms of crystal packing [47]. Pure triglycerides have sharply defined melting and boiling points. For each polymorphic state/form however, different lipids mix together well. Most familiar fatty foods, such as lard, butter and salad oil, are mixtures of different lipids, although often one or two types predominate [47]. On cooling, such mixtures do not readily pack together to form crystalline solids; instead, they freeze over more or less well defined temperature intervals. This has implication for the mouthfeel of foods that are high in lipids, e.g., chocolate. Natural triglycerides are common storage lipids and not usually viewed as major structural elements in food tissue. Phospholipids are key components of various cellular and subcellular membranes and occur widely in plants and animals. They have long, nonpolar tails and a small, highly polar head. In aqueous solution they disperse to form micelles, the non-polar tails clustering together in the middle and leaving the polar heads exposed to the aqueous environment [26]. Phospholipids also form bilayers, particularly at the interface between two aqueous surfaces, where the hydrocarbon tails cluster towards one another; such bilayers appear to form the fundamental framework of natural membranes [26].

Short-chain fatty acids result in softer fats, or lower melting points, than do long-chain fatty acids. Free short-chain fatty acids are highly volatile, generally giving rise to rancid odours [48]. Fatty acids can also include regions of unsaturation (double bonds), and the more unsaturated the fatty acids, the softer the fat at a given temperature, and the lower the melting point [14]. Double bonds between carbon atoms participate more readily in most chemical reactions than neighbouring single bonds [48]. When there is a considerable degree of unsaturation, the material will present as an oil. Unsaturated oils can be chemically hydrogenated, a process designed to harden such oils. Lipids containing a high proportion of saturated fatty acids are often called hard fats, because they are solid at room temperature. They readily solidify because their unbranched, regular chains pack together; but this tidy arrangement is disrupted in the presence of unsaturated fatty acids [48]. The double bonds bend the fatty acids so they no longer pack neatly side by side in triglycerides or phospholipids, and the melting point of the whole molecule is lowered. Unsaturated fatty acids, like many organic compounds, exhibit isomerism, i.e., different geometrical arrangements, which may result in different chemical, physical and physiological properties; most naturally occurring fatty acids occur in the *cis*-form, but they may be transformed into the *trans*-form under certain circumstances of processing [14]. The *cis* and *trans* formations bend the molecules in different ways, changing their overall shape. The hydrogen atoms associated with the C=C double bond lie on the same side in the *cis* formation and on opposite sides in the *trans* formation.

The trivial names of fatty acids frequently relate to one of the main source of the acids [48]. Myristic acid (C14:0) is found in the nutmeg tree *Myristica fragrans*; it has a chain of 14 carbon atoms and all of its carbon-carbon bonds are saturated. Palmitic acid (C16:0) was first isolated from the fruits and seeds of palms, and is the commonest saturated acid in mammalian triglycerides. Lauric acid (C12:0) was first found in the seeds of laurels, but is also present in substantial quantities in coconut and palm oils. Stearic acid (18:0) takes its name from the Greek word for hard fat. The commonest mono-unsaturated fatty acids are oleic and palmitoleic acids. The position of the double bond along the chain of carbon atoms is important to the identity, and often to the physiological role, of fatty acids. In modern nomenclature, bond positions n are identified counting from the methyl end [48]. For example, oleic acid (C18:1n-9) has a double bond at the ninth atom, whilst in palmitoleic acid (C16:1n-7) the double bond is between the seventh and the eighth carbons counting from the methyl end of the molecule. When the method of counting carbon atoms from the methyl end of fatty acids was first introduced, the pre-fix ω (omega) was used instead of n to indicate the location of the double bond. This convention still survives in some common-language context, e.g., typically in references to the health benefits of fish oils. Fatty acids with more than one double bond are known as polyunsaturates. Starting from C18:0 or C18:1, green plants can synthesise both linoleic acid (C18:2n-6), which has two double bonds, and α-linolenic acid (C18:3n-3), which has three double bonds. Here, n refers to the position of the double bond nearest the methyl end of the molecule. The 18-carbon fatty acids with two or three double bonds are usually the most abundant polyunsaturates in terrestrial green plants, but those of algae can have up to five double bonds [48]. Animals feeding on marine algae are able to incorporate algal fatty acids into their own tissues, and can also elongate and/or desaturate them further [48]. Thus, marine animals can elongate oleic acid to form gadoleic acid (C20:1n-9), common in many fish, and various 22- and 24-carbon fatty acids. Most marine fish readily convert α-linolenic acid into the long-chain polyunsaturates, eicosapentaenoic acid (C20:5n-3) and docosahexaenoic acid (C22:6n-3). The storage and structural lipids of almost all animals that feed in or from the sea are rich in n-3 polyun-saturated fatty acids, and mixtures of them are often referred to as fish oils.

Muscle lipids vary in quantity and composition. Phospholipids and choles-terol muscle lipids play important roles in the structure and function of muscle cells; they are mainly associated with membranes. Neutral lipid composed of triglycerides is present as microscopic droplets within the muscle cells or in the fat cells [12]. As these become more numerous, the fat becomes more visible in the muscle cross-section, exhibiting the phenomenon generally referred to as marbling. Foods like margarine, butter and chocolate consist of semi-solid fat in a continuous phase; semi-solid fat consists of high-melting fat crystals dispersed in an oil. The crystals in semi-solid fats form a network due to mutual adhesion, in which adhesion strength determines structure, consistency and stability [49]. The fat in margarine (80 % fat) consists of a mixture of liquid oil and crystallised fat. The structure is stabilised by sheet-like fat crystal aggregates, which inter-connect into three-dimensional networks, where they surround both water and oil droplets [7]. In butter, a limited number of fat globules are present in the fi-nal product, whilst an inter-globular phase is formed with a mixture of liquid oil, crystal aggregates and membrane residues [7]. The fracturing defect in butter is associated with the continuous butter-making process, and is due to the failure of adjacent layers of butter to coalesce during packing [50]. Structure develop-ment in low-fat spreads starts when surface-active agents or emulsifiers such as lecithins and monoglycerides form bilayers in response to exposure to an aque-ous phase [7]. To include more water, a second electrically charged surface-ac-tive agent is placed so as to stick out from the bilayers. This causes electric re-pulsion that increases the gaps between the bilayers. Finally, additional water structuring and spreadability may be achieved by immobilising free water through the formation of gelatine microgels [7].

Most confectionery products are made up of several phases, e.g., liquids, crys-tals and amorphous solids. When a temperature increase melts fat crystals and lowers viscosity, some liquid oil migrates from inside the product and badly af-fects the surface [51]. This becomes greasy and if the surrounding temperature is lowered, the liquid recrystallises into large white crystals, commonly known as fat bloom. It is not advisable to apply severe cooling to liquid chocolate, be-cause cocoa butter can set in six different crystalline forms [52]. Those with lower melting points are relatively unstable and will give neither the correct gloss nor the snap associated with quality chocolate products. Shortening, as used in baking, is so-called because it shortens the strands of starch that form when wheat flour is cooked. As a result of this, products such as shortbread turn out light and crumbly. Though most commercially available shortenings can be used for biscuit production, each type has a different effect on the spreading fac-tor of the dough and on the surface characteristics of the biscuit [12]. The three basic functions of fat in cake baking are the entrapment of air during the cream-ing process; lubrication of the protein and starch particles to break the continu-ity of the gluten and starch structure and to tenderise the crumb; and emulsifi-cation and liquid holding [12]. On baking, as the fat crystals melt, the air cells are transported into the aqueous phase where egg proteins stabilise the air/water in-terface [8]. During further heating, the air cells expand due to the formation of carbon dioxide and water vapour. The final development of the cake structure takes place in connection with starch gelatinisation and protein denaturation within the temperature range of 60–70°C [8]. The formula balance in cake batter systems varies with the product. Whilst sugar levels are consistently high, exceeding those of flour, shortening and milk fluctuate.

In living plants, the medium in which physiological activity takes place is the cytoplasm, a gel-like material in which most lipids are compartmentalised in discrete, homogeneous packets [7]. Here complex structures are formed as a result of the multiple interactions between polysaccharides and proteins. Food processing, the transformation of plant and animal tissues into man-made foods, tends to disrupt cell walls and membranes, releasing cell components and causing them to assemble into new patterns. Most foods are colloidal dispersions with gel-like flow properties, in which the liquid continuum consists of either oil or water, with macromolecules and particles in the continuous phase contributing to network formation [8]. Dispersions can be solid-in-liquid, liquid-in-liquid (emulsions) or gas-in-liquid (foams). In each case, a dispersed phase is distributed within a continuous phase, without the dispersed phase actually dissolving. Emulsions, which are mixtures of two immiscible liquids are, by definition, unstable. There are two tools for controlling emulsion breakdown, namely, the use of mechanical devices to disperse the system and the addition of emulsifiers [53].

Structure formation at the colloidal level is driven by the presence of interfaces and hydrophilic-hydrophobic balances [7]. Interactions of macromolecular aggregates of colloidal dimensions result in the formation of three-dimensional structural elements, e.g., networks in particulate gels, and stabilising surfaces in foams and emulsions [7]. When lipid and aqueous phases occur together, they form either oil-in-water (o/w) or water-in-oil (w/o) emulsions. Fat crystals are responsible for networks in the oil phase, proteins and carbohydrates in the aqueous phase; air bubbles are dispersed in a similar way to the dispersion of oil in an aqueous continuous phase [8]. When new interfaces are being established in complex systems such as foods, competitive adsorption takes place at oil/water interfaces. If both polar lipids and proteins are competing, the former are likely to succeed; in fact, the self-assembly of polar lipid molecules into infinite bilayers with water on each side is the mechanism behind the formation of biological membranes [8]. However, as proteins often adsorb irreversibly, films initially formed by proteins at the oil/water interface may remain there, even if lipids are present [8]. Proteins may change their conformation at interfaces, especially those lacking well-defined tertiary structures, which will be more or less unfolded after adsorption; proteins with well-defined globular tertiary structures, e.g., whey proteins, adsorb with a conformation close to that of their native state [8]. Protein foams can be destabilised by fatty acids, such as oleic acid; the protein-stabilised foam of beer can be controlled in this way when the brewing process has resulted in a beer with too much froth.

Droplets in ordinary emulsions, as opposed to micro-emulsions, have diameters from about 1 µm upwards [8]. Emulsion stability is limited by two phenomena, namely, flocculation and coalescence. During flocculation, intact emulsion droplets aggregate, leading to increased viscosity and sometimes, a gel network.

During coalescence, flocculated droplets fuse, leading to reduced viscosity. In the phenomenon known as creaming, an oil-rich layer forms on top of the emulsion [8]. Increased viscosity of the continuous phase of a dispersion results in reduced mobility of the dispersed droplets, and this means reduced flocculation and coalescence [8]. Milk is an example of a natural colloidal system. The milk fat globules of cows' milk are about 1–10 μm in diameter, and the globule membrane has an important role in stabilising the globules [8]. When milk is homogenised, globule size is reduced. Most dairy products owe their existence to changes at the colloidal level of milk [8]. Whipped cream requires the presence of fat crystals, and of a critical concentration of fat globules, to stabilise the foam bubbles. If whipping is continued after all the globules have aggregated at the gas/liquid interface, the globules will start to aggregate in three dimensions (churning). The resulting phase inversion (o/w to w/o) is the mechanism behind the formation of butter. Ordinary whipped cream is produced by beating air bubbles into cream. However, the foaming of aerosol whipping cream is due to a different mechanism. Due to the high pressure in the aerosol can, the propellant nitrous oxide is, for the most part, dissolved in the cream. When the cream leaves the can, the nitrous oxide comes out of solution, producing foam [54]. Ice cream has a complex physical and colloid chemistry. It contains four principal phases: ice crystals, fat droplets and air cells, all of which dispersed in a matrix of concentrated sugars, polymers and salts [55].

Margarine has traditionally been prepared with about 20 % water, and crystallisation of parts of the fat is the main factor responsible for stabilisation of the water droplets in the fat phase [8]. However, an emulsifier is usually added to the oil phase [8]. Compounded fats, such as margarine, are blended to achieve the desired properties [48]. In ice cream manufacture, it is crucial that the aqueous phase should go over to ice just before phase inversion towards butter takes place [8]. The whole structure, including the emulsion-stabilised gas cells, thus becomes immobilised. Ice cream should contain air cells to about half its volume, and the ice crystals should be below about 30 μm in diameter in order to prevent a so-called sandy texture from manifesting itself [8]. In chocolate, cocoa butter represents the continuous phase, the dispersed phase consisting of particles of sugar and cocoa powder, with mixtures of phospholipids being used as dispersing agents [8]. Mayonnaise is an o/w-emulsion, with egg yolk used for emulsification [8]. Dressings are o/w-emulsions with lower oil contents than mayonnaise, although the oil droplets should have a similar size distribution, i.e., about 1–5 μm [8].

4.5
Food Energy and Nutrients

Living tissues require nutrients for proper cell functioning, in particular, functions underlying growth, repair and work. Essential, or indispensable, nutrients cannot be synthesised by the human body and need to be obtained either via food, or directly from dietary supplements. All vitamins are, by definition, dietary essentials; the other essential nutrients are the essential fatty acids (EFA), the indispensable amino acids (IAA) and certain minerals [56]. Whilst there are no stated dietary requirement for carbohydrates, diets that are very low in these can cause keto-acids to be formed in the liver, and this may lead to a condition known as ketoacidosis [56]. Strictly speaking, individual metabolic fuels do not qualify as nutrients, because taken individually, they are not indispensable [1]. Nevertheless, nutrients are still commonly divided into macronutrients (fats, proteins, carbohydrates) and micronutrients (vitamins, minerals) [1, 57]. Fats, proteins and/or carbohydrates are required by the body in considerable quantity, as they are the main sources of dietary energy. Vitamins and minerals are required in much smaller amounts. The main feature that distinguishes essential nutrients from other beneficial dietary factors is the fact that there are recognised deficiency diseases for each of them. Essential nutrients therefore have both prophylactic and therapeutic functionality. The individual's need for a particular nutrient has been defined as the amount of that nutrient required in order to prevent clinical signs of deficiency [56]. Many nutritionists today take a broader perspective of nutrient requirement, one author citing "abundant energy and clarity of mind" as the outcome of satisfactory nutrition [58]. Certain essential nutrients, minerals in particular, are absorbed incompletely from foods, depending on the form and context in which they are present. Bioavailability is the portion of an ingested nutrient that, after being absorbed, is incorporated into a biologically active form [59]. As well as nutrients, human metabolism requires adequate amounts of water, oxygen and light. Some key nutrient-related terms and a list of essential nutrients are provided in Tables 4.1–4.2.

Dietary fibre is neither required as a player in metabolic processes, nor does it act as a dietary fuel. It is however increasingly regarded as an important dietary factor. Fibre acts primarily as a bulking and lubricating agent for food passing through the digestive tract. This is seen as helping to keep the tract in a good, healthy condition. It has also been demonstrated, in single meal studies using apples and oranges, that removing non-starch polysaccharides (NSP), i.e., the major fraction of dietary fibre, reduces satiety and increases post-prandial glucose and insulin levels [56]. A current definition of dietary fibre may be found in Table 4.1. The Panel on Dietary Reference Values of the Committee on Medical Aspects of Food Policy (COMA) proposed in 1991 that the concept of dietary fibre should become obsolete, and reviewed only specific evidence in relation to dietary NSP [56]. However, the term remains in common use, and the focus on NSP is controversial. In 1998, a COMA subgroup recommended a general re-

Table 4.1. Definitions of key nutrient-related terms

Macronutrients [1]	Nutrients needed in considerable amounts, i.e., fats, carbohydrates and proteins.
Micronutrients [1]	Nutrients needed in small amounts (μg or mg per day), i.e., vitamins and minerals.
Essential/indispensable nutrients [56]	Nutrients that cannot be synthesised in the body, or not fast enough for the body's needs.
Dietary fibre [62]	Food material, particularly plant material, that is not hydrolysed by enzymes secreted by the human digestive tract but that may be digested by the microflora in the gut.

Table 4.2. The essential/indispensable nutrients of foods

Essential fatty acids (EFA) [56]	Linoleic acid (C18:2, n-6); α-Linolenic acid (C18:3, n-3).
Indispensable amino acids (IAA) [16]	Isoleucine; Leucine; Lysine; Methionine; Phenylalanine; Threonine; Tryptophan; Valine (and, for infants only, Arginine and Histidine).
Vitamins [56]	Vitamin A; Thiamin; Riboflavin; Niacin; Vitamin B_6; Vitamin B_{12}; Folate; Pantothenic acid; Biotin; Vitamin C; Vitamin D; Vitamin E; Vitamin K.
Minerals [56]	Calcium; Magnesium; Phosphorus; Sodium; Potassium; Chloride; Iron; Zinc; Copper; Selenium; Molybdenum; Manganese; Chromium; Iodine; Fluoride; Silicon.
	Essentiality unproven: Aluminium; Arsenic; Antimony; Boron; Bromine; Cadmium; Caesium; Nickel; Silver; Strontium; Tin; Vanadium
	Cobalt is utilised only in the form of Vitamin B_{12}
	The active sulphate required by the body is probably derived from the amino acids cysteine and methionine.

classification of carbohydrates for labelling purposes [60]. Certain types of dietary fibre also have a secondary function, in that they serve as substrates for the growth of beneficial gut bacteria; these substances are referred to collectively as prebiotics. Related to the issue of fibre intake, the role of adequate consumption of water, either as part of a food, as in fruit, or as a beverage, is also increasingly being recognised [61]. Soluble fibre means that part of the dietary fibre which forms water-based gels. Soluble fibre is thought to increase the viscosity of the gut contents, whilst insoluble fibre appears to increase their bulk [1]. The COMA subgroup, aiming for clarity in terms of communication with the public, queried the usefulness of this distinction; the group was divided in its views on the status of so-called resistant starch [60]. There are three forms of resistant starch, i.e., protected starch molecules, unswollen granules, e.g., potato starch and retrograded starch [62]. Enzyme-resistant starch escapes digestion in the small intestine, but may undergo fermentation subsequently, in the large intestine [1]. Substances that cause gas production in the intestine by providing fermentable substrates for intestinal bacteria are termed flatogens; these include small oligosaccharides e.g., raffinose, stachyose and verbascose, which are present in various beans [1]. Many nutritionist now believe that the way in which foods are combined within a dish or a meal affects digestibility, and by extrapolation, colon health and the quality of nutrition [9, 58].

Energy may be defined as the ability to work. Nutritionally relevant amounts of energy are measured in kilojoules (kJ), the calorie also still being widely used [1]. There is a continuous requirement for energy to support cell functioning, with increased amounts being used during periods of heavy physical work. Carbohydrates, proteins and lipids all supply the body with energy, the gross energy provided per gram of carbohydrate being 17.2 kJ, per gram of protein 23.9 kJ and per gram of fat 39.5 kJ [1]. Those diets that are particularly high in fat have been associated with coronary heart disease and with some forms of cancer [56]. However in the modern world, individuals indulging in high-fat diets are also likely to be associated with overweight, a common risk factor in the development of a range of degenerative diseases. There is a popular argument that refined foods, sugar in particular, provide the body only with so-called empty calories. These foods are seen as lacking the nutrients required to metabolise the refined materials [57]. To that extent, such foods are also nutrient robbers, as they cause the nutrients for their digestion to be mobilised from body stores. After any eating event, as food is digested, glucose is released into the bloodstream. It has been suggested that the consumption of refined sugar produces surges of blood glucose. According to this argument, a few hours after a glucose "high", glucose falls to a very low level, which may cause brain function to be disrupted [57]. This is because the brain and nervous system, as well as red blood cells, have an obligatory requirement for glucose as their source of energy [57]. Sugars are soluble carbohydrates and occupy a key position in energy metabolism. They are classified, on the basis of their perceived availability for metabolic process, as intrinsic or extrinsic [56]. Intrinsic sugars are those sugars that are incorporated into the cellular structure; extrinsic sugars are found outside tissue cells, and include natural (honey) and processed (refined sugar) variants [56]. Extremely high intakes of sucrose, above 200 g/day, or about 30 % of food energy, may be associated in normal adults with elevations of cholesterol, blood glucose and insulin [56]. Complex carbohydrates have to be digested, i.e., they are broken down into units of glucose before they can be absorbed into the blood stream. Overweight and obesity basically result from a chronic excess of dietary energy intake over energy expenditure [56].

There are only two essential fatty acids (EFA), i.e., linoleic acid (C18:2n-6) and α-linolenic acid (C18:3n-3) [56] (Table 4.2). Once the human body has obtained these, its own enzymes can further desaturate and elongate the molecules as required [48]. Apart from the EFA, there is no absolute dietary requirement for fats. A number of longer-chain, physiologically important fatty acids can be synthesised in the body from linoleic acid and α-linolenic acid, e.g., arachidonic acid (C20:4n-6), eicosapentaenoic acid (C20:5n-3) and docosahexaenoic acid (C22:6n-3). If EFA deficiency is present, dietary intakes of these longer chain fatty acids may become critical [56]. EFA have a number of different physiological roles [56]. As components of phospholipids, they help to maintain cell membranes. They are also involved in the transport, breakdown and excretion of cholesterol, and they are precursors of prostaglandins and other physiologically active compounds. EFA deficiency is characterised by skin symptoms such as dermatosis and the skin becoming leaky to water, growth retardation, impaired reproduction and impairment of the function of many organs in the body [63]. EFA are present in all natural lipid structures, but are more common in the storage lipids of plants and of marine animals than in those of land animals [56]. Unsaturated fatty acids from all vegetable sources, and most animal sources, adopt the *cis*-configuration. However, increased consumption of hydrogenated vegetable and fish oils in margarines and other fat spreads has led to increased *trans*-unsaturated fatty acid (TFA) consumption. TFA are also known to be present in milk where their source is rumen bacteria, but do not generally possess

EFA activity [64]. As a general rule, the nutritional quality of fats and oils is compromised by heat processing [48]. Many deep-fried foods contain variable amounts of unsaturated fatty acids (both *trans-* and *cis-*) which are not encountered in native biological tissues [56]. Diets high in TFA (about 11% of energy) raise the concentration of damaging low density lipoprotein (LDL)-cholesterol, although to a lesser extent than is the case for diets high in C12- and C16- saturated fatty acids [56].

More than 20 amino acids have been obtained from food and body proteins; eight of these are indispensable in adult diets, and a further two (arginine and histidine) are indispensable for infants [16] (Table 4.2). The remainder can be synthesised by the human body. These non-essential amino acids are alanine, arginine, aspartic acid, asparagine, cysteine, cystine, glutamic acid, glutamine, glycine, hydroxyproline, proline, serine and tyrosine [1]. By definition, IAA cannot be synthesised in the body, at least, not fast enough for the body's needs [56]. Two of them, tryptophan and phenylalanine, stimulate the brain to produce the mood-enhancing endorphins serotonin and noradrenaline [61]. Phenylketonuria (PKU) is a disease which affects the metabolism of phenylalanine, where the amino acid becomes accumulated in the plasma and tissues, causing disruption of brain development [1]. Treatment is through strict limitation of phenylalanine intake. Animal-derived proteins tend to mirror human ones more closely than do plant-derived ones, and they tend to be more complete, or balanced, in terms of their amino acid composition. Plant-derived foods can be combined to achieve similar overall amino acid composition in the diet. There used to be an assumption that this sort of complementation, to achieve a complete protein, had to be achieved within the parameters of individual meals. One rule of thumb for this was that cereal grains would complement pulses. However, the performance of such intricate balancing acts in meal planning is no longer thought necessary [16, 65]. The limiting amino acid of a protein is the IAA present in the least amount relative to the requirement for that amino acid. The ratio between the amount of the limiting amino acid in a protein and the requirement for that amino acid provides a chemical estimation of the nutritional value, or protein quality [1]. Most cereal proteins are limited by lysine, and most animal- and other vegetable-derived proteins by the sulphur-containing amino acid methionine. Biological value is the proportion of absorbed protein retained in the body. A protein that is completely usable, e.g., egg and human milk, rates 0.9–1.0, meat and fish 0.75–0.8, wheat 0.5 and gelatine 0. Protein quality is a health issue only if the total intake of protein barely meets the requirements. In adequate, mixed diets, different proteins will complement one another [1]. Amino acids in excess of those needed for the synthesis of proteins and other biomolecules cannot be stored, in contrast with fatty acids and sucrose, nor are they excreted. Rather, surplus amino acids are used as metabolic fuels, mostly in the liver [25].

Vitamins, like EFA and IAA, are organic food components that are vital to human metabolic functioning. The Panel on Dietary Reference Values of COMA in 1991 reviewed all of the vitamins listed in Table 4.2 [56]. Among vitamins some are water-soluble, whilst others (A, D, E and K) are fat-soluble [66]. Vitamins perform a wide range of functions within the body. Some act as co-factors in enzyme activity, others are antioxidants and one (vitamin D) is a pro-hormone [66]. Hypervitaminosis means an overdosage with a vitamin, leading to intoxication. Hypervitaminoses A and D may result from eating certain (usually enriched) foods, but most problems encountered with high levels of intake of vitamins A, D and B_6, and of niacin are linked to the consumption of vitamin supplements [1]. Several foods supply the complex of B vitamins as a package. These include liver, kidneys and yeast extract, as well as meat, milk, eggs and fish, wholegrain cereals, wheatgerm, green vegetables, potatoes, nuts and pulses, bananas and dried fruit [61]. Thiamin (vitamin B_1) is required mainly in the metabolism of carbohydrates, fat and alcohol, so that symptoms of deficiency may arise in conjunction with diets that are rich in carbohydrates [56]. In the Far East, the initial popularity of polished rice became associated with memory problems in those who consumed large amounts of the product. The syndrome was referred to as beri-beri amnesia [57]. The body has only small stores of thiamin, so that a deficiency manifests itself relatively quickly [57]. Although thiamin is widely distributed in foods, it is readily destroyed during boiling [57]. Riboflavin (vitamin B_2) plays a key role in all oxidative processes [56]. Although there is no specific deficiency disease for this vitamin, low intakes lead to dryness and cracking of the skin around the mouth and nose [66]. Excess riboflavin is excreted in the urine [66]. Riboflavin is widely distributed in leafy vegetables, meat and fish, and intake has to be low for months before symptoms become apparent [57]. Vitamin B_6 is a mixture of pyrridoxal, pyrridoxine, pyrridoxamine, and their 5'-phosphates. Gross deficiency of vitamin B_6 may cause weakness, difficulty in walking, loss of a sense of responsibility, insomnia and depression, with elderly people particularly at risk [57]. Vitamin B_{12} is involved in the recycling of folate co-enzymes and is also needed for nerve myelination, so that prolonged deficiency leads to irreversible neurological damage [56]. Food sources of vitamin B_{12} include almost all animal products, and certain algae and bacteria; green plants do not contain it [56]. It is necessary for the formation of oxygen-carrying red blood cells [57]. As it is required in very small amounts, body stores will last several years [57]. In healthy individuals, large quantities of vitamin B_{12} are produced by beneficial colon bacteria, although there is some debate about how much of this is absorbed by the body [9]. Folate is involved in a number of single-carbon transfers, and rapidly regenerating tissues, such as the intestinal mucosa, are particularly badly affected if a deficiency is present [56]. The main dietary sources of folic acid are green leafy vegetables, orange juice and wheatgerm [61]. Dietary folate plays a central role in the proper development of the neural tube during the first month of pregnancy [67]. Spina bifida is an example of a neural tube defect.

The term vitamin C includes both ascorbic acid and dehydroascorbic acid, since both of them exhibit anti-scorbutic activity [68]. By virtue of its high reducing power, vitamin C functions physiologically as a water-soluble antioxidant; it is a co-factor for several enzymes [68]. The classic disease of severe vitamin C deficiency is scurvy, a connective tissue defect. Vitamin C aids wound healing and assists in the absorption of non-haem iron [56]. A combination of dose-dependent absorption and renal regulation allows the body to conserve vitamin C during periods of low intake, and to limit plasma levels during high intake [68]. Vitamin C is easily destroyed by oxygen, metal ions, increased pH, heat and light [56]. The main sources of vitamin C in the diet are fresh fruit and vegetables [61]. Biotin is required for lipogenesis, gluconeogenesis and the catabolism of branched-chain amino acids; it is widely distributed in foods, as well as being synthesised by the intestinal microflora [56]. The generic descriptor niacin includes nicotinic acid and nicotinamide; the latter can be synthesised from the amino acid tryptophan [56]. Niacin deficiency leads to pellagra. Pantothenic acid is a part of coenzyme A, and thus plays a key role in the catabolism of all the macronutrients to yield energy [56].

Vitamin A can be obtained as preformed vitamin A (retinol) and from some carotenoid pigments; however, of the over a hundred carotenoid pigments identified in plants, only a few are precursors [56]. Vitamin A deficiency can lead to permanent eye damage and eventually, death [56]. Foods rich in vitamin A include liver, kidneys, oily fish, milk and fatty cheeses, butter and egg yolk [61]. Foods rich in beta-carotene include carrots, dark green vegetables, tomatoes and peppers and many other fruits and vegetables [61]. The consumption during pregnancy of large amounts of vitamin-A rich foods, e.g., liver, is discouraged because excess intake can lead to fetal liver damage [61]. Ergocalciferol (vitamin D_2) is derived by ultraviolet (UV) irradiation of ergosterol, a substance that, although widely distributed in plants, fungi and lower life-forms, does not occur naturally in higher vertebrates [56]. Cholecalciferol (vitamin D_3) is derived from the action of UV irradiation on 7-dehydrocholesterol in the skin [56]. Vitamin D is involved in calcium homeostasis (plasma calcium is maintained within narrow limits); prolonged deficiency results in rickets, the main signs of which are skeletal deformity and muscle weakness [56]. In adults, hypovitaminosis D presents as osteomalacia, with muscle weakness and bone tenderness, or pain in the spine, shoulders, ribs or pelvis. There are a few dietary sources of vitamin D, e.g., oily fish [56]. Vitamin E activity is manifested by two series of compounds, the more important being the tocopherols, whilst the tocotrienols are less potent [56]. Vitamin E is thought to function primarily as a chain-breaking antioxidant that prevents the propagation of lipid peroxidation; deficiency syndromes include peripheral neuropathy [68]. Foods rich in vitamin E include avocados, blackberries, mangoes, tomatoes, sweet potatoes, spinach and watercress, nuts and seeds, wheatgerm, wholegrain cereal, and vegetable oils [61]. Vitamin K is required in the context of the γ-carboxylation of at least two coagulation inhibitors [56]. Vitamin K activity is exhibited both by plants (phylloquinone) and by related menaquinones, which are synthesised by intestinal bacteria [56].

Minerals are inorganic substances which are required by the body for a variety of functions [69]. These include the formation of bones and teeth, and roles as essential constituents of body fluids and tissues, as components of enzyme systems, in nerve function, and as part of hormones. Minerals are also involved in the maintenance of electrical potentials across membranes. Minerals are synthesised neither in the human body nor in plants, but must ultimately be derived from the soil. Plants which grow in mineral-depleted soils are therefore themselves likely to be mineral-depleted. Trace minerals – or trace elements – are essential minerals present in small amounts (parts per million); they are copper, chromium, iodine, manganese, molybdenum and selenium; although required in larger amounts, zinc and iron are sometimes included with the trace minerals [1]. The Panel on Dietary Reference Values of the COMA in 1991 reviewed the minerals listed in Table 4.2 [65]. Bioavailability of minerals may be influenced by a variety of factors [69]. Phytates in grains and pulses, and oxalate in spinach and rhubarb, reduce the absorption of calcium, iron and zinc [69]. Iodine absorption may be impeded by the presence of nitrates [69]. Any excess of one mineral may hinder absorption of another, e.g., excess iron reduces zinc absorption [69]. On the other hand, iron absorption may be increased when vitamin C is consumed in the same meal [69].

Body calcium amounts to just over 1 kg, with about 99% occurring in the bones and teeth, and the remainder in tissues and fluids [56]. Calcium it is essential for cellular structure, metabolic function and nerve signal transmission [56]. Children and pregnant women have a higher absorption than the standard absorption rate for adults [56]. Calcium plays an important role in the development and maintenance of bone tissues. Absorption is enhanced by a metabolite of vitamin D, which stimulates the biosynthesis of intestinal calcium binding protein [70]. Extreme calcium deficiency (tetany) causes muscle twitching and cramps, confusion, irritation and spasms of the throat [71]. Calcium may be obtained from dairy products, followed by cereal products and fruits and vegetables; tinned fish, such as sardines, are rich sources of calcium, as are citrus fruit, canned fish with edible bones, and pulses [72]. The human body contains about 25 g of magnesium [56], of which about 60% is located in the skeleton. Its physiological role lies in skeletal development and in the maintenance of electrical potential in nerve and muscle membranes [56]. Magnesium deficiency is characterised by progressive muscle weakness, failure to thrive, coma, and death [56]. Magnesium is widely distributed in plant- and animal-derived foods, especially nuts, legumes, green vegetables (chlorophyll), cereals and chocolate; tap water in hard-water areas also supplies significant amounts of magnesium [72]. About 80% of the 600–900 g of phosphorus in the human body is present as a calcium salt in the bones, where it imparts skeletal rigidity [56]. The energy that drives most metabolic processes is derived from the phosphate bonds of adenosine triphosphate (ATP) [56]. Phosphorus is present in all plant and animal cells [69]. Excess phosphorus may cause problems in babies fed unmodified cows' milk, as this can affect the calcium balance [69]. Fluoride forms calcium fluorapatite in teeth and bones; it may also have a role in bone mineralisation, and it protects against caries [56].

Sodium and potassium in the body are in a state of dynamic balance. Sodium occurs primarily in extracellular fluids, whilst 90% of potassium is found within cells [71]. If the body sodium burden is increased, water is retained and the volume of extracellular fluid increases, if the body sodium burden falls, the volume decreases [56]. The liberal consumption of salt may result in excessive pre-menstrual water retention; this presses on the body tissues, including the brain, typically resulting in tension and depression [71]. Hypertension, i.e., elevated pressure in the blood vessels, is often related to excessive sodium intake [71]. Potassium deficiency alters cell membranes, causing weakness of skeletal muscles, and it can cause arrythmias of the heart, as well as depression and confusion [56]. In infants and old people, sodium and fluid homeostasis are inefficient

[56]. Fresh fruits, vegetables and whole grains are rich in potassium [71]. Chloride is the major counter ion to both sodium and potassium.

Iron is a component of haemoglobin, myoglobin and many enzymes, and its stores can account for up to 30% of total body iron [56]. Whilst severe iron deficiency causes anaemia, less severe deficiency results in adverse effects on work capacity, intellectual performance and behaviour [56]. Iron from animal sources (haem iron) is absorbed more readily than iron from plant sources (non-haem iron) [69]. Absorption of non-haem iron is affected by anti-nutrient food factors, such as phytate, fibre and tannins, which can bind non-haem iron [69]. Zinc is involved in the major metabolic pathways which contribute to the metabolism of proteins, carbohydrates, energy, nucleic acids and lipids [56]. As it is a key component of several enzymes, it is present in all tissues [56]. Zinc also has a structural role in a number of non-enzymic proteins [56]. In legumes and animal products, zinc is associated with protein components; in meat products, the zinc content to some extent follows the colour of the meat, so that the highest content is found in lean red meat, where it is at least twice that in chicken [72]. In cereals, most of the zinc is found in the outer, fibre-rich part of the kernel [72]. Copper is a component of many enzymes [56]. It is involved in many mental mechanisms and acts as a brain stimulant [63]. Foods high in copper include liver, kidney, shellfish, wholegrain cereals and nuts [72].

The therapeutic and toxic levels of intake are much closer for selenium than for the other metals [73]. Selenium is part of the enzyme glutathione peroxidase, which is involved in one of the mechanisms that protect intracellular structures against oxidation [56]. Selenium deficiency has been linked to a form of heart disease (Keshan disease) in parts of China where soil levels of selenium are very low [69]. Alkaline soils favour plant uptake of selenites, whilst acidic soils, and areas rich in iron and calcium, restrict uptake by plants [72]. Animal tissues show smaller variations in selenium than vegetables [72]. Cereals, seafood and meat products are the richest sources of selenium, and are the main contributors to daily intake, whereas fruit and vegetables are generally low [72]. UK intakes of selenium have fallen over recent decades; this is thought to be due to the trend towards using selenium-poor European wheat rather than selenium-rich North American wheat for bread making [60].

Molybdenum is essential for several enzymes, as is manganese [56]. Relatively high concentrations of manganese have been reported in cereals, nuts, ginger and tea; concentrations of manganese in crops are dependent on soil acidity; animal tissues are low in manganese [72]. Chromium appears to function in an organic complex which potentiates the action of insulin, and may also participate in lipoprotein metabolism, in maintaining the structure of nucleic acids, and in gene expression [56]. The richest dietary sources of chromium are spices such as black pepper, brewer's yeast, mushrooms, prunes, raisins, nuts, asparagus, beer and wine [72]. Chromium occurs in green leafy vegetables [58]. Refining of cereals and sugars removes native chromium, but stainless steel vessels in contact with acidic foods may contribute additional chromium [72]. Iodine forms part of the hormones thyroxine and triiodothyronine, which are necessary for the maintenance of metabolic rate, cellular metabolism and the integrity of connective tissue [56]. In the fetus, iodine is necessary for the development of the nervous system during the first three months of gestation [56]. The concentration of iodine in plants and animals is greatly influenced by the soil; seafoods contain large amounts of iodine from seawater [72]. Whilst the bioavailability of minerals from some food sources may be poor due to an abundance of anti-nutrients, e.g., phytic acid and polyphenols, it may be improved through certain food processing steps. For example, the bioavailability in pearl millet of phosphorus, calcium, iron, zinc, copper and manganese was increased by germination of the grains [74]. Fermentation may have similar effects [75].

4.6
Bioactive Compounds

Edible biological tissue contains various substances which possess, or appear to possess, certain bioactive properties – not only in terms of the organism in which they occur naturally, but also in the human body. Clearly, the essential nutrients discussed earlier fall into this category. However, bioactivity in foods can refer to harmful, as well as to beneficial physiological effects. In many cases, dosage determines which of the two possible outcomes actually occurs. Few substances are either wholly beneficial or wholly harmful and besides dosage, context and frequency of consumption of a food may be important. This is clearly illustrated through the popular activities of alcohol and caffeine consumption. Intrinsic food toxins are traditionally removed by methods of food processing which may have developed over many generations. The cyanogenic glucosides in cassava exemplify a situation in which toxins are removed from a staple food through traditional processing [76]. The current section deals primarily with those food components which are believed to display beneficial physiological activity, but which are not currently defined as nutrients. Whilst nutrients have recognised deficiency syndromes associated with them, this is not the case with the so-called bioactive substances. The positive health benefits of these are more elusive, at least given the state of current knowledge about them. In discussing bioactive food components, there tends to be less of a focus on individual members of a group than on the group overall. For example, plant flavonoids are usually discussed as a group, and only rarely individually. The positive health effects exerted by a bioactive food component also tend to be thought of as long term, especially in relation to various degenerative diseases. This is an important difference between bioactive compounds and true nutrients. There is considerable interest in these compounds, not least in connection with the marketing of the so-called functional foods. In functional foods, an active principle has usually been isolated and added to a man-made base for a specific (health) purpose. These foods are discussed in Chapter 6. There is also considerable interest in these bioactive substances "in situ", particular among the nutrition education community [77]. Some believe that, once enough is known about them, individual bioactive compounds may be redefined as essential nutrients [78].

Dietary antioxidants appear to be particularly important in disease prevention, and include essential nutrients as well as bioactive compounds. As the body processes food and captures its energy potential, free radicals are generated which damage cell materials and membranes throughout the digestive, endocrine, circulatory and nervous systems, leading to common degenerative diseases such as heart disease, strokes, cancer, diabetes and osteoporosis [79]. Plant colour is mainly due to three groups of compounds, i.e., flavonoids, carotenoids and chlorophylls. Some of these are are now thought to have biochemical properties important to the physiology of man [76]. Glutathione, a tripeptide of glycine, glutamic acid and cysteine, is involved in oxidation-reduction reactions [1] and is emerging as an important factor in healthy ageing [79]. In addition to being a widespread antioxidant, it also binds pollutants (xenobiotics) [79]. Equally topical are the so-called prebiotics and probiotics. Bender and Bender define prebiotics as non-digestible dietary oligosaccharides that support the growth of certain bacteria in the colon, and probiotics as preparations of live microorganisms added to food, claimed to be beneficial to health by restoring microbial balance in the intestine [1]. Because this clearly refers to man-made foods, pre- and probiotics are discussed in Chapter 6, under the heading functional foods.

All dietary fats are not equal when it comes to health benefits that might be due to them. In particular, the polyunsaturated fatty acids (PUFA) linoleic acid and α-linolenic acid occupy key positions among fats. These are the only two essential fatty acids (EFA) required by the human body, which the body cannot synthesise for itself. Saturated (SaFA) and monounsaturated (MUFA) fatty acids serve mainly as metabolic fuels although, eaten in moderation, MUFA have beneficial effects on blood cholesterol levels [77]. MUFA do not cause cholesterol to accumulate as do saturated fats, as they do not deplete the blood of high-density lipoproteins (HDL); HDL pick up cholesterol from the arterial walls and transport it to the liver, where it is broken down into bile acids and flushed from the body [9]. Low-density lipoproteins (LDL) cause cholesterol to be deposited in the arteries, and MUFA reduce levels of LDL [9R]. Although PUFA too reduce LDL, they decrease HDL by an equal amount [9].

Almost all foods that contain fat have a balance of SaFA, MUFA and PUFA; a piece of meat will contain mainly SaFA and MUFA with little PUFA, olive oil has mainly MUFA, and sunflower seed oil mainly PUFA [65]. Fats and oils high in SaFA include most animal fats as well as coconut and palm oils [61]. Plants with abundant SaFA as storage lipids are usually natives of warm climates, e.g., coconut, oil-palm, nutmeg and cacao; they do not need PUFA to keep these lipids fluid and are able to maximise the energy obtainable from the smallest volume of storage organ [48]. Just a small number of fatty acids generally account for over 90% of the total composition of plant membrane lipids, varying little between different types of leaf and including palmitic, hexadecenoic, oleic, linoleic and α-linolenic acids [63]. The leaves of higher plants contain up to 7% of their dry weight as fats, some as surface lipids and others as components of leaf cells, especially the chloroplast membranes [63]. Seed oils vary widely in fatty acid composition and one fatty acid usually predominates [63].

It is the PUFA that have attracted special interest as bioactive dietary components, and in particular, the n-3 (or ω-3) and n-6 (or ω-6) families of PUFA. N-3 and n-6 fatty acids represent two series of long-chain PUFA derived respectively from α-linolenic and linoleic acids, with n being the position of the first double bond counting from the terminal methyl group [1]. The many physiological functions that these lipids perform in the body can be satisfied as long as there are adequate dietary supplies of the two EFA, as these are capable of undergoing the biochemical transformations necessary to produce the biologically active substances required. However, some of the fatty acids derived by this route may also be obtained directly from foods. The main *n*-3 derivatives are stearidonic acid (SDA), eicosapentaenoic acid (EPA) and docosahexaenoic acid (DHA) and the main *n*-6 derivatives, γ-linoleic acid (GLA) and arachidonic acid (AA) [80]. Highly unsaturated fatty acids, especially AA and EPA, are essential for the formation of the phospholipids in the nerve membranes; minor imbalances can distort the normal development of the brain and eyes [48]. This involves prostaglandins, short-lived, hormone-like chemicals that regulate cell metabolism and are essential for proper brain function, affecting vision, learning, coordination and mood [80]. Series 1 prostaglandins are produced in the body from DGLA and series 2 prostaglandins from AA, whilst series 3 prostaglandins derive from the *n*-3 family, with α-linolenic acid as the starting material [80]. Figure 4.1 provides a schematic representation of the key members of both families of PUFA and their transformations in the body.

Current government advice regarding dietary fat intakes holds that *cis*-PUFA should provide an average of 6% of total dietary energy and be derived from a mixture of *n*-6 and *n*-3 PUFA [56]. It is thought that the lipids from oily fish help to prevent heart disease and some types of cancer, improve skin conditions such as psoriasis and benefit individuals suffering from rheumatoid arthritis [61]. Organs and tissues that perform functions of storage (adipose tissue), chemical processing (liver), fuel utilisation (muscle) and excretion (kidneys), as well as the reproductive organs, tend to have membranes in which *n*-6 PUFA predominate, whilst in nervous tissue and the retina of the eye, *n*-3 PUFA are more widespread [63].

The best seed oils for *n*-3 PUFA are flax (linseed), hemp and pumpkin [65]. Most marine fish readily convert α-linolenic acid into EPA and DHA; these processes are much less efficient in humans who synthesise very little of them if they can obtain them by dietary means [48]. In green plants, PUFA of the *n*-6 family far outnumber those of the *n*-3 family; however, one of the best sources

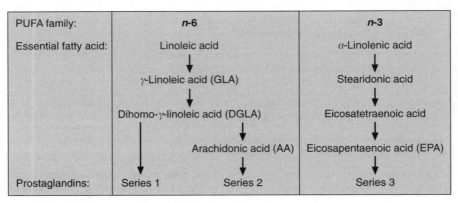

Fig. 4.1. Bioactive dietary fatty acids (adapted from [80])

of α-linolenic acid is purslane (*Portulaca oleracea*), a fast-growing, drought-tolerant plant that grows as a weed in open, semi-arid habitats throughout the world [48]. Plant oils rich in PUFA tend to be produced from crops grown in temperate climates and marine algae tend to have more highly unsaturated fatty acids than terrestrial plants [48]. The lipids of cold-blooded animals are generally richer in unsaturated fatty acids than those of warm-blooded animals, with species native to colder regions having the most PUFA [48]. The livers of gadoid fish such as cod and halibut are rich in oil, but their muscles contain almost no storage lipid; conversely, the lipids of fast-swimming fish such as mackerel and herring are stored in the muscle, and they can exceed 20% by weight of lipid [48]. Due to a combination of lifestyle and feeding regime, farmed fish such as salmon acquire a more terrestrial fatty acid composition than is typical of their wild ancestors [48]. In the last forty years or so, there has been increasing evidence about the beneficial effects of marine fats on plasma lipids and atherosclerosis, the major cardiovascular disease [81]. Eicosanoids are abundant in the metabolism of those consuming diets rich in n-3 PUFA; they probably mediate in reducing inflammation and (boosting) immune responses [81].

The phenolic compounds of plant tissues can be divided into simple phenols, with one aromatic ring; polyphenols, with two or more such rings; and tannins, large mass molecules [82]. Phenolic compounds are the major soluble constituents of tea, coffee and wine [83]. Chlorogenic acids are simple phenols and are found in coffee, maté, cider, blueberries and various fruits and vegetables [82]. Other examples of phenolic acids include gallic and salicylic acids [77]. Tannins are found in cranberries, certain pulses and red wine [77]. Among the polyphenols, flavonoids are the most widely researched in terms of their potential health benefits as well as their role in plant pigmentation. Food processing may generate derived phenols, e.g., theaflavins, polymeric flavonoids found in black (fermented) tea [82]. Table 4.3 gives an overview of dietary flavonoids for which possible health benefits have been identified.

Any diet which is rich in plant tissues is associated with considerable quantity and variety of intake of phenolic substances [83]. Flavonoids comprise one of the largest groups of secondary plant metabolites, occurring widely throughout the plant kingdom [84]. Their common structural feature is a three-ring skeleton (C_6-C_3-C_6) in which two benzene rings are joined with a heterocyclic, oxygen-containing, pyran group. Flavonoids commonly occur in leaves and fruits used as food, where they are present in the form of water-soluble glycosides, i.e., with sugars attached [83]. The sugar-free part of the flavonoid molecule is referred to as the aglycone [85]. Different classes and forms of flavonoids are present in plants in varying combinations; for example, onions contain quercetin as the aglycone as well as in the form of quercetin glucosides, diglucosides, rhamnosides, glucoronides and malonyl esters [86]. About 2000 flavonoids have been identified, among the most thoroughly studied being quercetin, myricetin and kaempferol, obtained from apples, onions and black tea [83]. Foods which are rich in phenolic compounds generally contain flavonoids too, so that population studies which show a link between high intakes of flavonoids and a reduced risk of heart disease and stroke typically hold true for other phenolic substances [77]. Much of the current evidence about pos-

Table 4.3. Bioactive flavonoids in foods

Flavonols	*Quercetin*: Onions, apples, lettuce (especially Lollo Rosso), kale, wine, broad beans, olives, potatoes, leeks, tea, etc.[77]; Onions, apples, broad beans [84] *Rutin*: similar to quercetin [77] *Kaempherol*: Strawberries, endive, black tea [84] *Myricetin*: Broad beans, red wine [84]
Flavones	*Luteolin*: Celery, olives, parsley, lemons, red peppers, artichokes, oranges, chilli peppers [77]; Red pepper [84] *Apigenin*: Celery, olives, parsley, lemons, red peppers, artichokes, oranges, chilli peppers [77]
Anthocyanidins	Red/blue plant pigments: Cherries, aubergines, cranberries, blueberries, blackberries, raspberries, black grapes, radishes, blackcurrants, etc. [77]
Flavanols = Catechins	Green tea, black tea, cocoa and chocolate, red and white wine, pears, apples [77]; Green tea, wine, chocolate [82]. Green tea [84]
Flavanones	*Naringenin*: Grapefruit [77] *Hesperidin*: Oranges [77]
Isoflavones	*Genistein;* Legumes, especially soya beans [77; 88] *Daidzein*: Legumes, especially soya beans [77; 88]

sible health benefits of flavonoids is due to such epidemiological studies; this type of research is inevitably affected by "noise" from other bioactive ingredients. A different kind of uncertainty attaches to *in vitro* studies where the active principle, removed from its normal context, may not perform in the same way as would be the case *in situ*/*in vivo*.

Phenols are antioxidants, i.e., they can inhibit the formation of free radicals; and once formed, they can also scavenge them [82]. Flavonols and flavones are especially important, both because of their ubiquity and because of their potential anticarcinogenic effect [84]. Quercetin scavenges active oxygen species and thereby inhibits free-radical dependent mechanisms of carcinogenesis [83]. Flavonols exhibit other biological effects that may enable them to inhibit later events that would lead to coronary heart disease [83]. The antioxidant and antitumour activities of many food-borne phenolics are the subject of much current research activity; but little is known about their absorption and mechanism in man [83]. The identity and concentrations of compounds reaching the target tissues via food intake is poorly understood in many cases, and it is therefore extremely difficult to interpret the significance of mechanisms identified *in vitro* [83]. Anthocyanins are flavonoids of the flavan type, the aglycones being known as anthocyanidins, and are extensively distributed in nature [86]. They are present in a large number of fruits and vegetables as well as in the petals of many red and blue flowers as well as the leaves of some plants. The six common ones in foods are pelargonidin, peonidin, oetunidin, cyanidin, delphinidin and malvidin [86]. Anthocyanins gradually appear with ripening as chlorophylls fade away. Anthocyanins differ from other plant flavonoids by strongly absorbing visible light [85].

Redbush, or rooibos, a tea long popular in South Africa, is currently being introduced into the UK as a health drink. Flavonoids such as quercetin and luteolin have been isolated from rooibos tea [87]. Flavanols, also known as catechins, occur in cocoa beans [77]. Epicatechin flavanols, the main phytochemicals in green tea, have antioxidant, antithrombic, antibacterial, antiviral, anticancer and immune system regulating effects; flavanols are the main phytochemicals in black tea [77]. The flavonol quercetin, which is found in apples, is one of the most potent antioxidant flavonoids; it is also said to possess anti-inflammatory and anti-cancer actions; quercetin also occurs in onions and Lollo Rosso lettuce [77]. The flavone apigenin is a key phytochemical in celery; like all flavonoids it has antioxidant properties [77]. Flavones occur in olives among a range of polyphenols; these, along with vitamin E, give olive oil its well-known antioxidant power [77]. Flavanones, such as hesperidin and naringenin, are mostly found in citrus fruit [77].

Phytoestrogens are phenolic compounds in plants that are structurally similar to mammalian oestrogens [88]. Research into the possible benefits of phytoestrogens has focused on cancer (breast and prostate in particular), menopause, osteoporosis and heart disease (antioxidant acitivity), as well as diabetes and cognitive function [88]. There is strong epidemiological evidence that phytoestrogens have a protective effect against hormone-dependent cancers [83]. The main classes of phytoestrogens are the isoflavones, which are flavonoids; as well as cumestans and lignans [88]. Isoflavones are found almost exclusively in legumes, the soy bean being the most abundant source; the most important soy isoflavones are genistein and daidzein [88]. Lignans are also an important source of phytoestrogens in the UK diet as they are present in most fibre-rich foods [83]. The main sources of coumestrol are mung bean sprouts, alfalfa sprouts and soy bean sprouts [77]. The oestrogenic activity of phytoestrogens in humans is much lower than that of human oestrogen; phytoestrogens may inhibit human oestrogens by blocking oestrogen receptors and may have desirable effects, for example to reduce the risk of breast cancer [88]. The outcome of increased phytoestrogen intake is unpredictable due, in part, to a poor understanding of the mode of action; there is the possibility of both adverse and beneficial effects in some individuals in different organs [88]. The best food sources for phytoestrogens include soy beans, lentils, chick peas, other pulses, bean sprouts, linseed and wholegrains, with much smaller amounts found in fruit and vegetables [77]. Antibiotic treatment can block the proper absorption of phytoestrogens, indicating a role for gut bacteria in their absorption [77].

Carotenoids are synthesized by bacteria, fungi and higher plants; they contribute to the colour of many yellow, red or orange tissues [89]. They are a diverse group of lipid-soluble pigments, with some 500 known to exist in the human food chain [83]. Their molecular structure includes an extended chain of double bonds, which enables carotenoids to function as antioxidants [83]. Carotenoids occur universally in leaves and other green plant tissues and are widespread in yellow-orange flowers, orange and red fruit such as orange, tomato and peppers, roots, notably carrots, and seeds like sweetcorn [90]. Carrots and other yellow root vegetables accumulate carotenes and small amounts of xanthophylls, whilst the carotenoids in fruits are more diverse [89]. The carotenoids in ripe fruit usually differ from those that are present in unripe, green fruit [89]. Bioavailability varies between carotenoids and types of food matrix; β-carotene in the form of supplements that are solubilised with emulsifiers has a much higher bioavailability than β-carotene from foods, in particular, from raw vegetables [68]. Blood concentrations of carotenoids are the best biological markers for consumption of fruit and vegetables; high blood concentrations of β-carotene and other carotenoids are associated with lower risk of several chronic diseases [68]. There are two classes of carotenoids, the carotenoid hydrocarbons or carotenes, and the oxygenated derivatives or xanthophylls [90]. Carotenoids that can be converted into vitamin A include α-carotene, β-carotene and β-cryptoxanthin; by contrast, lycopene, lutein and zeaxanthin have no vitamin A activity [68]. In UK diets, only β-carotene is of major importance as a precursor of vitamin A [56].

Carotenoids are thought to have various actions, including antioxidant activity, immune system enhancement, and inhibition of mutagenesis and of pre-malignant lesions [68]. They have been associated with a range of health effects, i.e., decreased risk of contracting certain eye diseases, some cancers and of cardiovascular disease [68]. However, there is no proof that it is the carotenoids specifically that constitute the protective factor in fruit and vegetables [56, 68]. It has been suggested that carotenoids influence levels of DNA damage in a manner that does not rely simply on a direct antioxidant effect but via novel mechanism, including the regulation of DNA repair processes [91]. The main carotenoid present in tomatoes is lycopene; up to 85 % of lycopene in the diet comes from tomatoes and tomato products [77]. Lycopene is also present in pink grapefruit, watermelons, rosehips, guava and apricots [77]. Carotenoids vary in their antioxidant potency, with lycopene significantly outpacing both β-carotene and lutein [92]. Epidemiological studies suggest that lycopene may protect against cancer and cardiovascular disease; high intake of tomatoes and tomato sauce was associated with decreased risk of prostate cancer, and high adipose tissue levels of lycopene were associated with decreased risk of myocardial infarction [93]. Lycopene is the most efficient carotenoid singlet oxygen quencher [93]. Spinach is an important source of lutein; mango, papaya and red peppers contain β-cryptoxanthin [77]. Cos or romaine lettuce are an important source of zeaxanthin and in addition, contain lutein [77]. Sweetcorn is a prime source of zeaxanthin [77].

There are many thousands of secondary plant metabolites whose potentially beneficial effects on human health have not yet been investigated. To conclude this section, just two more beneficial types of phytochemical will be mentioned: glucosinolates and α-lipoic acid. Glucosinates are potentially anticarcinogenic, secondary plant metabolites, with brassicas, e.g., cabbages, broccoli, Brussels sprouts and cauliflower, the main contributors to intake [84]. The main ones are sinigrin, progoitrin and glucobrassicin [94]. Many are also flavour active. Glucosinolates are usually broken down through hydrolysis catalysed by myrosinase, an enzyme that is released from damaged plant cells, resulting in an aglycone, glucose and sulfate [84]. The aglycones are unstable and undergo further reactions and among these, isothiocyanates and indoles possess anticarcinogenic properties [84]. Epidemiological studies suggest inverse associations between the consumption of brassica vegetables and cancer, in particular, lung, stomach, colonic and rectal cancers; however, the association between intake of glucosinolates or their hydrolysis products and risk of cancer has as yet not been investigated specifically [84]. Cooking usually reduces glucosinolate levels by 30–60%, depending upon the vegetable [94]. α-Lipoic acid is a naturally occurring potent antioxidant, with particularly high levels found in spinach and lower concentrations in garden peas, Brussels sprouts and rice bran; it is an integral component of mammalian cells [95].

Many endogenous food components are potentially harmful to health; some may be beneficial in small quantities and in certain contexts, others are toxic under any circumstances. The phytochemicals in foods that are commonly eaten may be beneficial in moderate amounts only, with negative effects if consumed in excess; examples include phytoestrogens, glucosinolates, phytic acid, furanocoumarins and xanthine alkaloids [77]. Isoflavones have been found to influence not only sex hormone metabolism but also malignant cell proliferation [94]. Very high intakes of glucosinolates, e.g., through diets rich in raw cabbage, may cause reduced iodine absorption leading to swelling of the thyroid gland (goitre) [77]. Phytates are found in all plants as they are associated with the storage of phosphates and other minerals [77]. About 90% of dietary phytates come from cereal products, especially bran; pulses, seeds and nuts are the other main sources [77]. Phytic acid can bind with minerals such as iron, zinc and calcium, thereby reducing their absorption in the body [77]. Xanthine alkaloids have a stimulatory effect on the central nervous system (CNS); they include caffeine in tea and coffee, theophylline in tea, and theobromine in cocoa [77]. Bananas are sources of two biogenic amines, serotonin and dopamine [77]; such amines are also made in the body and influence mood and appetite (see Chapter 7). Chocolate too contains biogenic amines [77]. Alcohol is a CNS depressant which, consumed in large quantities, leads to intoxication and possibly, addiction. A range of pharmacologically active compounds are present in herbal products, e.g., coumarin, safrole, isosafrole and pulegone in mint [96]. Pyrollizidine alkaloids, which can damage liver cells and trigger cancer, are present in comfrey tea [77]. Very high intakes of oxalic acid can be toxic; in addition, oxalic acid may bind to and reduce the absorption of minerals, especially calcium [77]. Plants rich in oxalic acid include rhubarb, spinach and beetroot, whilst chocolate, cocoa, nuts and tea are moderate sources [77].

Lectins are proteins with a characteristic affinity for certain sugar residues or glycoproteins present in the membranes of animal cells, including those of the intestinal mucosa [94]. Since lectins react with red blood cells *in vitro* to induce agglutination, the alternative term haemagglutins is often used; they are absorbed intact into the systemic circulation to elicit profound effects on organs, in addition to which an immunological response may occur [94]. Lectins are present in the seeds of a variety of plants, inluding kidney beans and lima beans, which must be cooked in order to inactivate the toxins [94]. Small amounts of glycoalkaloids are found in potatoes and other members of the nightshade family (tomatoes, aubergine, red and green peppers); these can be toxic if present in high concentrations, e.g., in green and sprouted potatoes and in green tomatoes [77]. Potatoes also produce glycoalkaloids (such as solanine) in response to bruising, cutting and other forms of physical damage, and during rotting [97]. Cassava and lima beans contain cyanogenic compounds which are broken down to form cyanide [77]. Cassava is a staple, rich in starch, for millions of consumers in the tropics; it contains cyanogenic glycosides in the form of linamarin and, to a lesser extent, lotaustralin [98]. These glycosides produce hydrocyanic acid when the action of an endogenous enzyme, linamarase, is initiated by crushing or otherwise damaging the cellular structure of the plant; cassava roots therefore require processing to remove the toxins [98]. Small amounts of hydrazine are found in a wide range of mushrooms, including cultivated ones [77]. Some varieties of wild mushroom contain highly toxic metabolites, making them unsuitable for food use. The ovaries, intestines and liver of puffer fish (fugu) are rich in tetrodotoxin, a substance said to be 10,000 times more poisonous than cyanide [99]. Despite the precaution of allowing only licensed chefs to prepare the fish, several persons die every year from fugu poisoning.

5 Origins and Nature of Sensory and other Performance Attributes in Foods

5.1
Introduction

Each quality food individually must "perform" in accordance with a whole host of quality criteria. Food components and ingredients, once consumed, are involved in physiological processes, which may impact health, either for good or bad. However, even before a food is swallowed, the human organism responds to the sensory stimuli which characterise that foods. Many food components act as sensory stimuli. The first three sections of the current chapter therefore describe the perceptions that result from a food interacting with a consumer's special senses. Three key sensory properties of foods are introduced from an objective, value-neutral point of view. These are the appearance, texture and flavour of foods. Since the current part of the book is designed to provide a focus on foods rather than on the consumer, affective aspects are dealt with in Part 3. Affective aspects of the sensory perception and evaluation of foods involves many different motivational factors, of which expectations are especially important. Food components, and the manner in which they are arranged in a particular food, also determine the performance of a food in a slightly different, but equally important, context. This is to do with the suitability of a raw material for different food technologies, and of a finished food product for different storage and distribution systems. All food processing is designed in some way to render some edible material more fit for some particular purpose, one of the most common goals in food processing being the extension of shelf life. The domestic preparation and cooking of foods generates different quality requirements, particularly on the selection of raw materials, than does industrial manufacturing. Other aspects of fitness for purpose include use context and convenience, although these tend to be more susceptible to individual consumer perspectives. Nevertheless, there are aspects of so-called appropriateness-in-use which are relatively untouched by subjective considerations. Pasteurised milk in the UK is commonly used both as a beverage and for pouring over breakfast cereals, and the appropriateness of both of these contexts is generally acknowledged. However, there will be some people, who would consider one or the other, or both, of these uses as inappropriate. On the other hand, once it is accepted that milk and dairy products are suitable items for human consumption, milk, as a liquid, is clearly used differently from butter, or cheese. Homing in on the cheese offer, clearly, different types of cheese differ in terms of their suitability for different types of application. Some cheeses are soft and can be spread on bread, but cannot be sliced; others are sliceable but not spreadable; yet others are crumbly and cannot readily be used in either of these applications. Some cheeses perform well in cooked dishes, others do not. Another way of looking at the concept of fitness for purpose is in relation to different stakeholders in food supply chains. Whilst the producer and consumer will share certain requirements, they will differ in others. For example, commercial growers of peas will plant varieties that mature simultaneously, as this will facilitate harvest. In contrast, the domestic gardener is likely to require continuity of supply, and will therefore select varieties of pea with extended cropping seasons.

The sensory attributes of foods and beverages are broadly categorised as appearance, texture and flavour, reflecting the special senses primarily involved in the determination of such attributes (see Chapter 7). The objective description of a food in terms of its sensory attributes, and the intensity of each of these attributes, depend on suitably trained assessors and the elimination of any affective interference. Nevertheless, it has to be recognised that even a strictly objective methodology, e.g., sensory profiling by quantitative descriptive analysis [100], incorporates at least one subjective component, namely, the *a priori* selection of quality attributes which the taste panel will be required to score. This is in accordance with the arguments about the nature of quality developed in Chapter 1, i.e., that it is not the totality of attributes of a product that determine quality, but the selection of attributes important to the user. Food is perceived initially either via the sense of vision or the sense of smell, or via both of these senses simultaneously. Both appearance and aroma provide cues about the identity of a food and generate expectations about its eating quality. Appearance plays an increasingly important role in the way in which consumers select food. Most food is now bought in supermarkets, where it is rarely possible to sample a product before purchasing it for the first time. Consequently, packaging design often has to stand as a proxy for appearance of the food itself. Both appearance and aroma of foods are powerful forces in appetite, appearance being especially important for foods with low aroma intensity, e.g., many cold dishes. Taste as one of the special senses and taste as a sensory attribute share a common terminology, and the term flavour refers to the aroma and the taste of a food combined (see Chapter 7). In traditional cuisines throughout the world, it is typical for both individual dishes and complete courses to be put together in order to maximise sensory responses and enhance the eating experience. This highlights the importance of hedonic consumption value, especially in formal and social dining contexts. For example, a so-called full Italian meal would typically consist of a starter, first course, main course, cheese and finally, fruit and desserts [101]. The main role of the starter will be to stimulate appetite, making appearance, and an attractive presentation, a key requirement for this course. The first course, which follows the starter, will serve as a filler. The main course of meat or fish and vegetables, and the dessert, contribute flavour complexity and impact to the meal, and are therefore its most indulgent part. Whilst elaborate meals must be especially carefully designed to engage all of the senses positively, the same basic principles also apply to simple dishes. A range of salad leafs combined with nuts, cheese, vegetables, fruit, sprouted seeds and beans, rice and other grains is an example of such a dish, as it presents an interesting range of colours, flavours and textures to its target consumer [102].

Measurement of the sensory attributes of foods and beverages forms the basis for determining their contribution to consumer evaluation of these products. Accurate characterisation of such attributes depends on appropriate sensory vocabularies; these are specific for each product category, and sensory panel training therefore needs to be product-specific. Sensory profiling is relatively easy with homogeneous, formulated foods of known recipe, e.g., tomato ketchup [103], and it is for these types of product that consumer expectations about batch-to-batch conformity will be the greatest. On the other hand, sensory profiling of animal- and plant-derived foods that have undergone little or no processing is more difficult; this is largely due to sampling difficulties. For example, where a taste panel is required to profile fish, such as salmon, samples will vary, particularly in appearance, depending on which area of the fish the sample was taken from. The food industry has traditionally considered sensory evaluation to be the responsibility of a so-called company expert, who, through years of accumulated experience, was able to describe a company's products, as well as setting sensory quality standards for them. In the modern food industry, such experts still work as flavourists, wine experts, brewing experts, and coffee and tea tasters. Although these roles still exist, in terms of food quality overall, the impact of expert tasters has been eroded as taste panels have become increasingly professional [104]. In the past, when dealing solely with the preservation of basic agricultural crops, it was more feasible for product experts to understand a particular product category. These experts tended to create score cards and unique terminology [104], and scores soon became established as targets or standards. Examples of this approach include the 100-point butter score card, the 10-point oil quality scale, and the 20-point wine score card, all of which had specific numbers that connoted levels of product acceptance [105]. With the evolution of processed and formulated foods, experts faced increasing difficulty in maintaining a detailed knowledge of all products [104]. First, the Arthur D. Little Company introduced the Flavor Profile® method, a qualitative form of descriptive analysis, which replaced the individual with a panel of experts responsible for obtaining a consensus decision [106]. The next descriptive method of importance was the Texture Profile® method developed by General Foods [107, 108]. The Quantitative Descriptive Analysis (QDA) Method® [109] is characterised by a consensus approach to vocabulary development, use of replication for assessing subject and attribute sensitivity, and defined statistical analyses [104]. A considerable body of literature has been developed in which lists of words are published for specific types of products [e.g., 110, 111]. Aroma terminology may be based either on individual chemicals, on chemical categories, source materials (e.g., vanilla) or aroma associations. These are all legitimate provided that the communication that results is unambiguous.

5.2
Appearance

The appearance of all objects, including foods, is based primarily on their structure and pigmentation. Appearance includes all visible attributes and derives from the interactions between a substance or object and its environment as perceived by the human observer [112]. Appearance combines the visual information contained in reflected, transmitted and scattered light, and the colour of that light [113]. Shape, pattern, size and colour provide means of identification of both basic produce and man-made foods; and where a product is concealed from visual inspection at the point of purchase, it is the task of food packaging to accurately represent the contents of a pack. Other important appearance attributes in foods include properties such as translucency (as opposed to opacity) and gloss. Appearance is examined here both in terms of the product itself and in relation to the context and environment in which the product is presented to the consumer. An in-depth discussion of individual perceptions and meanings in relation to appearance is provided in Chapter 7. Appearance cues are widely used by consumers to infer food product quality; frequently, these are the only cues available, especially at the point of purchase. Whilst many sensory product attributes do not necessarily qualify as quality attributes, i.e., they are not particularly important in terms of acceptance, appearance attributes largely do. One of the reasons for this may be the fact that appearance constitutes the very first screening device in the evaluation of any substance or material as a potential food. For example, consumers' attention is readily drawn to just noticeable differences (JND) in the colour of branded products, or in the colour of their packaging, especially when individual packs are seen next to each other on the supermarket shelf. Visible product variation is likely to be interpreted by consumers as some lack of control in the manufacturing, handling or distribution of a product. In fact, one of the key considerations determining the use of colours in processed foods is the perceived need for consistency (see Chapter 6). Visible order and harmony, such as uniformity of display in boxes of fruit, oysters etc., further communicate both competence and care on the part of the supply chain.

Appearance cues are significant firstly in the identification of a food and secondly, in the assessment of the condition of that food. Foods must look normal, i.e., conform to expectation. They must be undamaged and of normal colour. Chocolate bloom, wilted lettuce, sweaty cheeses and delicatessen meats are abnormal. Foods must look clean and be seen to be free from extraneous matter. Visible abnormalities, such as blemishes in fruit and vegetables, abnormal colour in meat, bloom of chocolate and cloudiness in beer will cause a product to be rejected. Appearance may also provide cues on the degree of excellence, or grade, of a product. This applies to tea, where evenness of leaf size denotes teas true to their grade, with raggedness describing an uneven, badly manufactured and graded tea [114]. Surface dryness or wetness of meat, turgor of fruit and vegetables, gloss on apples, and polish of wine and other clear beverages all are appearance attributes capable of providing cues to overall quality. Appearance tends to be the first milestone in food product quality evaluation as conducted by a prospective consumer, where full confidence in the consequences of consumption is still being built up. Once this initial hurdle has been overcome, subsequent evaluations will be carried out with progressively greater confidence, with the likelihood of a satisfactory outcome increasing at each stage.

The physical dimensions of a food are likely to be related to its internal structure. This is obvious for fruit and vegetables, and cuts of meat, but also applies to man-made foods. The form of a food provides the initial means for its identification, thus differentiating apple from pear, lemon from grapefruit, and bass from bream. Shape and pattern also play an important role in areas of food quality such as the conformation of animal carcasses, where different classes are related to yields in specific cuts [115]. The curvature of cucumbers may make them more or less suitable for pickling and subsequent presentation in small glass jars. This is a typical illustration of the fitness-for-purpose concept. Pore size, and the size distribution of these pores, are important appearance attributes in baked goods, e.g., bread, as they signal the level of process control exerted during the manufacture of products. In mass produced, man-made foods, in particular, size and dimensions need to be controlled very accurately, because of (raw material) costing issues and because of the need to fit product into its packaging. This latter aspect takes on an added importance as the role of the pack as part of product value increases, e.g., the fancy packaging employed in the presentation of fancy chocolate confectionery. In contrast in so-called artisanal foods, somewhat less regular appearance of the product may in fact be desirable. Appearance can be a good predictor of the texture and mouthfeel of a food. For example, soups may appear as either light, clear, frothy, viscous, thick, creamy, smooth, grainy or lumpy, or as a combinations of several such appearance attributes, all of them clearly related to texture. Visual texture further includes the so-called tears, or legs, that form on the side of a wine glass after wine has been swirled around the glass. These tears are indicative of the viscosity, or body, of the wine in question, attributes that are likely to be reflected in mouthfeel. Similarly, the thickness of a food portion, e.g., of a slice of smoked salmon or of a bar of chocolate, is likely to be utilised by the consumer as a texture cue. Food photography and labelling can be pressed into service in order to communicate textural and flavour character of a food to consumers. For example, a photograph of a bottle of white wine covered in condensation suggests a light and refreshing product.

Colour is the perception that results from the impact of the wavelengths of light in the visual spectrum (390–760 nm) on the retina in the human eye [116]. It is often considered as the most important appearance attribute in food, and this is reflected in the widespread use of colours as additives in man-made foods. The colour wheel is a simplified spectrum showing the arrangement of the primary (red, yellow and blue) and secondary (orange, green and violet) colours, from which all other colours can be mixed; the wheel should be consulted when assembling colours for either harmonious or contrast effects [117]. If too many colours are present, a picture becomes confused. An effective way of creating colour harmony is to focus on a small selection of colours which lie next to each other on the colour wheel, i.e., analogous colours. On the other hand, too much harmony can produce visual blandness, and a small area of contrasting colour composition can be included to avoid this. Like form, colour serves initially as an identification tool, especially where processing has destroyed, or obscured, form, as is the case with fruit juices and juice drinks. For example, raspberry flavour(ed) products will be expected to be red, banana flavour(ed) products yellow, and so on. If this basic requirement is ignored, the consumer may even be unable to attribute flavour accurately, able only to identify fruitiness rather than specific fruit character. As well as colour hue or shade, i.e., red, yellow green etc., brightness or luminance (from light to dark), and purity or saturation (from vivid to dull) are important in colour expression [118]. However for many foods, colour specification only partially describes appearance, and an indication of their light scatter and its relation to the pigment absorption is also required [119]. The colour of milk is due to light scattering by the dispersed phase, combined with feed and processing effects [120]. The overall whiteness of milk arises from light scattering by the casein micelles and, to a lesser extent, the fat globules. Coloured milk components also contribute to the colour of milk, in particular, greenish-yellow riboflavin in the aqueous phase and yellow carotenoids in the lipid phase. The green, almost fluorescent colour imparted by riboflavin can be clearly seen in skimmilk. The amounts of riboflavin and carotenoids in milk vary seasonally, and according to the feeding regime for the cow. Homogenisation of milk increases its whiteness, because of the increased light scattering by the now more numerous, as well as smaller, fat globules. On heating, whiteness is similarly increased due to damage to the casein micelles. However, on further heating, the Maillard reaction gives rise to brown compounds. Whilst some basic foods happen to exhibit strong or attractive colouring in addition to other quality attributes, other foods and food ingredients are used in recipes specifically because of their colour. Plants that yield stable yellow colours include saffron, marigold petals, annatto and turmeric. Vegetables such as spinach may be used to add colour to colourless foods such as pasta.

The colour of fruits and vegetables, as well as of meat and fish, is due largely to the presence of certain types of biological pigment. Individually, colours vary depending on the type and quantity of pigment present. The distribution of chlorophylls and carotenoids in fruits and vegetables changes with the stage of maturity. In general, as colour changes from green to yellow-orange-red, chlorophylls decrease and carotenoids increase. However, it is not always possible to generalise, even in terms of the cultivars within a particular type of vegetable. Capsicums do not follow the general pattern of pigment composition [121]. Giant *Cavendish* bananas remain green on ripening at tropical temperatures 30–34 °C due to incomplete chlorophyll degradation, while at 20 °C complete degreening occurs and fruits turn yellow, with attendant improvements in both eating quality and shelf life [122]. Besides carotenoids, anthocyanins are important sources of plant pigments, providing plant colour mainly in the red-blue range. Anthocyanins are susceptible to chemical and photochemical degradation, but complexes of phenolic compounds increase anthocyanidin stability [123]. Natural pigments in biological systems are synthesised in living cells; others, e.g., oxidised phenols and simple phenolic compounds, such as coumarins, may be formed in the dying cell [124]. Fruits and vegetables may undergo browning reactions as a result of tissue disruption. For example, in pear juice or purée manufacture, colour deteriorates rapidly due to the action of the endogenous enzyme, polyphenol oxidase [125]. Thermal inactivation of the enzyme directly after cutting or maceration of plant tissues is an effective means of controlling this type of browning reaction. In heat-processed fruit and vegetable products, non-enzymic browning (the Maillard reaction) is also common [126]. Enzymic browning is not always undesirable as the quality of apple juice, dates, prunes, raisins, soy sauce and black tea depend on it; the reactions also contribute to the quality of fermented drinks such as beer, cider and perry [90]. Non-enzymic browning includes caramelisation, dextrinisation and sugar-amine browning. This type of browning of food is often welcome as it goes along with distinctive and attractive flavour development, e.g., loaves of bread, slices of toast, and grilled and roasted meat.

The metalloproteins are a group of pigments with important biological function, and include the haems and chlorophylls, which are categorised as metalloporphyrins [124]. Haems are ubiquitous in nature and perform a central role in cellular energy metabolism [127]. Haems are either pink or red. Myoglobin is the haem responsible for the colour of red meat. Major changes can involve the iron atom at the centre of haem, which can be present either in the oxidised (Fe^{3+}), or in the reduced (Fe^{2+}) form, and the protein globin, which can be either native or denatured [90]. In raw meat, the complex occurs in three forms: purple myoglobin; bright-red, oxygenated oxymyoglobin; and brown, oxidised metmyoglobin (Fe^{3+}) [90]. Among these, metmyoglobin is unable to bind oxygen [90]. Muscles contain varying amounts of myoglobin, with muscles designed to carry out more work containing more. For example chicken leg meat contains more myoglobin than breast meat. In fresh meat, myoglobin occurs together with haemoglobin, the oxygen carrier of blood [90]. Both pigments undergo similar colour change [90]. Good quality beef needs to be well hung, and as it matures, its colour changes from bright red to a dark, purplish red [128]. However, high concentrations of oxygen in meat packs incorporating a so-called modified atmosphere, cause meat to retain its original, bright cherry-red colour [90]. The appearance of fresh meat is therefore due to the concentration and oxidative state of myoglobin, any desiccation (including freezer burn), and the oxidative state of the pigment just below the meat surface. Cooking changes the red colour of raw meat to grey or brown. In cured meats, nitrates from the curing salts are reduced to nitric acid, which complexes with myoglobin to yield nitrosomyoglobin, the stable red colour typical of such meats [90]. PSE (pale, soft, exudative) and DFD (dark, firm, dry) meats provide visual cues towards diminished eating quality, both in terms of texture and of flavour [129]. The colour of fish flesh is governed by blood, melanin derivatives (produced for skin colouration), or arises directly from feed. The pink flesh colour of salmon arises from the carotenoid astaxanthin, which is contained in the diets of fish which naturally feed on crustacea [90]. The colour of farmed fish can be varied through the types, and amounts, of natural, or synthetic, pigment fed to them. Synthetic pigment is also added to formula feeds for layers feed to achieve deeper yolk colour (see Chapter 6).

5.3
Texture

The texture of foods and beverages derives, in the main, from their structure and other mechanical properties. However, as with all sensory perception and measurement, the perceiver is an integral part of the process. This is particularly evident in astringency, a drawing, or puckering, of the lining of the mouth, which is caused by substances such as tannins [112, 130]. This is typical of full-bodied red wines. A recent explanation of the mechanism underlying astringency is based on the formation of complexes between proteins contained in saliva and tannins and other astringent substances in foods and beverages [131]. The astringency of sterilised milk is produced by the aggregation of certain milk components, in particular, whey proteins and salts [120]. Astringency in sterilised milks is commonly referred to as chalkiness. Astringency lingers in the mouth, and has a tendency to become more pronounced on repeat exposure, e.g., when sipping wine [130]. Texture means "all the mechanical, geometrical and surface attributes of a product perceptible by means of mechanical, tactile and, where appropriate, visual and auditory receptors" [112]. Visual texture has already been referred to. The role of sound in texture perception is familiar from the crunching sound of apples, the snap of chocolate and the sizzle of a steak when it is being grilled. Texture may be perceived in the mouth and through the skin, and it may be transmitted to skin receptors via cutlery and other eating implements. For example, a tough steak resists the action of a knife, a floury potato can be easily mashed with a fork, and so on. The ease with which cooked, or cured, meat and fish products can be cut by knife and manipulated in the mouth, depends both on the overall tenderness of the lean tissue and on the nature and distribution of connective tissue and fat. Intra-muscular fat in meat is associated with succulence. Succulence is an example of a mechanical food attribute, i.e, a group of attributes which express the response of a food to a force being applied to it. The nature of mechanical attributes may be further illustrated by comparing the characteristics of some popular hard cheeses. For example, Appenzeller is hard and elastic; Grana is very hard and grainy, making it suitable for grating; Cheshire is firm, open-textured and crumbly; Caerphilly is flaky; and so on. Possible mechanical textural attributes of foods are categorised, with typical examples, in Table 5.1

The geometrical component of texture relates to the size, shape and arrangement of particles within a product [112]. Smoothness, grittiness, graininess and coarseness are categorised as granularity, and fibrousness (as in celery), cellularity (as in citrus fruit) and crystallinity are examples of conformation. Surface attributes are textural attributes that are related to the sensations produced by moisture and/or fat content [112]. Moisture describes the perception of water absorbed by, or released from, a product. Related textures range from dry, as in cream crackers, to juicy, as in fruit and meat, to watery. Fatness attributes range from oily, as in salad dressings, to fatty, as in lard. Bakers blend lipids into cake recipes as softeners and lubricants, to tenderise the cake [14]. Overall, biologically grown foods exhibit more complex textural properties than do man-made foods, and they respond differently to mastication. All fruits and vegetables have cellular structures. However, they differ in the sizes, shapes and arrangement of

Table 5.1. Mechanical textural attributes of foods (adapted from [112])

	Attribute	Scale points and food examples	
Hardness:	Force required for deformation or penetration of food	Soft: Moderate: High:	Cream cheese Olive Boiled sweets
Fracturability:	Force required to break food into crumbs or pieces	Crumbly: Crunchy: Brittle: Crispy: Crusty	Corn muffin cake Apple Brandy snaps Cornflakes Fresh French bread
Chewiness:	Length of time, or number of chews, required to masticate food ready for swallowing	Tender: Chewy: Tough:	Young peas Fruit gums Old cow meat
Gumminess:	Effort required to disintegrate a tender product ready for swallowing	Short: Mealy, powdery: Pasty: Gummy:	Shortbread Mealy potatoes Chestnut puree Edible gelatine
Viscosity = Resistance to flow:	Force required to draw a liquid from a spoon over the tongue, or to spread it over a substrate	Fluid: Thin: Unctuous: Viscous:	Water Sauce Double cream Runny honey
Springiness:	(a) Speed of recovery of food from a deforming force; (b) Degree to which a deformed food returns to its original condition after the deforming force is removed	Plastic: Malleable: Elastic, springy, rubbery:	Margarine Marshmallow Calamary
Adhesiveness:	Force required to remove material that adheres to the mouth or a substrate	Sticky: Tacky: Gooey, gluey:	Marshmallow topping Cream toffee Tapioca

the cells, as well as in the substances which hold these cells together [132]. Texture is an important attribute of both raw and processed fruits and vegetables and is often used as a major criterion in selecting such foods for processing [132]. The main determinant of texture in plant-derived foods is the plant cell walls, i.e., the chemical nature of the cell wall and the interrelationships between the component polymers, together with the structural arrangement of tissues [19]. Such variations exist between different fruits and vegetables as well as between different varieties of a particular type of fruit or vegetable. For example, waxy and floury varieties of potato exhibit distinct texture differences when cooked, reflecting differences in their cell and tissue structure. On the other hand, mealy texture in apples tends to reflect senescence and storage-related changes. The rupturing of plant cells does not merely consist of brittle fracturing, but is accentuated by the sudden release of fluid. Fruits and vegetables therefore do not respond to stress in a purely elastic manner, but instead, exhibit simultaneous viscous behaviour. They may therefore be described as viscoelastic [132]. Meat products, such as ham, are also viscoelastic [133].

Oral evaluation of food texture is dynamic with respect to mechanical, thermal and salivary action. Solid and semi-solid foods, e.g., ice cream, gelatine gels and chocolate, exhibit significant textural changes with temperature during consumption, providing a kind of textural contrast generally valued by consumers [134]. The mouthfeel of chocolate is perhaps its most unique characteristic. Cocoa butter has s very narrow melting interval, which happens to be close to human body temperatures. Therefore, when chocolate is eaten, it is transformed from a solid with some snap into a smooth, mouth-coating liquid [135]. This melting behaviour also affects the evolution of the perceived chocolate flavour. The particle size of chocolate is a factor in smoothness; in premium chocolates, particle size can be as low as 10–12 µm, whilst ordinary chocolate may have particle sizes of 40–80 µm [135]. High contents of solid fat in a food result in a cooling effect in the mouth during eating. This is due to the heat needed for melting; however, if the fat does not melt completely during eating, a sticky sensation results [63]. Access of tastants to receptors in the mouth decreases with increasing structural complexity of a food, meaning that flavour perception depends on structure, texture and oral manipulation of the food [136]. Oral manipulation includes frequency and pattern of chew, the flow of saliva and breathing pattern. Texture-flavour interactions can be measured using time-intensity methodology [136]. This includes the rate of onset, duration of maximum intensity and rate of decline of perception of sensory attributes [136]. Difficult-to-chew and difficult-to-swallow textures may represent significant barriers to the consumption of certain foods. Tough pieces of leaf, e.g., flat-leafed parsley can catch at the back of the throat, as can fish bones. It has been suggested that fear of choking and gagging during mastication, due to difficult-to-manipulate textures, could be one of the causes of children's aversion to certain cooked vegetables [137].

5.4
Flavour

Flavour, unlike appearance and texture, derives from specific, chemically defined food components. The British Standard on terms relating to sensory analysis provides a definition of the term flavour [112]. This makes some reference to the special senses involved, a topic developed at some length in Chapter 7. The definition of flavour is given in Table 5.2.

Tastants dissolve in saliva and are transported to receptor sites in the mouth; odorants are volatile chemicals that can gain access to receptors in the olfactory mucosa either through the nose, or through the mouth via the retronasal route. The retronasal route tends to generate more complex flavour impressions. One reason for this is that the temperature of a cold food is raised in the mouth, which leads to more volatile material reaching the olfactory mucosa. In addition as indicated earlier, flavour release from a food matrix, and therefore flavour perception, are dynamic processes. The term taste refers both to the sense of taste and to product attributes capable of inducing taste sensations; similarly, smell describes both the odour attributes of a product and the sense of smell [112]. Four basic tastes are commonly recognised, i.e., sour, bitter, salty and

Table 5.2. Definition of "flavour" [112]

Flavour	Complex combination of the olfactory[1], gustatory[2] and trigeminal[3] sensations perceived during tasting The flavour may be influenced by tactile, thermal, painful and/or kinaesthetic[4] effects.

[1] pertaining to the sense of smell.
[2] pertaining to the sense of taste.
[3] irritating or aggressive (sensations) perceived in the mouth or in the throat.
[4] (sensations) resulting from pressure on the sample, produced by a muscle movement.

sweet. By contrast, there are thousands of aroma chemicals in foods, each with its own particular smell. Some of these are already present in live biological tissues. Others develop during food processing from precursors present in the raw materials, or they are degradation products which accumulate during the storage of a food. Trigeminal sensations (Table 5.2) include the burning sensation of chilli peppers and the pungency of vinegar and mustard. The view expressed in the British Standard about what should be included in the basic tastes is somewhat broader than normal [112]. Here, alkaline, umami and metallic are included as well as sour, bitter, salty and sweet. The Standard also proposes that the term acid should be used instead of the term sour to describe a basic taste, and that the word sour should be reserved to describe the complex smells or tastes that characterise certain organic acids. A range of food examples for the basic tastes are shown in Table 5.3.

Fruit acids are a common source of acid, or sour, taste in foods, as are fermentation acids. Lactic acid is responsible for the sour taste of fermented dairy products and for that of lactic acid-fermented vegetables, e.g., sauerkraut. Acetic acid fermentation produces vinegar, which is used to acidify many man-made

Table 5.3. Basic tastes of foods and beverages

Taste [112]	Reference substances [112]	Food and food ingredient examples
Acid	Citric acid	Acid fruits; Fermented milks and vegetables; Vinegar pickles
Bitter	Quinine, Caffeine	Naringenin in grapefruit [77]; Chicory; Marmalade (bitter oranges); Vermouth (wormwood); Beer (hop bitter acids) [139]; Cocoa extract (caffeine, theobromine)
Salty	Sodium chloride	Table salt
Sweet	Sucrose	Ripe tropical fruits; Certain fruits from temperate regions; Honey; Sweeteners
Alkaline	*None given*	
Umami	*None given*	Monosodium glutamate and 5' nucleosides [143]; Peptides [144]
Metallic	*None given*	cis-3-Octenol (also has mushroom-type flavour)

foods, e.g., pickled vegetables. Acidulants are widely used in man-made foods and beverages, citric acid being commonly used in soft drinks.

The original bitterness of their wild ancestor has mostly been bred out of cultivated food plants. Recipes using aubergines still occasionally recommend pre-salting, ostensibly to facilitate the removal of bitter juices; but in fact, modern commercial cultivars of aubergine are very low in bitterness. Chicory is a slightly bitter vegetable, but in this case bitterness is a desired attribute. Cucumbers may develop bitterness if affected by water shortages during fruiting. Different types of food chemicals cause bitterness in foods. Citrus bitterness arises from flavanones [138]. The main compounds in hops are bitter resins called α-acids or humulones and β-acids or lupulones; different beers contain different concentrations of the major bitter acids [139]. The major contributions to the bitterness of cocoa come from the xanthines, caffeine and theobromine [140]. Part-digestion of protein leads to bitter peptides. Foods for special medical purposes are provided for individuals who are unable to take, digest or absorb adequate amounts of normal foods, and may include bitter tasting amino acids or peptides [141]. Bitterness in standard foods is more common in adult beverages, confectionery etc. than in staples. Bitterness tends to be an acquired taste, and a liking for bitterness may be associated with pleasant physiological effects, e.g., from caffeine.

Natural sweetness is widespread, especially in fruits, as is sweetness in man-made foods. Sweetness in fruits and vegetables is due to native sugars and can be enhanced through processing, e.g., caramelisation in fried onions. Equi-sweet concentrations of nine sugars and sweeteners have been established [142]. With sucrose as the benchmark (100), relative sweetness of fructose is 120, of glucose 61, of sorbitol 57, of lactitol 35, of Acesulfame K 19,500, of Aspartame 12,800 and of Saccharin 36,000. Umami is the name given by Japanese researchers to the taste sensation produced by certain substances present in seaweed, traditionally used to enhance the flavour of meat, fish, certain vegetables and mushrooms [143]. Umami is particularly associated with monosodium glutamate (MSG, the salt of a common amino acid) and certain 5'-nucleosides (IMP, inosine monophosphate and GMP, guanosine monophosphate). Isolated from their natural sources, these substances are widely used as flavour enhancers for savoury foods. Some peptides obtained through proteolytic processes, e.g., in cheese, meat, gravy and from soy proteins are taste active, albeit typically in an undesired manner, i.e., as bitter components. Glutamyl oligopeptides isolated from modified soy bean protein were the first known to have savoury taste character [144].

The particular odour sensation a chemical creates depends, to a large degree, on its molecular structure and volatility. As there are thousands of odorants associated with foods, these are much more difficult to categorise than the tastants. The most accurate way of naming an odorant is through its chemical name. Professional flavourists will be able to recognise and name a large proportion of odorant chemicals. However, naming an odour becomes more difficult where complex mixtures of odour chemicals are concerned. In either case, most ordinary consumers find it extraordinarily difficult to recognise, identify and name specific odours. Communications in this area therefore tend to rely on associative terms and descriptions. Over a number of years, there have been many attempts to assign individual chemicals to broad odour categories. For example,

Table 5.4. Aroma note categories [145]

Aroma note	Present in food/flavouring	Aroma note	Present in food/flavouring
Fruity	Peach (ethyl caproate, ..), strawberry (ethyl butyrate, ..), peach (γ-undecalactone, ..) ...	Green	Grass (hexenyl acetate, ..), tropical fruit (2-isopropyl-4-methyl thiazole, ..) ...
Citrus	Lemon (citral, ..), orange (limonene, ..) ...	Coniferous	Pine, juniper (terpinolene, ..) ...
Orange flower	Neroli (methyl anthranilate, ..), petitgrain oil (linalool) ...	Medicinal	Eucalyptus (eucalyptol, ..), camphor oil ...
Vanilla	Vanilla beans (vanillin, ..), Peru balsam ...	Tobacco	*for flavoured cigarettes*
Dairy	Milk (δ-dodecalactone, ..), butter (diacetyl, butyric acid, ..), cheese (isovaleric and caprylic acids, ..) ...	Honey-sugar	Honey (ethyl phenyl acetate, ..),
Roasted	Hazelnut (2-methyl-3-methoxy pyrazine, ..), caramel (maltol, ..) ...	Floral	Geranium rose (phenylethyl alcohol, ..), jasmine (benzyl acetate, ..), violet (α-ionone, ..) ...
Spicy	Cinnamon (cinnamic aldehyde, ..), Cloves (eugenol, ..) ...	Smoke	*Rectified smoke*
Anisic	Anise (anethol, ..), tarragon, basil (estragole) ...	Anamalic	Civet (scatol, ..) ...
Wild-herbaceous	Thyme (thymol), cumin (cumin aldehyde, ..) ...	Woody	Cedarwood (cedrol, ..), orris (β-ionone, ..) ...
Marine	Sea-like (nerol, ..), fish (trimethyl amine, ..) ...	Aromatic	Thyme (thymol, ..) ...
Minty	Mint (menthol, ..), spearmint (carvone, ..), pennyroyal (pulegone, ..) ...	Fungal-earthy	Patchouli ...
Citronella-vervain	Citronella (citronellal), lemongrass (citral, ..) ...	Aldehydic	Aldehyde C12 ...
Alliaceous	Onion (dipropyl disulfide, ..), garlic (diallyl sulfide, ..) ...		

the *Perfumer & Flavorist* published 25 training charts [145] based on distinctive aroma notes. These categories are shown, with a small number of examples, in Table 5.4.

In aromas and flavourings, top notes represent the most volatile chemicals, e.g., the alcohols and aldehydes typically found in "green" notes; floral volatiles typify so-called middle notes, and odour components with low volatility are known as base notes. Base notes include vanilla, maltol and the longer chain esters and alcohols. The roasted flavour of nuts and of certain savoury foods is due largely to volatile alkylpyrazines. Pyrazines are generated in bread crusts during baking and in beer malt upon extrusion cooking, and they can be extracted from roasted oats [146]. There are a number of so-called signature volatiles, also known as character-impact volatiles. These, although rarely representing an overall rounded flavour impression of a particular food product, serve as a key chemicals allowing recognition of that food. One substance, for which this is eminently the case, is *cis*-3-Octenol, which very clearly signals "mushroom" without providing a rounded mushroom flavour. Similarly, benzaldehyde is characteristic of almond and cherry flavour and *trans*-2, *trans*-4 decadienal of cucumber. However, many foods do not possess signature volatiles. This makes it difficult to recognise complex, but subtle aromas, such as the aroma of strawberries if there are few supportive visual cues. Dilution profiling is a method that can be used to deconstruct complex flavours, and allow individual components to be identified [103]. Fat-soluble molecules are partitioned between the food lipid phase and the vapour phase above the food. Short-chain fatty acids are more volatile, and have more intense odour, than longer chain acids; those with chain lengths greater than ten carbon atoms have little or no odour [63].

Flavour profiles have been established for a large range of common foods and beverages. In the whisky tasting wheel, whisky aromas are related either to production or to maturation [147]. Production aromas include aldehydes (leafy, floral), esters (fruity, fragrant) and phenols (wood-smoke, iodine); flavours associated with whisky maturation depend on the types of cask used. Sherry-type aromas are the most obvious, whilst new wood (not commonly used) confers resinous, pine-like aromas. Bourbon wood is the most commonly used, and it typically bestows a rounded vanilla-like, nutty, cigar-box aroma characteristic of well-matured malt whiskies [147]. Twenty-two descriptive terms were used to profile 36 soy sauces; this enabled individual sauces to be allocated to distinct categories [148]. Twenty-four wines from Bordeaux were profiled and the major aroma differences established [149]. The largest differences among the wines were those attributed to variations in green bean/green olive and blackcurrant aromas. Individual wines also differed in their bitterness and astringency. Sensory profiling scales have also been published for Finnish rye breads, where the odour component is represented by terms such as musty, burnt, rye-like, malty, sweet and sour [150].

Raw meat has little aroma. However, when meat is cooked, a complex series of reactions occurs between non-volatile components of the lean and fatty tissues. This results in over 1000 volatile compounds, many of which contribute to the aroma typical of the cooked meat [151]. All cooked meat has a characteristic, meaty savouriness. The aroma of a fat-free meat bouillon typifies this generic meaty flavour, which does not carry any associations with any particular species of meat. The species-specific components of meat aroma derive from the fat of animals [151], and it is those that allow beef flavour to be distinguished from pork flavour. Grilled, or roasted, meats exhibit characteristic roast aroma notes, which differ from the aromas associated with boiled or stewed meat. The Maillard reaction and lipid degradations are the most important routes to flavour compounds in meat, but the initial products of these reactions may undergo further interactions. Heterocyclic compounds, such as pyrazines and thiazoles, derived from the Maillard reaction, contribute to roast and grilled aromas, while certain aliphatic and heterocyclic sulphur compounds provide some of the characters of boiled meat [151]. Maturation of meat too is associated with flavour development. Maillard reaction describes a category of chemical reactions that occurs at high temperatures between amino compounds and reducing sugar. It occurs in roasted foods generally, e.g., baked goods and coffee, as well as meat. The high temperatures required for sugar caramelisation are not needed for the reaction to occur but low moisture conditions are required.

Cereal products may gain their distinctive flavours enzymically, e.g., during dough-kneading and fermentation, and non-enzymically, e.g., during heat treatment [146]. Microbiological processes are started by adding yeast, starter cultures or sour dough, as in bread dough fermentation, or by inducing mould and bacterial growth, as in the manufacture of tempeh. Fermentation by yeast, or lactic acid bacteria and yeast (sour dough), gives rise to typical flavour compounds in bread, e.g., ethanol and an array of organic acids such as acetic, propionic, butyric, lactic and malic, as well as citric and succinic acids [146]. The uniqueness of cocoa flavour arises from the fact that cocoa is one of the few foods that are both fermented and roasted. The flavour is the result of the combination of a variety of enzymic, microbial and thermal pathways [140]. Citrus volatile flavour components can be split into two broad categories, i.e., oil-soluble constituents in peel oil and juice oil, and water-soluble juice constituents [138]. Most of the characteristic flavour of fresh citrus fruit derives from water-soluble components found in the juice sacs; the oil in these sacs differs slightly in composition from the peel oil [138]. Peel oil from each citrus cultivar provides much of the characteristic aroma of that cultivar [138].

Grape variety, and factors affecting wine development and berry composition exert major influences on the distinctive flavours of different wines, whereas fermentation has very little overall effect [152]. In two cases, character-impact volatiles have been identified in grapes; these have been shown to respond to climatic and/or geographical influences. In *Muscat of Alexandria* and other "aromatic" varieties, terpenes contribute to the distinctive floral aromas; and 2-methoxy-3-isobutylpyrazine has been correlated with the intensity of "vegetative" *Sauvignon blanc* and *Cabernet Sauvignon* wines [152]. The specific cultivar, or variety, of grape used will influence the sensory properties of a wine, but the effects of geography, e.g., climate and soil, are also important. Vineyard management, which includes trellising, irrigation, vine age and grape maturity, also affects the character of the wine. Finally, the winemaker influences the flavour of a wine, e.g., the length of time during which the skins of the crushed grapes remain in contact with the fermenting grapes, and length of time of holding the wines in contact with the yeast lees [152].

Cheese maturation is a complex process, involving the breakdown of both proteins and fats. Enzymes involved in these reactions can be indigenous to the milk, added with the rennet, produced by starter bacteria cultures and cheese ripening cultures, and to contamination. Proteolysis influences the texture, but also leads to the formation of flavour peptides and free amino acids, which form the precursors for the development of aroma compounds [144]. Lipolysis is important in the development of the typical flavour of mould-ripened cheeses, while the starter bacteria have a direct impact on the flavour of fresh cheeses such as cottage cheese [144]. As cheeses mature, during their so-called affinage, their savouriness increases together with the depth and complexity of the flavour. This is also the case with other fermented or ripened foods, e.g., Parma ham. Cheeses such as Camembert, Manchego and Taleggio must be made from raw milk in order to develop their characteristic flavour [153].

Capsaicinoids are responsible for the pungent taste of chillies, and a widely-used measure of pungency is the Scoville index [154]. Chillies range from 100–500 Scoville Units for mild bell peppers and pimentos to over 100,000 units for habaneros and Scotch bonnets [155]. However, chillies are not just pungent but, depending on variety, also provide a wide variety of complex aroma notes, ranging from fruity to chocolate.

Undesirable flavours can either be inherent, or their presence may be due to degradations, e.g., rancidity in oils, or to biological or physical contamination, e.g., adsorbed flavours. Environmental odours are readily transmitted via animal feed, as in the case of milk [120]. Even the air that a dairy cow breathes can transmit foreign odours to milk [120]. There is a conceptual difference between an off-flavour and a taint. An off-odour is an atypical odour and is often associated with deterioration or transformation of a product, whereas a taint is a taste or odour which is foreign to a product [112]. Taints include the baggy flavour of tea that has been wrapped in unlined hessian bags, earthy odour, which is caused by damp storage of tea and musty odour, which is associated with mould growth [115]. On the other hand, the burned odour of certain poor-quality teas is due to extreme over-drying, and would have to be classified as an off-flavour [115]. One of the most common sources of off-odours and -flavours in foods is the degradation of lipids, either through autoxidation or through lipolysis [156]. The so-called warmed-over flavour in reheated meat [157] is an example of a mild, lipid-related off-flavour. Rancidity is a more serious defect. Oxidation of other food components can also result in off-flavours, e.g., a floral note in carrots [156]. Aspects of plant or animal metabolism can also give rise to off-flavours, e.g. boar taint from entire (un-castrated) male pigs [156]. Green/grassy and beany notes in products made from soy beans also fall within this category [158]. Food packaging materials, especially plastics and paper-based packaging, are often found to be responsible for tainting food. Chlorophenols and chloroanisols are frequent sources of disinfectant taints, including in milk [120] and in biscuits, chickens and eggs [156]. Microbial growth and activity in food is another common source of taints [159].

5.5
Keeping Quality

Durability, or shelf life, is one of the most important quality attributes of many foods, both traditional and modern. On the one hand, shelf life is closely associated with food safety, in particular, the potential emergence of a pathogenic microflora. On the other hand, a strong requirement for extended-shelf life foods arises from current food distribution and shopping practices and patterns. One aspect of the complexity of modern food supply chains is the large number of individual stages, in handling, transport and storage, and so on, which it contains. The modern food supply system can cope with highly perishable foods only if the shelf life of these foods is maximised. Foods that are perceived to be both fresh and minimally processed, and that adapt themselves to a once-a-week shopping pattern, are precisely the foods that modern consumers most desire. Consumers also look for other aspects of efficiency, and these are usually considered under the broader concept of convenience. Biological tissues that have been cut off from their life support systems deteriorate rapidly. This is obvious in the case of milk, meat, fish and mushrooms, and in the case of certain plant-based foods, e.g., lettuce. Generally as far as fruits and vegetables are concerned, spoilage rates depend on whether there has been any major tissue damage. Thus, apples, potatoes and onions have good storability in principle; but when they are cut or bruised, each will deteriorate rapidly. Man-made foods have limited shelf lives too, although food processing frequently aims expressly to improve keeping quality. The primary goal in this historically would have been to allow the supply of seasonal, or slaughter, products to be evened out over the course of the year. Traditional food preservation methods, e.g., drying, salting and heat sterilisation, are severe by contemporary standards. Nevertheless, many of the products concerned have maintained their popularity simply because consumers have acquired a taste for them, e.g., Parma ham. The modern, global food trade has tackled the seasonal supply problems of old by developing systems of counter-seasonal supplies. The sophisticated logistics of today's food supply chain have meant that required shelf life for many foods is of the order of weeks rather than months. This is reflected in the fact that chilled distribution, often used in combination with modified headspace gases (modified atmosphere packaging, MAP), are of particular interest at present. In defining keeping quality and establishing the shelf life for individual products, the quality of a food at the time it is consumed must be as originally intended. A suitable definition is available from the Institute of Food Science and Technology (UK) (IFST) [160], and is shown in Table 5.5.

The maintenance of quality throughout the stated shelf life of a food relies on the integrity of the food supply chain [160]. Temperature variations in chilled distribution chains are likely to reduce shelf life, but do not necessarily alter the type of spoilage. However, in certain cases, temperature abuse of a food during storage and distribution can bring about radical changes in the spoilage mechanism. For example, if bacterial spores are allowed to germinate, their vegetative stages may eventually outgrow the normal microflora, as is the case with bitty cream [120]. During temperature abuse, pathogens may gain an opportunity for

Table 5.5. Definition of shelf life [160]

Shelf life is the period of time during which the food product will

(i)   remain safe,
(ii)   be certain to retain desired sensory, chemical, physical and microbiological characteristics,
(iii)   comply with any label declarations of nutrition data,
when stored under the recommended conditions.

significant growth, outgrowing the normal spoilage microflora of a food, and presenting a risk of food poisoning. The UK Food Labelling Regulations 1996 make provisions for minimum durability to be indicated on food labelling [161]. The shelf life of a food must be indicated by the words "best before" in conjunction with the expiry date and conditions for optimum storage of the food. "Use by" dates are similarly applied to perishable foods, i.e., where the expiry date represents a food safety threshold. The clearest perspective on food spoilage undoubtedly is that whereby spoilage defines the point at which a food becomes unsafe to eat, as in the presence of significant numbers of pathogenic microorganisms. Spoilage may also mean altered sensory characteristics, in the absence of any safety issues. Altered sensory attributes of a food do not necessarily equate with spoilage of that food. In fact, one consumer's perceived spoilage may be another's perceived excellence, e.g., ripened cheeses exhibiting strong off-flavours or rich aromas, respectively. In the case of a number of food products, post-processing storage constitutes an integral part of their manufacture and is necessary to achieve optimum condition of the product. The hanging of meat and maturing of certain (vintage) wines are other familiar examples where storage is required to develop the desired flavour and texture of products. As highlighted in Table 5.5, the shelf life of a food also expires when it ceases to comply with labelling claims, e.g., contents of certain vitamins that may be subject to breakdown during storage.

Spoilage processes in foods occur due to either intrinsic or extrinsic factors, or, most likely, both. Biological tissues and fluids are rich in enzymes, which break down food components, in particular fat (lipolysis) and proteins (proteolysis). The disruption or destruction of cell and tissue structures allows cell contents to mix freely, and thus promotes enzymic activity. This is also the cause of other chemical spoilage reactions, e.g., non-enzymic browning, and creates opportunities for microorganisms to become active. Normally, several spoilage mechanisms proceed concurrently within a food, but amongst these, it is the shelf life-limiting mechanisms which are of the greatest practical interest. The rate of spoilage, and the type of spoilage that eventually predominates in a given food depend on the type of food, its vulnerability to spoilage, any preservation processing it might have undergone, and the storage and distribution conditions the food is held under. Intrinsic factors affecting shelf life include the raw material(s) used, the product formulation and composition, the product make-up, the water activity (α_W) and pH value of the product, and the availability of oxygen [160]. Extrinsic factors include processing, hygiene procedures, packaging materials and systems, storage and distribution systems, and the nature of the retail display [160]. Water activity refers to the amount of water in a food which is available to participate in spoilage reactions [162]. Enzymic changes and microbial growth occur readily in foods with high α_W when not restricted by environmental factors such as pH or temperature. This is generally the case with unprocessed foods, e.g., fruits and vegetables immediately after harvest, and meat immediately after slaughter. However, chemical reactions, such as oxidation and enzymic acitivity, may occur at considerably lower α_W than those required by microorganisms [11]. These reactions do however require certain level of molecular mobility and proximity of reactants. Both enzymic changes and non-enzymic browning have been found to occur above a critical α_W, to show maximum reaction rates at intermediate moisture levels, and to slow down at higher α_W.

Temperature plays an important role in food spoilage. All microorganisms have defined growth ranges and optima, whilst chemical reactions and physical spoilage phenomena generally accelerate as temperature increases. An obvious exception concerns certain fruits and vegetables, in particular tropical fruits. Here very-low temperature storage alters normal physiological processes, leading to cell damage and accelerated spoilage due both to intrinsic and extrinsic factors. Mangoes [163] and bananas [164] are examples of a fruits that are highly susceptible to chill injury. Post-harvest fruit and vegetables are metabolically active. Flavour and colour development often carry on, as it does in tomatoes and bananas. Tomatoes genetically modified to achieve extended shelf life have been place on the UK market (Flavr Savr and Calgene); the additional shelf life has been attributed to decreased polygalacturonase activity and/or an inhibition of ethylene production [165].

Depending on the temperature and amount of free water present, many foods will undergo phase and state transition upon storage. This includes crystallisation and melting phenomena and the phase change that occurs at the glass transition of an amorphous material into a crystalline one, or vice versa [11]. Whilst pure compounds have exact phase transition temperatures, complex biological materials tend to exist in amorphous states. Processes resulting in amorphous states in foods include baking, evaporation, extrusion, dehydration and freezing; they have in common the production of a melt of mixtures of food solids and water at a high temperature; this is followed either by rapid cooling or removal of water [11]. The glass transition temperature of amorphous biological materials is one of the main issues in the shelf life of low-moisture and frozen foods [11]. Other transitions include protein denaturation and starch gelatinisation [11]. Above the glass transition temperature, molecular mobility is sufficient to allow reorganisation of individual molecules; below it, molecular mobility is low. Time-dependent changes do also occur below the glass transition temperature. These are often referred to as physical ageing. [11]. Crystallisation occurs during the storage of a number of amorphous foods containing sugars, polysaccharides and proteins. It is the major cause of the common structural changes that occur in bread and other bakery products during storage, in particular, staling [11]. The staling of bread is characterised by increased firmness of the crumb and crust, by reduced crust crispness, and by substitution of fresh bread flavour by stale flavour. The recrystallisation of cocoa fat, and its leading to fat bloom in temperature-abused chocolate, is another well-researched spoilage phenomenon. Amorphous sugars may also crystallise from freeze-concentrated solute matrices, known to be the primary cause of the development of sandiness in ice cream [11]. The stickiness and the caking of powders, including lactose powders, are governed by the glass transition temperature.

Water activity plays a crucial role in the keeping quality of foods. Foods are most stable when water that is strongly bound to specific sites in a particular food occupies all of these sites; example are 11 % for gelatine, 11 % for starch, 6 % for amorphous lactose and 3 % for spray-dried milk. At moisture contents below the full occupation rate of adsorption sites, rates of lipid oxidation increase [46]. Spoilage due to physical changes in a food is often connected with migration and phase transitions of water. Man-made foods may not be as complex in their structure as biological tissues, but they too frequently are multi-component systems. Components are not usually separated from each other by physical barriers, but instead, one phase may migrate into another. For example in stored coleslaw, water may migrate from the cabbage into the mayonnaise. Other examples are sogginess of sandwiches and other bread-, pastry-, and crust-based products such as pies and pizzas, which may result from moisture migration out of fillings or toppings. Butter spread on sandwiches, and cheese toppings on pizzas, may be used as a barrier to counteract moisture migration from other components. Physical spoilage due to phase transition of water in foods leads to the familiar phenomena of drying out of unwrapped delicatessen products, weight loss from fruits and vegetables and wilting of lettuce leaves. Freezer burn is surface drying of frozen foods during storage, which occurs when foods are held in packaging with insufficient moisture-barrier properties. Enzyme activity is related to water activity, especially in terms of the mobility afforded to both water-soluble enzymes and their substrates; this obviously does not apply to lipase mobility.

By far the most common deterioration associated with products containing an appreciable amount of oils and fats is rancidity; this can be either lipolytic or oxidative in nature [160]. Lipolytic rancidity is due largely to the presence of free fatty acids from hydrolysed lipids. Oxidative rancidity develops either via metal catalysis or by an enzyme-initiated oxidative degradation [160]. There are potentially two types of oxidised flavour in milk, one catalysed by copper ions, and the other light-induced. The two processes differ from each other in several respects, including the exact nature of the substrate that is being attacked, the level of oxygen consumption involved, and the precise sensory character of the resultant off-flavours [166]. Oxidative rancidity can occur in baby milks due to their relatively high levels of PUFA and metal ions, both of which are added to the product for nutritional purposes [160]. The defect is also found in spray-dried fats, in chocolate confectionery, and in comminuted meat products. Buttermilk has a high proportion of phospholipids and can bind to metal ions in a pro-oxidative fashion. Fish contains a high proportion of PUFA which are vulnerable to oxidation. Coffee is well known to start oxidising soon after roasting, but can be protected through suitable packaging technology, e.g., vacuum packing. Several vitamins are subject to oxidation and are lost from foods via this mechanism.

Minimally processed fruits and vegetables are foods that are pre-prepared by the food industry for the convenience of consumers. Minimal processing of fruits and vegetables entails some degree of wounding, and wounding is a powerful morphogenic stimulus [167]. The consumption of oxygen, and the production of carbon dioxide and ethylene are stimulated, as is the synthesis of metabolically active compounds. The wound signal and responses probably arose to thwart attacking pests and to coordinate healing [167]. MAP designed on the basis of normal respiration will be unable to maintain the desired gaseous composition during the burst of respiration following wounding [167]. Polyphenol oxidase, an enzyme found in most fruits and vegetables, is responsible for enzymic browning following bruising, cutting or other damage to the cell [168]. The most important factors that determine the rate of enzymic browning are the concentrations of the active enzyme and the food content of phenolic compounds, pH, temperature and oxygen availability. The control of enzymic browning through inactivation of polyphenol oxidase is one of the key motivations of food processing and can involve heat, pH, oxygen and/or additives [168]. Non-enzymic browning is an important reaction that produces typical flavours of several categories of foods, but it also decreases food quality during storage in products such as dehydrated fruits and vegetables, instant potato and dried egg white [160].

MAP technology has the potential to extend the shelf life of foods, and has brought about major changes in the storage and distribution of raw materials, and in the marketing of food products [169]. MAP foods include raw and cooked red meats, poultry, fruit, fresh pasta, crisps, coffee, tea, vegetables, cheese, bread, fish and crustaceans; ca. 95 % of fresh pasta sold in the UK is marketed in MAP [169]. Fish spoils relatively quickly and, except for tropical fish, chilling is less effective in decelerating spoilage than it is for red meat and poultry [169]. In general, aerobic microorganisms are sensitive to carbon dioxide and it is this, along with their requirement for oxygen that is utilised to delay the spoilage of fish and other foods [169]. A detailed discussion of microbial growth in foods is found in Chapter 6. In most oily fish, particularly the Atlantic mackerel (*Scomber scombrus*), lipids readily undergo oxidation on exposure to the air, even in frozen storage. However, rapid development of rancidity in subcutaneous fat of these fish at $-15\,°C$ was effectively inhibited at $-40\,°C$ [170]. Odour is a good indicator of freshness in fish. However, not all spoilage mechanisms in all foods are accompanied by readily detectable sensory changes, and this includes the growth of pathogens. This re-enforces the important role of the statutory date marking requirement.

5.6
Fitness for Purpose

Whilst many writers on quality theory view the concept of fitness for purpose as an all-encompassing definition of quality itself, the approach taken here is more specific. The approach is two-pronged. All food processing may be conceptualised as value-adding activities. The first consideration therefore concerns the matching of food processing technologies to goals, or types of added value. For examples, food dehydration and freezing may have some shared goals or none, but can never share all goals. Secondly, the suitability of foods and food ingredients for specific uses (and users) is examined, including aspects of production, manufacture and consumer (domestic) use. For example as has been mentioned earlier, certain varieties of peas are more suitable for commercial, others for domestic, growing. The commercial grower is likely to require varieties that are highly resistant to disease, that produce high crop yields and that are suitable for mechanical harvesting. This means sturdy compact plants and simultaneous ripening of the pods. The domestic grower on the other hand requires a prolonged cropping season combined with what they would consider to be excellent sensory quality. In addition, different varieties of pea will be required for different types of manufacture, e.g., canned mushy peas as opposed to frozen garden peas. Similarly as is well understood, different cuts of meat require different culinary treatment, e.g., frying, stewing or roasting. Commercial specifications for food ingredients, where mass manufacture is involved, are likely to be considerably tighter than requirements for either domestic or artisanal application.

There are many possible motivations for processing foods. Foremost among them is the desire to preserve a food in a state as close as possible to the original, with shelf life extension therefore the single goal. They further include transformation of raw materials into a product with more desirable properties overall, e.g., the creation of new forms and structures. However, process engineering means finding the balance that ensures minimising detrimental and maximising beneficial effects of the process. The sensory attributes of food materials such as strawberries are severely affected by processes such as canning. Bright red colour turns into brownish-grey, texture weakens and cooked flavour notes take the place of fresh, green ones. Nevertheless, even though fresh strawberries are now available from supermarkets throughout the year, there remains a small market for the canned type, indicating a difference in the type of (use) value. Much modern food processing is concerned with transferring aspects of traditional, domestic food preparation to the factory, e.g., ready-to-eat prepared fruit and vegetables, ready-to-wok chopped vegetables, meal kits and pre-cooked ready meals. Table 5.6 presents an overview of today's popular food processing technologies, categorised by processing goal.

Packaging is an integral part of most contemporary commercial foods. Amongst all food processing operations, packaging has the least negative impact on the nature of a food. This is assuming that the packaging material is compatible with the food it is meant to protect, i.e., it does not permit migration of low molecular weight chemicals into each other. Packaging must hold a food securely, preventing spillage of solids, leakage of liquids and the unwanted escape of gases, e.g. water vapour and aroma chemicals. It must also act as a barrier towards undesired environmental influences, protecting the product from physical impact (including tampering) and barring access to dirt and foreign materials, pests and microorganisms, gases such as oxygen, water vapour and taints, and light. The ease with which packaging can be opened and, where appropriate, re-closed during domestic use are relevant here.

However, whether a strictly hermetical seal is required depends both on the type of product and on the length of its expected shelf life. For example, packaging of pasteurised milk requires a liquid seal, but does not require aseptic filling techniques (which prevent microorganisms from gaining access), nor are protection against the effects of oxygen and light essential. Although this will lead to some losses of quality from the product, these loss are minor when seen in the context of the shelf life of pasteurised milk. In fact, it is well recognised that conventional pasteurised milk is subject to post-pasteurisation contamination [171]. The situation is very different where sterilised milks are concerned; these have to be protected against any ingress of microorganisms, and the packaging should also protect the milk from oxygen and light [166]. However, there

Table 5.6. Goals in food processing and examples of technologies

Purpose	Technology	Food examples
Shelf life extension without significant food transformation	Packaging	Universal
	Modified atmosphere packaging	Chilled foods
	Refrigeration	Universal
	Food irradiation	Licensed applications only
	Temporal additives	(see Chapter 6)
	Freezing	Universal
Shelf life extension with food transformation	Thermal processing	(see Table 5.7)
	Drying	Universal
	Concentration	Juices
	Reduction of water activity	Salt and sugar preservation
	Fermentation	Alcohol, acids, biomass
	Enzyme treatment	Processing aids, cheese flavouring
Creating form, structure and sensory character	Moulding	Burgers
	Baking	Bread, flour confectionery
	Extrusion	Snack foods, noodles
	Grating	Parmesan cheese
	Slicing	Ham, cheese
Multi-component foods and recipe dishes		Prepared salads
		Ready-to-cook vegetables
		Pizzas, sausages, meal kits
		Pre-cooked ready meals

are products commercially available in the UK where the latter requirements appear not to have been addressed. There are two basic mechanisms in achieving hermetic packaging of sterilised foods, i.e., in-container sterilisation and aseptic packaging [14]. A basic requirement of aseptic packaging is a packaging material that is entirely free from microorganisms. This can be done by recontamination-free extrusion of plastics, application of hydrogen peroxide to the packaging, UV- and γ-irradiation and application of superheated steam, or combinations of such treatments [172]. In-container heating of a food has more severe consequences in terms of composition and attendant sensory and nutritional attributes than does heating a food separately and then filling it aseptically. This is due to the longer heating and cooling times of a food when processed in its container. An example of an in-container heat treatment of a food, where there are no negative side effects, is the baking of cakes in hermetically sealed containers which double as display units at the point of sale [173].

Smart, or active packaging does more than simply provide a barrier; it can control, and even react to, events taking place inside a pack [174]. The metabolism of freshly harvested, or slaughtered, foods continues, using up oxygen and generating carbon dioxide within the pack; humidity tends to build up at the same time. Each of these foods has its own, particular, optimal gas composition for maximising shelf life, as do man-made foods. In MAP, the initial composition of the storage atmosphere is adjusted from ordinary air (21% oxygen, 79% nitrogen, 0.03% carbon dioxide) to give a gas mix designed to optimise shelf life [175]; this also requires packaging materials with closely defined barrier properties. In order to accommodate an adequate reservoir of gas, trays for MAP foods are deeper than conventional packs. A typical MAP atmosphere for red meat is 80% oxygen, 0% nitrogen and 20% carbon dioxide, with corresponding percentages for poultry 0, 75, 25, white fish 30, 30, 40 and oily fish 0, 40 and 60% [175]. In the case of prepared fresh fruit and vegetables the atmosphere is not injected into the pack. Instead, careful selection of breathing film allows the living produce to generate a suitable carbon dioxide-rich, low-oxygen atmosphere in the pack [175]. MAP technology is almost always combined with chilled distribution.

Table 5.6 also refers to refrigeration, or chilled distribution, as a process. Most chilled foods are highly perishable foods, whose keeping quality is improved by holding them within a specific, and closely controlled, temperature range. This normally lies between the freezing point and 8°C [176]. Chilled foods are very popular at present due, in part, to the high level of sophistication of chilled food processing and distribution, which reduces failure to a minimum. Chilling may be combined with other "hurdles" against deteriorative mechanisms in foods; MAP has already been mentioned. Other examples include the submersion in oil, or oil-vinegar dressings of vegetables (e.g., sun-dried tomatoes, herbs, spices) and cheeses (e.g., Feta). Chilling slows down most spoilage reactions in most foods and prevents others, e.g., the growth of non-psychrotrophic microorganisms. However, as indicated previously, certain fruits and vegetables may be damaged by low storage temperature [46]. Prepared fruits and vegetables range from ready-to-eat washed, sliced, chopped or shredded single commodity packs to complex mixed salads and ready-to-stir-fry-vegetable mixes. These products are highly perishable and subject to all kinds of deteriorative mechanisms, including biochemical, physical and microbiological changes. Chilled storage and distribution is therefore a major factor in the stability of these products [177]. The perishability of meat similarly increases with tissue damage, with minced meat particularly susceptible to rapid spoilage. Sometimes, foods are chilled for non-preservative purposes, i.e., to facilitate in-process operations such as cutting (meat), slicing (bread) and pitting (cherries). Strict temperature controls are especially important for cook-chilled foods, whether or not they are designed for re-heating, and this is reflected in the issuing of guidelines for the process by the UK Department of Health [178].

Low-dosage food irradiation is a process that can be applied to many foods to sterilise them, but without modifying their sensory properties. From a technical point of view, this would seem to make it the ideal food technology. However in practice, the concept of irradiating food has proved highly unpopular with UK consumers. Food irradiation entails the exposure of a food to a carefully controlled dose of ionising energy from either machine-generated electron beams, or from cobalt-60-derived γ-rays [179]. This generates short-lived free radicals, which kill microorganims and inhibit processes such as sprouting and ripening. Suitable applications include poultry and red meats and products made from them, dried herbs and spices, seafood, certain fruits and vegetables, bulbs and tubers, cereals and grains, and sterile ready meals [179]. So-called temporal additives, e.g., preservatives and antioxidants, also have the advantage of leaving the sensory properties of a food more or less intact. In this they differ from the more traditional chemical preservation methods, e.g., salting, sugar preservation and pickling. The use of temporal additives in foods is discussed Chapter 6.

Freezing, like chilling, is designed to leave the compositional and sensory nature of a food unchanged, but for much longer. Nevertheless, the consequences of freezing for product quality can be severe, because freezing involves more than just a drop in the product temperature. For example, associated with the freezing of most foods is a volume increase of around 10% [180]. Freezing is accompanied by localised dehydration and concentration phenomena, which are the direct results of water crystallisation. In frozen products in which enzyme systems are not fully inactivated, the freeze concentration of enzymes and other substances can cause damage to texture and flavour [176]. The freezing interval of most foods extends to about $-2\,^{\circ}$C [180]. Temperatures just below freezing are the most damaging to quality, due to the continuous freezing and thawing of individual ice crystals. Speed of freezing plays an important part in quality and shelf life. Very low rates encourage formation of large ice crystals causing physical damage, especially to biological tissues and cells, where osmosis is another important issue. During thawing, this may manifest itself in excessive drip. However, very fast freezing speeds, as in so-called cryogenic freezing, may also damage product structure, in this case because of the almost instant build-up of pressure as the food expands [180]. This can shatter the external layers of already frozen, and therefore brittle, foods. Much of the frozen meat in international trade is carried today on ships at $-18\,^{\circ}$C. This, although relatively high, reflects the impracticability of providing different temperatures for different parts of a cargo [180]. Although microbiological growth can be discounted at frozen food storage temperatures, microbial and endogenous enzymes continue to contribute to spoilage. Oxygen is a problem in most frozen foods, leading to oxidative rancidity and loss of colour [180]. Moisture loss by sublimation from the surface of products leads to so-called freezer burn, white patches on the surface of meat that can be mistaken for mould [180].

Thermal food preservation has a longer history than low-temperature processing, but leads to fundamental changes in product structure and the sensory quality of foods. There are many types of heat processing, both in commercial and domestic settings, each with slightly different goals, but all transforming a food in terms of its sensory qualities. The extent of such transformation varies and is more obvious in the case of biological tissues than for fluids such as milk. Heating alters or destroys native structures, but can also be used to generate new structures, as in baking, extrusion cooking and the formation of gels. The main categories of thermal food processes are shown in Table 5.7.

Thermal effects on foods are expressed as time-temperature relationships. For example, milk may be pasteurised using either the holder, or batch, process (62.8–$65.6\,^{\circ}$C for 30 min), or the high-temperature-short-time (HTST), continuous, process ($71.7\,^{\circ}$C for 15 s) [181]. In each case, microbial death rates are the same. At the same time, combinations of higher temperatures with shorter times tend to be less severe in terms of thermal damage to the food itself, e.g., sensory properties and nutrient losses. This is most clearly demonstrated by comparing

Table 5.7. Thermal food processing

Process goal	
Blanching	In-process operation to inactivate enzymes and reduce microbial loads
Pasteurisation	Thermal death of pathogens (milk, eggs) and shelf life extension
Sterilisation	Thermal death of all viable microorganisms to allow ambient storage of product
Process technology	
Conventional heat transfer	Hot air, hot water, heat exchange surfaces, infrared radiation
Volumetric heating	Microwave technology
	Ohmic heating
Packaging system	In-container processing
	Continuous-flow processing with aseptic filling

the quality of in-container-sterilised milks ($>100\,°C$) with that of UHT milk ($135\,°C$ for 1 s) [182]. Again, the microbial kill for both processes is the same; but nutrient and flavour retention are far superior in UHT milk.

There are various degrees of preservation by heat. Blanching is an in-process operation, generally applied to fruits and vegetables intended for further processing (e.g., freezing, drying) to inactivate natural food enzymes [14], especially polyphenol oxidase. Blanching also causes some of the tissue gases to be expelled from biological tissues and softens them for containerisation and further treatment, e.g., canning. Pasteurisation has two primary objectives, which vary with the application. In the case of milk and liquid egg, it is specifically designed to eliminate pathogenic microorganims [14]. Pasteurisation also extends product shelf life both from a microbiological and an enzymic point of view, where pathogens do not pose a threat, e.g., in beer, wine and fruit juices [14]. Pasteurisation of fruit juices inactivates pectinesterase which, if left intact, might lead to cloudiness and related spoilage phenomena; it further destroys spoilage yeasts and moulds [46]. Pasteurised products contain a residual microflora, which is initially mainly heat resistant in nature. However, post-pasteurisation contaminants are likely to predominate eventually, as is the case with pasteurised milk [171]. Many processes combine pasteurisation with other treatments, leading to products such as cook-chill dishes; ambient in-pack products such as pickles, sauces and some canned fruit; and cured products such as ham [183].

Low acid foods having in any part of them a pH value of 4.5 or above, and intended for storage under non-refrigerated conditions, must be subjected to the minimum botulinum process – unless the formulation, or water activity, or both are such that *Clostridium botulinum* growth cannot occur [176]. The toxin that is produced by *Cl botulinum* is destroyed by exposure of the organism to moist heat at $100\,°C$ for at least 10 min [14]. Cooking a potentially contaminated food immediately before eating it therefore provides a safety net against poisoning [14]. A major problem with in-container sterilisation of solid or viscous foods is the low rate of heat penetration to the centre, which causes partial overheating and therefore unnecessary sensory and nutritional damage [46]. In continuous-flow sterilisation, product depth for thermal treatment is dramatically reduced, but the technology is more complex than that of in-container sterilisation. Inflow sterilisation can be achieved indirectly, in heat exchangers, directly with steam injection and using novel processes, in particular microwave and Ohmic heating [46]. These novel processes are examples of electro-technologies and represent so-called volumetric heating rather than conventional heat transfer. The dielectric energy of microwaves can penetrate to the centre of a food, whilst in Ohmic heating an electric current is passed directly through the body of the food [184]. In both cases, rapid uniform heating of a food is achieved, whilst avoiding any contact with hot heat transfer surfaces. Microwave technology is now well established both for commercial and for domestic application. Microwaves form part of the electromagnetic spectrum close to the section allocated for radar and radio communications. Two frequencies are reserved for the industrial use of microwaves, i.e., 2450 MHz and 896 MHz, the latter exhibiting the greater depth of penetration [185]. Microwave technology is used to temper frozen foods during thawing, to dry pasta and onions, in baking and the pasteurisation of bread, and to sterilise foods in flexible pouches [186].

The purpose of drying a food is preservation through elimination of spoilage mechanisms involving water, and this means microbial activity in particular. Like thermal processing, drying alters the essential nature of a food. Reconstitution will be more or less successful in achieving the original state of a food, depending, in part, on the drying technology employed. However, reconstitution is never perfect. Drying is achieved either through heat application, causing water to evaporate, or by freezing followed by sublimation of ice (freeze-drying) [46]. Freeze-drying is considered the gentlest among the various drying technologies. In conventional (thermal) drying, water migrates from the inside of a food to its periphery, taking with it solutes, and thus changing the local composition of the food. In freeze-dried foods, migration and osmosis also occur, but are more localised. Although dried foods are microbiologically safe as long as they remain dry, they often carry a heavy microbial load. This is notoriously the case with onion powders. Such potential microbial time bombs become a practical hazard as soon as they regain access to some water reservoir. This is common in foods because foods tend to have complex structures. Water removal may also be desired for reasons other than food preservation, e.g., fruit juice concentrates where a portion of the water is removed from the juice largely to reduce transport and distribution costs. Water can also be removed from foods through osmotic drying, e.g., immersion in liquids with water activities lower than that of the food (e.g., sugar or salt solutions). Traditional curing methods are still popular, e.g., that of pork to produce bacon. Superficial surface drying of meat is an important aspect of hanging, as the dry surface constitutes a barrier for microbial access. Cold smoking has a similar effect but here, in addition, anti-microbial substances are deposited on the food [14]. Meat hanging is a process during which both the colour and the texture of the product change, the most important outcome of which will be tender eating quality. It is interesting to note that in MAP meat, there is no such surface drying.

Traditional fermented foods [14, 46], such as alcoholic beverages, bread, fermented milks including cheese, fermented meat including salami, and fermented vegetables including sauerkraut, remain popular. Modern applications include the production of biomass, e.g., yeast and yeast extracts, the popular vegetarian meat substitute Quorn [35], and of food chemicals such as citric acid [46]. Enzyme technology is widely used to break down macromolecules, e.g., the aforementioned pectinesterase [46]. Lipases made to act on cheese slurries produce enzyme-modified cheeses. These have a more intensive flavour than the original cheeses and are therefore suitable for application as a flavouring preparation in soups, dips and snacks. Commercial processing may simply involve adding value by transferring common, domestic food preparation techniques to the food factory, possible types of added value generated being convenience, competence and variety. In the case of ambient temperature processing of fruit and vegetables this may mean washing, sorting, and slicing or chopping, mixing together different kinds of fruits and vegetables, and packaging. Another example is grated cheese, e.g., Parmesan and pizza toppings. The validity of this approach to modern food processing is also recognised in the meat supply chain. An example of this is Hazelwood Foods plc with its "Meat Plus" range of meal kits for in-store butcher's counters, e.g., "Paupiette of Beef with Wild Mushroom Risotto" [187]. So-called recipe dishes may therefore appear on the market in the guise either of meal kits, meal components (e.g., pasta, bread, sausages) or pre-cooked ready meals. The latter include traditional tinned soups and stews and chilled, frozen or ambient ready meals, where meal components may or may not be presented separately in the pack. Recipe dishes may also be partially pre-cooked, e.g., pizza.

The intrinsic quality of foods influences their suitability for processing and domestic food preparation. An alternative interpretation of the concept of fitness for purpose can thus be readily illustrated by revisiting one of the processing technologies outlined above, in this case, freezing. Foods suitable for freezing include meat, fish, most fruit and vegetables, and cooked ready meals; raw materials unsuitable for freezing include salads, bananas and eggs. It is therefore important to select appropriate raw materials and processes in order to achieve desired outcome. In domestic settings, potatoes may be used for boiling, mashing, chipping, baking or slicing into salads. An important commercial process using potatoes is the manufacture of crisps. Clearly, one particular potato cultivar is unlikely to meet all these requirements to optimum effect. For example, in order to minimise waste, potatoes for crisp making should be round to oval, with shallow eyes. Secondly, the dry matter content of the potato influences both the crisp yield and the uptake of frying oil. Potatoes for crisp manufacture should also be low in reducing sugars, in order to minimise browning (Maillard reaction) during frying. Similarly, different types of cheese are suitable for different uses, such as slicing for use in sandwiches (Edam), slicing for use in salads (Mozzarella), chunking or grating for use in salads (Cheddar, Feta), melting for sauces and pizza toppings, and frying (Halloumi). Among the more familiar examples of grading of foods or food category in terms of fitness for purpose are wheat flour, edible oils and fats, and rice. These will be examined in the following.

There are soft, hard and durum wheat varieties. Hardness refers to the resistance of the endosperm to grinding and is often associated with higher protein concentrations than are found in soft wheat varieties [15]. However, it is not inevitable that protein concentration parallels hardness [15]. Flour made from hard wheats is especially suited for breadmaking, whereas soft wheat flour is used for cakes, pastry and biscuits [15]. The superiority of hard wheat flour for bread reflects the large proportion, and good quality, of gluten, which it forms when it has been mixed with water. However, the use of UK-grown wheat is often limited by excessive α-amylase activity, which decreases the amount of water that can be held by the starch both in the dough and in the bread as it is baked [188]. Durum wheat varieties, although hard and high in protein, do not make good bread flours, but are used in the manufacture pasta [15]. The desired properties here are bite (al dente) and stickiness [15]. So-called strong flours enable a dough to retain gas during fermentation; they are higher in protein content, have greater elasticity and resistance to extension, and greater ability to absorb water [1]. This is particularly important where bread is baked free standing, without the use of a baking tin, as in the traditional Scottish batch loaf. French bread, e.g., baguettes, are made mainly with soft flour (80% soft wheat, 20% hard wheat); this gives a dough which rises less, as well as a drier loaf [189].

The functionality of edible oils and fats relates to five main areas, i.e., as a frying medium, for emulsification of batters (shortening), as a spread, to dress salads and to flavour cooked dishes. The maximum temperature to which a fat or oil can be heated is determined by its smoke point, i.e., the temperature at which degradation is sufficient to result in the evolution of smoke [15]. Soybean oil is the major frying oil in the USA, having replaced cottonseed oil, but with interest in rapeseed oil expanding [15]. Both corn oil and peanut oil are popular in Chinese stir- as well as deep-frying [190]. Suet is the hard white fat surrounding beef kidneys, shredded and floured ready for domestic use; it is needed for making traditional British suet pastries and puddings [191]. Monoglycerides added to hydrogenated shortenings increase the extent of emulsification of batters [15]. Mono- and diglycerides in fat lower the smoke point; therefore, a fat that is ideal in a cake batter cannot also be the best frying fat [15]. Sesame oil is golden or dark brown, thick and strongly flavoured; like walnut and other aromatic culinary oils, it is used as a flavouring [190]. Extra Virgin olive oil is the result of the first cold pressing of olives and has a pungent taste; it is therefore used as a flavouring rather than a cooking oil [192]. Successive heated pressings of the olives are used to produce refined, blended olive oils that can be used in all kinds of cooking and food preparation [192].

The favoured variety of rice in China is the white, long-grain type, which cooks up relatively dry, and into easily separated grains [190]. Short-grain white rice is stickier than long-grain white rice, making it easier to pick up using chopsticks. This, non-glutinous, rice is more frequently eaten in Japan than in China, though it is useful for making rice porridge, a popular southern Chinese morning dish [190]. *Arborio* and *Vialone* are popular rice varieties for making Italian risottos. The grains are rounded, and their starchiness enables them to swell to at least three times their original volume, absorbing cooking liquids whilst retaining a firm, al dente, texture [192]. Glutinous rice is rich in soluble starch, dextrin and maltose, and on boiling, the grains adhere in a sticky mass [1]. This type of rice is used for sweetmeats and cakes [1].

6 Food Additives, Functional Food Ingredients and Food Contaminants

6.1
Introduction

The processing of edible biological tissues into contemporary foods frequently brings such foods into contact with defined chemical or biological entities. This is often deliberate, as with food additives, food ingredients with closely defined technological functionality and microbial starter and ripening cultures; or it may be incidental. Incidental exposure results in contaminants and residues being present in the foods. For example, crop protection and disease control agents are applied to crops deliberately, but usually have neither function nor desired presence in a finished food. This is also true of processing aids, e.g., filtering aids and release (de-moulding) agents, and of so-called carry-over additives. These are food additives with functionality in an ingredient of a food, but with no functionality after that ingredient has been incorporated into the finished product. An example would be the stabilisation of citrus oils using antioxidants, and the subsequent use of such oils in the manufacture of a flavouring, where the antioxidant effect may have been lost. On the other hand, ingress into a food of foreign matter, chemicals, micoorganisms and pests is categorically undesirable, and can only be described as contamination. Food contamination is, for the most part, a safety issue, with microorganisms and microbial toxins the main concerns. Foreign body contamination is important too. For example, metal shavings from machinery represent a constant hazard in all food factories, although this is readily managed by integrating metal detectors into processing streams. Glass however is much more difficult to deal with. Throughout the history of the food industry, there have been unscrupulous traders who have subjected food to adulteration, e.g., by substituting poorer grade ingredients for ingredients that consumers would have expected to be present in a food. Latterly, food tampering has become an important issue, both at the start of food supply chains, where malicious interference with a food may be due to an employee, and at the point of sale, where members of the public can be causing a problem.

The large-scale, commercial manufacture of food products demands stricter specifications for the ingredients to be used than does domestic or restaurant food preparation. Firstly, it lacks the scope for intervention and adjustment by a skilled cook; secondly the consumer, deprived of direct communication with the maker of the food, shows less tolerance towards variations in the final product. Accordingly, in most industrial food processing, chemically defined food ingre-

Table 6.1. Definition of food additive [4] and categorisation of additives for food labelling purposes [161]

Food additive

(a) any substance not normally consumed as a food in itself and not normally used as a characteristic ingredient of food, whether or not it has nutritive value, is intentionally added to food for a technological purpose in manufacture, processing, preparation, treatment, packaging, transport or storage, and results in it or its by-products becoming directly or indirectly a component of the food; or

(b) a carrier or carrier solvent

Categories of additives which must be identified in a list of ingredients by their category name (SCHEDULE 4)

Acid	Colour	Gelling agent	Raising agent
Acidity regulator	Emulsifier	Glazing agent	Stabiliser
Anti-caking agent	Emulsifying salt	Humectant	Sweetener
Anti-foaming agent	Firming agent	Modified starch	Thickener
Antioxidant	Flavour enhancer	Preservative	
Bulking agent	Flour treatment agent	Propellant gas	

dients and additives are preferred to more complex, traditional materials. In fact, additives often represent the purified, active principle which has been isolated from some naturally occurring material. Sucrose, sugar, is a familiar example of a functional food ingredient – but not an additive – which is commonly used both in industrial and domestic settings. The material is highly refined, and its characteristics closely defined. Its functionality in a food includes sweetness, flavour modification, bulking, structure and mouthfeel, surface appearance, preservation, control of crystallisation, colour enhancement, antioxidant, fermentability, decorative finishes and freezing point depression [193]. Food additives are categorised by their functionality, and for label declaration purposes in the Food Labelling Regulations 1996 [161]. Table 6.1 gives a definition of the term food additive [4], together with a listing of food additive categories [161]. Omitted from this list are flavourings, although the labelling format for these is dealt with extensively in the regulations themselves.

For the purpose of this book, food additives, and where appropriate, functional ingredients, are conceptualised as either structural, temporal or sensory, with each category being discussed separately. It is however recognised that each such material may have multiple functionality. For example, a structural additive may also affect the stability and/or the mouthfeel of a food. Additives are regulated both through horizontal and vertical legislative mechanisms. This means that in order for a particular additive to be able to be used, it must firstly be permitted for food use per se; but it must also be permitted for use in the food and/or target market in question. The chapter covers the addition of micronutrients to a food, as well as the related issues of functional foods and dietary supplements. The final two sections deal with biological contamination, in particular, food spoilage by microorganisms and their metabolic products, and with chemical residues in, and foreign body contamination of foods.

6.2
Food Structuring Additives and Ingredients

Man-made foods lack the complex cell structures of fruit, vegetables and meat. Except for certain minimal processing technologies, MAP especially, primary food processing either alters food structures or destroys them altogether. At the extreme of the processing spectrum, processing aims to extract specific food fractions or ingredients from the raw material, e.g., flour from cereal grains and pulses, and cream and butter from milk. An important objective of much secondary processing is to combine a number of more or less refined food components into structures with desired quality attributes, e.g., cake. Alongside the creation of different sorts of new food structures arises the need for structuring additives and/or ingredients to provide the necessary control in manufacturing, and during distribution and storage. Structural and temporal aspects in the design of made-up foods are therefore closely related. The main structure-related additive categories, as defined in the Miscellaneous Additives Regulations 1995 [4], are outlined in Table 6.2. This list is not complete, but space considerations demand selectivity.

Edible biological tissues hold large amounts of water. In man-made foods, the main alternative to this function of tissue consists of the creation of gels, e.g., jellies and jam, yoghurt, processed meats and fish such as frankfurters and surimi [7, 194]. Gels may be conceptualised as materials which fix water into semi-solid, three-dimensional structures at room temperature; they are formed from polymer solutions (sols) in which a continuous network of material spanning the whole volume becomes swollen with liquid [7]. Gels allow the creation of uniform textures in foods and permit stable and uniform distribution of food ingredients throughout a food. The process whereby water may be expelled from a gel, i.e., syneresis, is commonly observed with commercial yoghurts. There is no universally suitable gelling agent, which means that different ones are selected for different applications. Alginates (E400-E404) form heat-stable, irre-

Table 6.2. Classes of structural food additives [4]

Emulsifier	Any substance which makes it possible to form or maintain a homogeneous mixture of two or more immiscible phases, such as oil and water, in a food
Foaming agent	Any substance which makes it possible to form a homogeneous dispersion of a gaseous phase in a liquid or solid food
Gelling agent	Any substance which gives a food texture through the formation of a gel
Modified starch	Any substance obtained by one or more chemical treatments of edible starch, which may have undergone a physical or enzymatic treatment, and may be acid or alkali thinned or bleached
Stabiliser	Any substance which makes it possible to maintain the physico-chemical state of a food, including any substance which enables homogeneous dispersion of two or more immiscible substances in a food to be maintained, and any substance which stabilises, retains or intensifies an existing colour of a food
Thickener	Any substance which increases the viscosity of a food

versible gels in cold foods; gelling takes place via a calcium-bridging mechanism, although acidic gels may also be formed [7]. Alginate jellies lend themselves to the preparation of specialised confectionery such as mock glacé cherries as, unlike other confectionery jellies, they do not liquefy upon heating [195]. Alginates are also used as gelling agents in icings and toppings, fruit pie fillings and table jellies, but are incompatible with milk, except in the presence of calcium sequestrants [195]. Carrageenans (E407) are structurally related to alginates and are widely used because of their ability to provide a wide range of food textures [195]. They dissolve when heated in the presence of water, and form rigid gels upon cooling [7]. There are three types, i.e., kappa-, iota- and lambda-carrageenan [195]. Both kappa- and iota-carrageenans form thermally reversible gels. In the case of kappa-carrageenan, these provide a strong, rigid texture with potassium ions, whilst iota-carrageenan gels with calcium ions are of an elastic nature. Kappa-carrageenan at low concentrations weakly gels ice cream mixes, whilst at slightly higher levels, it stabilises cocoa particles and fat suspensions in chocolate milk [195]. Uniquely, iota-carrageenan gels are freeze-thaw stable and reform after mechanical breakup [195]. This allows ready-to-eat water jellies to be filled below gelling temperature, making possible products such as multi-layer desserts and jellies topped with whipped cream [195].

Pectin (E440) is divided into two main categories, i.e., low methoxyl- (LM) and high methoxyl- (HM) [7]. LM pectin is structurally similar to alginates and, like alginates, gels via calcium-bridges. It does not require added solids (sugar) and/or acids, as does HM pectin, and is suitable for use in non-acidic gels, e.g., mint-flavoured ones. The thickening effect of pectin is exploited mainly where food regulations prevent the use of cheaper gums, or where a more natural image is desired [196]. LM pectin is used in fruit preparations for yoghurt to create soft textures which are firm enough to ensure uniform fruit distribution, whilst still allowing the fruit preparation to be stirred into the yoghurt [196]. HM pectin is used in jams and jellies, and in fruit drink concentrates, where it stabilises emulsified oils and suspended fruit particles. Compared with other gelling agents for confectionery products, pectin requires strict observance of the recipe and process parameters [196]. The extent of esterification of HM pectin determines the speed of gelation, i.e., slow-, medium- or rapid-set, whilst in LM pectin esterification determines calcium reactivity [196]. The functionality of LM pectin is variable and may require performance testing before use [196]. Various gums also play a part in the gelation of foods, e.g., locust bean gum, xanthan gum, cellulose gum and guar gum [195].

Gelatin is a traditional and commonly used gelling agent. Obtained by purification of collagen, it is not classified as a food additive [197]. Gelatin swells due to hydration when in contact with cold or warm water; gelatin dissolves as the temperature is raised above its the melting point. On cooling, gelatin forms clear, transparent, highly elastic and reversible gels. Gelatin has a melting range of 27–34 °C, which means that gelatin gels melt in the mouth, accounting for its desirable characteristics in terms of mouthfeel and flavour release [7]. Gelatin is used in jelly babies, pastilles and wine gums, in desserts and jelly-covered fruit tarts, and in meat products, e.g., the aspic on pâtés and garnishes [195, 197]. Gelatin counteracts sugar crystallisation in confectionery, and syneresis in dairy products, and its foaming action makes it suitable for structuring aerated confectionery such as marshmallows, meringues and nougat [195]. Different gel strengths of gelatin are available for specific applications, whereas other hydrocolloids will usually need to be blended with additional ingredients, such as sugars and salts, in order to modify the strength of a gel [197].

Starch and starch derivatives are also used to create food gels, e.g., for partial substitution of gelatin in wine gums and jellies [195]. When starch granules are heated in the presence of excess water, they gelatinise, causing amylose to leach out [7]. On cooling, this free amylose then forms a gel. Other functions include binding of hams and sausages, and thickening of instant bakery cream powders and toppings, and of instant pudding and soup mixes [195]. Native starches have limited use in commercial food processing as they lack process tolerance and distribution stability and only perform a narrow range of functions [198]. Acetylated, cross-linked waxy maize starch, e.g., Colflo 67, is freeze-thaw stable with improved texture and appearance characteristics, i.e, smoothness and sheen [198]. Pre-gelatinised, instant starches allow food manufacturers to dispense with in-house starch preparation, but at the cost of grainy texture and reduced sheen compared with cook-up starches; however, ultra starches combine both advantages [198]. Textra creates cling in very thin liquids, such as dipping sauces, helps suspend the pulp in fruit juices, gives extra body to egg washes, and increases adhesion of glazes for meat and baked goods [198]. Novation starches are so-called functional native starches, said to offer the technical advantages of modified starches, whilst circumventing the requirement to declare them on the food label [198].

Skimmilk based chocolate milks behave like fluids with a weak network structure, with cocoa particles being incorporated into that structure [199]. Carrageenans stabilise such products by promoting the interaction between milk proteins and protein-covered cocoa particles, and this also counteracts the sedimentation of cocoa particles. Pectin stabilisers react with positively charged macromolecules, e.g., proteins, at pH values below their isoelectric point (pH). At low pH, pectin may therefore combine with casein particles to produce stable acidified milk [196]. Low-fat substitutes for traditional foods tend not to share their stability. In the absence of a stabiliser, low-fat spreads may break down during spreading. Stabilising agents for such products fall into four categories: viscous (polysaccharides, proteins), gelling (based on gelatin or carrageenan), phase separating (using mixed biopolymers) and synergistic (e.g., caseinates with starch) [200].

Foods are multi-phase systems in which solid, liquid and gaseous phases may coexist and furthermore, any of these phases may be either aqueous or lipid-based. In dispersions, two or more immiscible phases are distributed throughout one another. In colloids, the dispersed phase consists of either droplets (emulsion) or particles (colloidal dispersion) of about 1μm in diameter and suspended in a continuous phase [201]. In foods, the continuous phase usually consists of either water or fat, or of solutions based on either, e.g., an alcoholic, aqueous solution. In ordinary suspensions, the dispersed particles are observable by optical microscopy, but in colloids the particles are too small to be seen [15]. A gel is a liquid-in-solid dispersion. A sol is a solid-in liquid colloidal dispersion in the liquid state, e.g., gelatin hydrated with cold water, followed by dispersal in hot water [15]. A foam may be either a gas-in-liquid dispersion, e.g., whipped cream, or a gas-in-solid-dispersion, e.g., marshmallow [15]. In foams, gas bubbles are dispersed through a liquid or solid continuous phase. In emulsions, both the continuous and the dispersed (droplet) phases are liquids. Milk is an oil-in-water (o/w) emulsion, whilst butter and margarine are water-in-oil (w/o) emulsions. Food emulsions tend to be complex colloidal systems, containing other phases as well as oil and water, e.g., gas cells, protein micelles or aggregates, fat crystals, dissolved carbohydrates, and starch granules or gels [202]. Hydrophilic polymers are sometimes referred to as hydrocolloids, and some of these form gels as already discussed. Most do not, but all have the ability to thicken aqueous systems. They have an impact on the viscosity of aqueous foods, their texture and flavour release.

Emulsifiers are substances that facilitate the formation of emulsions because the molecule carries both hydrophilic and lipophilic regions. This surface activity causes films, micelles and similar ordered surface structures to form. There are two distinct approaches to the stabilisation of emulsions, as stability depends both on the viscosity of the continuous phase and on the degree of dispersion of the dispersed phase. In aerated desserts, such as ice cream, the primary function of emulsifiers is to destabilise fat globule membranes during homogenisation [203]. They thereby promote partial churning-out of milk fat during freezing [203]. In butter, most of the milk fat globules are disrupted during churning, and this inter-globular phase contains a mixture of liquid oil, crystal aggregates and membrane residues [7]. Margarine on the other hand is characterised by a fine network of crystallised fat [7]. In the manufacture of low-fat spreads, these approaches are reversed as it is the aqueous phase, rather than the lipid one, that is being structured here. This structure development commences with emulsifiers forming bi-layers in response to exposure to the aqueous phase [7]. Emulsifiers influence, and allow the standardisation of, product attributes such as viscosity, consistency, texture and mouthfeel and the wetting ability of (instant) powders. Meat comminution releases and solubilises muscle proteins which, upon heating, set to form a fixed matrix which entraps both fat and other desired particles [7]. In this way, the tougher cuts of meat can be transformed into meat products such as frankfurters. Here it is the raw meat preparation requiring emulsification and stabilisation, because the finished product is structurally stable due to the heat-set protein network it contains. Lecithin (E322) is a familiar, and widely used, emulsifier consisting of a mixture of phospholipids and triglycerides [204]. It is traditionally used in the manufacture of chocolate to achieve improved fat distribution as well as in instant milk powders, ice cream, yoghurt and mousse, baked products and snack foods [204]. In instant powders it promotes controlled hydration [205]. Lecithin is also used as a release agent, to prevent adherence of foods to oven belts, drying and smoking racks, moulds, meat casings and other food contact surfaces [205]. Whilst emulsifiers are surfactant in nature, stabilisers usually form part of the continuous phase of a dispersion, and must therefore be compatible with that phase [206]. Emulsifiers and stabilisers play an important role in ice cream manufacture, as they counteract recrystallisation phenomena that can lead to sandy textures [7].

Many polysaccharide gums are used in this, including locust bean gum, guar gum, carboxymethyl cellulose and carrageenan [7]. Xanthan gum forms gel-like networks in sauces and similar products thereby stabilising them; however, when subjected to sheer, the molecules dissociate instantly allowing the sauce to flow again [207]. This is a unique feature among thickeners and stabilisers.

Many of the above additives and functional food ingredients find application as so-called fat replacers in low-fat food products, the exact type required depending on which of the properties of fat need to be duplicated. Carbohydrate-based fat replacers are commonly used in formulated foods but, in common with protein-based ones, are unsuitable for frying. Fatty acids which have been chemically altered to make them less absorbable are suitable for frying foods [208, 209].

6.3
Temporal Food Additives

As a key quality attribute, the shelf life of all foods needs to be carefully managed. There are a few man-made foods that are inherently stable, resulting from a combination of composition and design, as are those storage organs of plants which serve to nourish the developing germ, e.g., cereal grains and nuts. Most unprocessed foods deteriorate rapidly due to both chemical and microbial processes and reactions. Other spoilage mechanisms relate to moisture migration and other physical phenomena. Biological tissues are especially vulnerable once their cell integrity has been compromised, and this is true both of microbial growth and of chemical reactions between food components, and between food components and oxygen. Temporal additives describes those food additives that improve keeping quality, and although this definition might be said to include certain of the structural additives discussed earlier, only preservatives and antioxidants will be discussed here. Table 6.3 gives the definitions of preservatives and antioxidants [4]. Many food preservation treatments, including traditional chemical preservation using salt and sugar, alter the nature of foods substantially. In contrast, preservatives and antioxidants are applied in small concentrations and have little impact, if any, on the sensory attributes of foods.

The term preservative has a precise meaning and covers specific, named chemicals listed in the Miscellaneous Food Additives Regulations 1995 [4]. Unfortunately, this is not always remembered, even where public information brochures are concerned. For example, Ministry of Agriculture, Fisheries and Foods' (MAFF) booklet "Food Sense. About Food Additives" refers to salt, sugar and vinegar preservation, food ingredients which clearly do not fulfil the crite-

Table 6.3. Definition of preservatives and antioxidants [4]

Preservative	Any substance which prolongs the shelf-life of a food by protecting it against deterioration caused by microoganisms
Antioxidant	Any substance which prolongs the shelf-life of a food by protecting it against deterioration caused by oxidation, including fat rancidity and colour changes

ria for food additives [210]. Fruit-, vegetable- and meat-based foods are characterised by high moisture levels and an abundance of nutrients, i.e., nutrients not only for humans, but also for microorganisms. They therefore constitute favourable habitats for many microbial contaminants. However, because of certain differences between individual foods, e.g., pH and moisture level, different foods allow the development of different types of microflora to take place. Different preservatives are required to control the activities of bacteria, moulds and yeast in different foods.

There are a number of additives with preservative and antioxidant functionality that are generally permitted for use in foods [4]. Examples are acetic acid (E260), lactic acid (E270) and carbon dioxide (E290), and ascorbic acid (E300) and the tocopherols (E306-E309), respectively. However, most preservatives and antioxidants are permitted conditionally, i.e., they are permitted for specific applications and at specified levels. Sulphur dioxide (E220) is one of the oldest preservatives and in addition, has multiple functionality relating to keeping quality, as it participates in reactions with a wide range of food components [211]. In addition to inhibiting microbial growth, it acts as an enzyme inhibitor and antioxidant; as such, it is effective against enzymic browning in fruit and vegetables [212]. It also counteracts non-enzymic browning in foods [213]. Sulphur dioxide is highly reactive, and its anti-microbial activity is attributed to its ability to pass through cell membranes, and the subsequent extensive reactions with cell components, in particular, enzyme systems [214]. Major applications include fruit products, e.g., fruit-based desserts, sauces, spreads, juices and jams; other uses include pickles, salad cream and mayonnaise, sauces, and dried and frozen vegetables [215].

Benzoic acid (E210) acts as a preservative, and is normally applied in the form of its sodium, potassium or calcium salt. However, the preservative effect derives from the undissociated benzoic acid, i.e., at low pH (optimum pH 2.5–4.0) [214]. Benzoic acid is therefore best suited as a preservative in acid foods. Among the benzoates, the sodium salt has the greatest solubility [214]. Sodium benzoate is effective against yeasts and bacteria, but less so against moulds [214]. Benzoic acid is less detrimental to food colours in soft drinks than sulphur dioxide, and another application is in cooked pre-packed beetroot [215]. Benzoic acid occurs naturally in a number of fruits and spices, e.g., cranberries, prunes, greengage plums, cinnamon and ripe cloves [214]. The esters of p-hydroxybenzoic acid (parabens) are permitted for use in a wide range of foods and food ingredients, their antimicrobial activity being directly proportional to their chain length [215].

By far the biggest use of propionic acid (E280) and the propionates is in baking; they find use as antifungal agents in breads, cakes, buns, flan cases, puddings, frozen pizza and dairy products [215]. Calcium propionate is normally used just for bread, while the neutral taste of sodium propionate is taken advantage of in the perservation of cakes [215]. Sorbic acid (E200) and its sodium, potassium and calcium salts are used principally as fungistats; they have a broad spectrum activity against yeasts and moulds, but are less effective against bacteria [214]. Although their activity increases with decreasing pH, their upper limit of effectiveness, at around pH 6.5, is considerably higher than that of benzoates and propionates. Sorbates find far-reaching use in low-fat emulsions, on hard cheese surfaces and in baking powder-raised baked goods (in yeast-raised goods it may inhibit fermentation) [215]. Other applications include marzipan and fillings in confectionery, fruit preserves and semi-dried fruit, glacé fruit, tomato concentrate, soft drinks, sauces and dressings [215]. Sorbic acid occurs naturally in the berries of mountain ash, or rowan. Both sodium hydrogen acetate and sodium diacetate (E262) are particularly effective against spore-forming bacteria; applications include bread, cheese, frozen vegetables, fruit pastilles, packet soups, quick-set gels, crisps, salad cream and pickles [215].

Sodium and potassium nitrates and nitrites (E249-E252) are permitted for use in cured meat products, e.g., bacon and salami, and certain cheeses [214]. Nitrate must be reduced to nitrite by bacteria in the meat to bring about both a preservative and a colour effect, and the preservative effect persists only as long as residual nitrite remains in the meat [215]. The stable pink colour, typical of many cured meats, is due to a reaction between nitrate and myoglobin. Interaction between nitrate and protein upon heating results in strong action against spores, so that food is made safe against botulism even after mild heat processing [215]. Nitrates and nitrites have been used for many years in conjunction with salt as components of dry mixes and brines in the preparation of cured meats. Nisin (E234) is a short polypeptide antibiotic produced by certain strains of *Streptococcus lactis*, which has been found to delay, or prevent, germination of bacterial spores, in particular, clostridia [214]. It has little or no effect on the vegetative forms of most food spoilage bacteria, and is approved for use in cheese, clotted cream and canned foods [214]. Biphenyl (E230) is a fungistat which is used on citrus fruit, where it is generally impregnated into wraps or containers [214]. From there, because of its high vapour pressure, it manages to penetrate into the peel. The preservatives discussed here are perhaps the most important ones, but the list is far from complete [4].

Antioxidants are used to inhibit or delay the oxidation of food components, in particular unsaturated fatty acids and pigments. Oxidative rancidity is a cause not only of off-flavours but in addition, consumption of foods so affected is undesirable in nutritional safety terms [216]. The following is a discussion of some of the antioxidants commonly used in foods. As mentioned previously, two groups of generally permitted additives have antioxidant activity, i.e., ascorbic acid and tocopherol. The conditionally approved antioxidants are the gallates (E310-E312), erythorbic acid (E315) and sodium erythorbate (E316), butylated hydroxyanisole (BHA) (E320) and butylated hydroxytoluene (BHT) (E321) [4]. The gallates, BHA and BHT are used in fats and oils used for thermal food processing, in frying oils and fats, cake mixes, cereal-based snack foods, milk powders for vending machines, dehydrated soups, sauces, dehydrated meat, processed nuts, seasonings and condiments and chewing gum [4]. Erythorbic acid and sodium erythorbate are permitted for use in semi-preserved and preserved meat and fish products [4]. BHA is insoluble in water, but is highly fat soluble and provides good carry through potency, especially in animal fats used in baked foods [217]. This refers to its ability to be added to a food component, survive a processing step such as frying or baking, and impart stability to the finished food product. BHT has similar properties to BHA [217]. Propyl gallate imparts good stability to vegetable oils and provides good synergism with BHA and BHT [217]. It has poor carry through properties, and decomposes at 148 °C, its melting point [217]. Propyl gallate can form coloured complexes with metal ions, which can bring about discolouration in fats and oils [217]. Gallates remove any free peroxides that have been generated within a lipid material; they exhibit good stability in fats at room temperature [215]. Unlike the other gallates, propyl gallate is soluble in water as well as fats, and it also works better in anhydrous foods [215]. Antioxidants are often applied to foods as preparations consisting of several chemical components, e.g., tocopherols, ascorbyl palmitate, citric acid and lecithin; such mixtures are permitted for general use in most countries [216]. Here the tocopherols react directly with fatty acid radicals, thereby disrupting the oxidative chain reaction. Ascorbyl palmitate does not act directly as an antioxidant but boosts the action of the tocopherols. Citric acid complexes pro-oxidant metals and the antioxidant mechanism of lecithin are not currently well understood.

6.4
Sensory Food Additives and Industrial Ingredients

The food additives covered under the current heading are colours, sweeteners and flavourings; this means substances regulated under three sets of regulations, i.e., the Colours in Food Regulations 1995 [218], the Sweeteners in Food Regulations 1995 [219] and the Flavourings in Food Regulations1992 [220]. The industrial ingredients also discussed here are materials which, whilst possessing closely specified quality attributes suitable for large-scale commercial application, are not regulated as food additives. The fact that food colours, sweeteners and flavourings are regulated individually, outside the scope of the Miscellaneous Additives Regulations 1995 [4], may reflect their potential to mislead consumers about the true nature of a food. In fact, colours in particular have often been referred to as cosmetic additives. Colours can potentially be abused to disguise the signs and symptoms of processing faults and spoilage, or to mislead about the nature, condition or degree of excellence of a food. Colour use in foods is particularly controversial. Flavourings can be similarly abused, e.g., to camouflage spoilage taints, and to mislead consumers about the true nature of a food.

The appearance of an unprocessed food is largely determined by its external structure, e.g., shape and size, and by its colour; these are therefore important factors both in product recognition and in giving a first impression of quality. In man-made foods, colour takes on added importance as an identifier and as a signal of quality. Colours may be added to soft drinks and other colourless foods to aid orientation and perception, starting from when the product is viewed by a potential purchaser at the point of sale. Thus, orange-flavour(ed) drinks will be, and will be expected to be, orange in colour. In another scenario, the purpose of colouring a food may be to draw attention, or generate interest, without in any way attempting to imitate nature. This is the case with fantasy-flavour products, e.g., the deep-orange coloured, very popular Scottish soft drink called Irn-Bru. Another example of this kind of colour use is that of sugar-craft-type celebra-

Table 6.4. Goals of colour application in foods and beverages

Goal	Food examples
Restoring colour impression lost during food processing	Strawberries turning grey-brown, peas turning pale, on canning
Visual batch-to-batch consistency	Soft drinks and alcoholic drinks, boiled sweets
Boosting weak colour impressions	Fruit yoghurt
Giving identity to colourless products	Soft drinks, boiled sweets, e.g., red raspberry., providing consonance between flavour and colour impressions,
Providing product character	Soft drinks and boiled sweets with fantasy names; sugar craft products such as novelty cakes

tion cakes, e.g., wedding cakes and novelty cakes. A very important motivation for adding colours to foods and beverages is for visual consistency across packs and batches. Visible colour variation raises doubts in the observer's mind about the condition of a product, and probably also about the competence of its maker. Some typical goals of colour application in foods and beverages are summarised in Table 6.4.

A colour is a food additive "used or intended to be used primarily for adding or restoring colour to a food" [218]. Nevertheless, some colours possess additional functionality, e.g., caramel, may impart some flavour impact as well as colour. The Regulations list the colours that are permitted in food and drink products, identify those foods that may not be coloured, and set maximum levels for named applications. There are no separate categories of natural and artificial colours. Although these terms are frequently used colloquially, in the context of colours they have no legal meaning. Permitted colours occupy the E-No. range from E100 (Curcumin) to E180 (Litholrubine BK). An example of restricted colour use is beer, which may only be coloured with plain caramel (E150a). As well as food colours, a number of other additives are used to improve the appearance of foods. For example, glazing agents are used to achieve surface polish/sheen in foods such as confectionery products and citrus fruit. To avoid use of, and therefore having to declare, colours, coloured foods may be used as ingredients for finished food products, e.g., products obtained from red grapes, elderberries, blackcurrants and red cabbage [221]. Colours come in two physical forms, i.e., straight colours, which are usually water-soluble dyes, and lakes [222]. Lakes are generally insoluble in water, and are used in products where leaching, or bleeding, would cause problems, e.g., biscuit fillings. In fact, suitable choice of colours depends on a series of factors, some of which are listed in Table 6.5.

Whilst the law does not distinguish between natural and artificial food colours, consumers may well attach related meanings to individual colours. Since the discovery of the coal tar dyes, most of the so-called (colloquially) artificial colours have fallen into that particular category. These dyes are stable and very effective, even in very small amounts, which is why they are favoured by industrial users [189]. Artificial in this context seems to suggest synthetic materi-

Table 6.5. Factors to be considered when selecting food colour

Statutory compliance in relation to the application being considered

Compatibility with the food matrix, e.g., phase solubility (aqueous/fatty/alcohol etc.)

Stability of colour and colour shade in terms of interactions with food ingredients, e.g., proteins, metals, acids (pH) etc.

Stability in terms of the external environment, e.g., towards light

Table 6.6. Coal tar dyes [218]

E102 Tartrazine*	E127 Erythrosine	E154 Brown FK*
E104 Quinoline Yellow	E128 Red 2G	E155 Brown HT
E107 Yellow 2G	E129 Allura Red	E180 Litholrubine BK
E110 Sunset Yellow FCF,	E131 Patent Blue V	
Orange Yellow S*	E132 Indigotine, Indigo	
E122 Carmoisine*	Carmine	
E123 Amaranth	E133 Brilliant Blue FCF	
E124 Ponceau 4R,	E151 Brilliant Black BN,	
Cochineal Red A*	Black BN*	* "Azo dyes"

als that do not have equivalents in edible biological tissues. In some of these, the chromophore is an azo group, and they have therefore been referred to as azo dyes. Permitted food colours that do not appear to have equivalents in edible biological tissues are listed in Table 6.6.

Some of the colours in Table 6.6 are only permitted for specific uses, e.g., Brown FK (E154) in kippers [218]. Other colours that do not have equivalents in edible biological tissues are Vegetable carbon (E153), Calcium carbonate (E170), Titanium dioxide (E171), Iron oxides and hydroxides (E172), Aluminium (E173), Silver (E174) and Gold (175). All of these are inorganic pigments. Many colours are extracted from natural sources, some of which are foods in their own right, and some of which are not; or they are synthetic copies of such compounds. Table 6.7 lists those permitted colours that are related to natural food pigments [218].

Naturally occurring pigments include carotenoids, flavonoids, porphyrins (chlorophyll), betalaines (beetroot), as well as some natural, but non-food pigments, e.g., quinonoids [223]. Carotenoids are widespread and several can be extracted, particularly lutein, bixin, beta-carotene, crocin and the pigments contained in paprika [223]. The seeds of the annatto plant have long been used as a source of food colour, i.e., Annatto (E160b). Their principal pigments are bixin, which is oil soluble, and norbixin, which is water soluble [223]. Like all carotenoids, annatto is susceptible to oxidation. It is important in the development of products containing annatto and protein that conditioning time is allowed, as norbixin has a tendency to bind to proteins [223]. Curcumin (E100) is the principal pigment of turmeric. Although curcumin is relatively heat stable, it is susceptible to bleaching by sulphur dioxide [223]. However, the main limitation of its use is its poor stability when exposed to light [223]. The skins of

Table 6.7. Permitted food colours related to naturally occurring food pigments [218]

E100 Curcumin	E160d Lycopene
E101 (i) Riboflavin, Riboflavin- 5′-phosphate	E160e Beta-apo-8′-carotenal
E140 Chlorophylls and chlorophyllins	E160f Ethyl ester of beta-apo-8′-carotenic acid
E141 Copper complexes of cholorophylls and chlorophyllins	E161b Lutein
	E161 g Canthaxanthin
E160a Carotenes	E162 Beetroot Red, betanin
E160c Paprika extract, capsorubin	E163 Anthocyanins

black grapes are a major commercial source of anthocyanins (E163). The reactions and colours of individual anthocyanins are dependent on a number of factors, including the degree of polymerisation and complexing [223]. They are relatively heat sensitive. Betanin (E162) is extracted from red table beet. The major limitations of betanin use are sensitivity to heat and oxygen [223]. Metal ions promote discolouration [223]. Beetroot Red is therefore limited to use in short shelf life, or low water activity, products that only undergo mild heat treatment, and with limited exposure to air, e.g., dairy desserts and dry mixes [223]. Chlorophyll (E140) and its greener copper complexes (E141) are commercially derived from three main sources, i.e., grass, lucerne (alfalfa) and nettles. Stability of the copper complexes is superior to the natural magnesium varieties, both in respect of heat and light exposure [223]. The major pigment of wild salmonid muscle, e.g., trout, is astaxanthin, which derives form food carotenoids [224]. To simulate this effect, beadlets of synthetic canthaxanthin are added to the feed of farmed salmonids [224]. Canthaxanthin (E161 g) is an approved food additive but, transmitted to food via animal feed, its presence in fish, as well as eggs, is not usually declared to the consumer.

A sweetener is a food additive "used or intended to be used to impart a sweet taste to food, or as table-top sweetener" [219]. These Regulations list the authorised sweeteners, the food and drink products in which they may be used, and the corresponding maximum usable doses. The permitted sweeteners are listed in Table 6.8.

In terms of relative sweetness, sweeteners can be divided into two categories, namely, intense (E-No. range E950-E959) and bulk sweeteners [225]. The sweetness intensity of the former is many times that of sugar, and as they are used at very low levels, they are usually supplied in combination with texture-providing polyols, or fillers [226]. Different sweeteners may be blended together for optimum sweetness performance. For example, saccharin boosts the sweetening power of cyclamate, and cyclamate reduces the aftertaste of saccharin [227]. Su-

Table 6.8. Permitted sweeteners [219]

E420 Sorbitol	E967 Xylitol	E954 Sacharin and its Na, K and
E421 Mannitol	E950 Acesulfame K	Ca salts
E953 Isomalt	E951 Aspartame	E957 Thaumatin
E965 Maltitol	E952 Cyclamic acid and its	E959 Neohesperidine NC
E966 Lactitol	Na and Ca salts	

crose is usually regarded as the reference material against which other sweetening agents are compared. Sucrose and other refined sugars, e.g., fructose, are energy-rich carbohydrates. Although they are chemically defined, pure substances, they are not included in Table 6.8 and are not, strictly speaking, sweeteners, as they are not classified as food additives. True bulk sweeteners are sugar alcohols, or polyols, i.e., sugar derivatives that are slowly, or incompletely, metabolised so that they are particularly suitable for diabetic consumers [1]. They are also less cariogenic than sugar [1].

A flavouring is a "material used or intended for use in or on food to impart odour, taste or both" [220]. Flavourings can either be so-called flavouring substances (chemically defined chemical compounds), flavouring preparations (obtained by physical, enzymic or microbial processes from suitable materials, e.g., citrus oils), process flavourings or smoke flavourings; or they can consist of combinations of two or more of these categories. So-called flavour modifiers, or flavour enhancers, e.g., monosodium glutamate (MSG, E621) are regulated by means of the Miscellaneous Food Additives Regulations 1995 [4]. Whilst MSG is widely used as a flavour enhancer in savoury foods, maltol and ethyl maltol (an artificial flavouring substance) perform a similar function in sweet foods. Food flavourings, unlike other additives, are self-limiting in terms of dosage. For a preservative to function effectively, it will typically be present in a food at a level of around 0.2%, whilst a stabiliser will be present at higher levels [228]. In contrast, the vast majority of flavouring substances are used in quantities that are measured as ppb [228]. In order to achieve the desired flavour impact, most commercial flavourings contain at least some flavouring substances. Depending on their origin and manner of preparation, flavouring substances are categorised as natural, nature-identical or artificial [220]. If a flavouring formulation contains any artificial flavouring substances, the whole product becomes artificial. Similarly, natural flavourings contain only natural components. Natural flavouring substances are "obtained by physical, enzymatic or microbiological processes from appropriate material of vegetable or animal origin" [220]. Interestingly, what is an appropriate material is left open to interpretation. Nature-identical flavouring substances are "either obtained by chemical synthesis or isolated by chemical processes" and are "chemically identical to a substance naturally present in appropriate material of vegetable and animal origin" [220]. Finally, artificial flavouring substances are obtained by chemical synthesis, but have not been shown to be present in "appropriate material of vegetable or animal origin" [220]. Process flavourings are obtained "by heating to a temperature not exceeding 180°C for a continuous period not exceeding 15 minutes a mixture of ingredients ... of which at least one contains nitrogen (amino) and another is a reducing sugar" [220]. This is a controlled application of the Maillard reaction. Smoke flavourings are extracts from smoke of a type normally used in food smoking processes. If a food contains a level of 3,4-benzopyrene, a flavouring which contains the material can also be added, as long as the flavouring does not exceed a specified maximum amount. The Flavourings in Food Regulations 1992 also list, and set permitted levels in specified foods for, several naturally occurring, bioactive substances. These include coumarine, which is permitted for use in chewing gum, alcoholic drinks and caramel confectionery, and pulegone,

Table 6.9. Flavouring categories [220] illustrated using vanilla flavour

Alcoholic extract of vanilla pods	Flavouring preparation
Naturally extracted vanillin	Natural flavouring substance
Synthetic vanillin	Nature-identical flavouring substance
Ethyl vanillin	Artificial flavouring substance

which is permitted in mint confectionery, mint or peppermint flavoured drinks and other drinks. Table 6.9 provides an illustration of some of the flavouring categories defined in the Regulations [220] using the example of vanilla flavour.

Like colours, flavourings and other flavour-active food ingredients are added to different foods for various reasons. Some of these are summarised in Table 6.10.

Flavour is a major brand differentiator and consequently, product differentiation is a major motivation for flavouring use. Use of flavourings also protects a brand in that it will be hard for competitors to match a particular flavour exactly. Food manufacturers may use a mixture of different flavourings, from different suppliers, to further protect the confidentiality of product formulations. Flavourings, especially those whose character is strongly defined by flavouring substances, can be manufactured to very narrow flavour specifications. This helps to ensure consistency between batches of the flavouring and, more importantly, between batches of the finished food. Some flavour impressions in respect

Table 6.10. Goals of flavouring application in foods and beverages

Goal	Food examples
Imparting flavour to flavourless food	Boiled sweets, chewing gum, Quorn™
Imparting character, identity, novelty to flavourless food	Boiled sweets, chewing gum, certain "fantasy" flavour soft drinks, e.g., Irn-bru™ and Vimto™
Strengthening flavour impact	Yoghurt, fruit juice drinks
Modifying existing flavour profile	Adding back top notes which may have been lost in processing; In general, differentiation within product categories
Masking undesirable (but not spoiled!) flavour	Diet formulae
Providing cheap versions of desired foods/flavour notes	From vanilla extract to ethyl vanillin (Table 6.9)
Ensuring batch-to-batch consistency	Ingredients subject to seasonal or varietal variation, e.g., fruit-based products
Enhancing or reinforcing weak intrinsic flavour	Fruit flavoured yoghurts
Imparting flavour where use of ordinary aromatic materials is technologically infeasible	Dry powder mixes

Table 6.11. Complexity of flavourings

Flavour	Signature chemical
Pear	isoamyl acetate
Cucumber	*trans*-2,*trans*-4 decadienal
Cherry (USA-style)	benzaldehyde
Mushroom	*cis*-3-octenol
Raspberry	raspberry ketone (para-methoxy phenyl butanone)
Lemon	citral
Grapefruit	nootkatone
Peach	γ- and δ-lactones
Strawberry	complex mixture of acids, alcohols, esters, lactones and carbonyl compounds

of particular foods can be imitated much more readily than others. Table 6.11 provides an overview of some signature chemicals.

Ready-to-use flavouring products are manufactured by compounding one or more categories of flavourings as defined in the Regulations [220], with additives such as flavour carriers, stabilisers and antioxidants. For example, in a fruit-type flavouring, fruit extract may provide the base notes, whilst impact is achieved by the use of flavouring substances. Several thousand flavouring substances are known, and up to a hundred may be used in the formulation for a single flavouring [229]. Under USA law, the definition of the term food additive exempts any substances "generally recognised, among experts qualified by scientific training and experience to evaluate its safety, to be safe under the conditions of its extended use". This provision has come to be designated by the acronym GRAS [229]. The US Flavor and Extract Manufacturers' Association (FEMA) organised a panel of experts not affiliated with the industry to evaluate the GRAS status of flavouring substances. The Food and Drug Administration subsequently adopted virtually the entire FEMA GRAS list [229]. Another difference between the EU and the USA is that in the USA, all synthetic flavourings, including those that would be considered nature-identical in the EU, are classified as artificial. This has created a perceived need to extend the range of "naturals" available to the industry and encouraged research into the manufacture of such substances. Whilst the definition of natural flavouring substances under UK/EU law already leaves room for interpretation [220], it should not be assumed that all US "naturals" will necessarily be considered natural in the EU. Flavouring preparations do not have to be declared as flavourings, but instead, can be referred to more specifically. Many will be able to flavour a food without the addition of flavouring substances, e.g. essential oils (steam volatile odour chemicals) and essence oils (alcoholic extracts). Citrus essence oils contain more volatile, fruity topnotes than the essential oils [230]. Standardised herbs and spices, spice essential oils and spice oleoresins are similarly used for flavouring foods and beverages [231]. Condiments and relishes are examples of highly flavoured food items, which are typically eaten with a plainer food to add flavour and interest, e.g., a small serving of pickled vegetables taken with plain bread and cheese [189].

Protein hydrolysates may be obtained from various foods, in particular vegetables and yeast. Hydrolysed vegetable protein (HVP), whether as part of a complex flavouring, or on its own, has for many years been the mainstay of flavouring vegetarian products [232]. Soy sauce and miso are related products, one already being familiar in the West and the other becoming increasingly so [189]. Meat extracts are soluble substances extracted from meat when it is put into water. Bouillon bases are available both for vegetarian and non-vegetarian products [232]. Yeast extracts are made both from brewer's yeast and from baker's yeast. This usually involves autolysis, i.e., hydrolysis by intrinsic proteolytic enzymes [233]. As with meat extracts, the flavour enhancing properties of yeast extracts derive from their high contents in free amino acids (e.g., glutamic acid) and 5'-ribonuleotides (e.g., inosin 5'-monophosphate) [233]. These substances are available in purified form, where they are controlled as additives (E653) and are chemically related to the more familiar MSG (E621). Yeast extracts may be able to be used in place of MSG to potentiate or enhance savoury flavours, depending on the application. Fish sauces, primarily autolytic products, are typical foods of South East Asia and embrace a range of traditional products, including shrimp pastes [234].

6.5
Micronutrient Additions, Functional Foods and Dietary Supplements

In Chapter 4, nutrients and bioactive food substances were discussed as natural components of biological tissues and fluids. Some of these are now examined in their role as additions to man-made foods, and to diets. Conceptually, and from a potential legislative point of view, there are overlaps between nutritionally enhanced foods, functional foods and dietary supplements [235].

The UK legal definition of additives excludes vitamins, minerals and other nutrients, if they are used solely for the purpose of fortifying or enriching food, or of restoring the constituents of food [235]. Fortification usually refers to the addition of nutrients to foods in which they are naturally present in insignificant amounts, or from which they are absent [235]. The rationale behind fortification is the existence of actual or potential deficiencies among specific populations; this would also include nutrient addition to a food for it to fulfil a specified dietary role. For example, margarine is fortified with vitamin A in many countries to replace that which is "lost" when margarine is substituted for butter. In this particular context, fortification may also be referred to as substitution [235]. Enrichment is a more general term and means an increase in the level of some nutrient to make the food a richer source of that nutrient [235]. Restoration refers to the replacement, in full or in part, of losses incurred in processing, e.g., loss of B vitamins and iron in milling of cereals to low extraction rates, and loss of vitamin C in the manufacture of instant potato products [235].

UK food law requires the compulsory addition of certain micronutrients to bread and flour (except wholemeal), and to margarine and spreadable fats [235]. The Bread and Flour Regulations 1995 require the addition of not less than 1.65 mg iron to 100 g flour and the source of the iron can be any of ferric ammonium citrate, green ferric ammonium citrate, ferrous sulphate, or iron powder

[236]. The Regulations also make it compulsory to fortify extracted flour with calcium, thiamin and niacin. The Spreadable Fats (Marketing Standards) Regulations 1995 require margarine to be enriched with vitamins A and D to a level to make it comparable with butter [237]. The addition of vitamins A and D to skimmilk powder, of vitamin D to evaporated milk, of vitamin C to the preparation of instant potato and to some juices and nectars are examples of voluntary vitamin restoration in food products in the UK [238]. Other UK foods that are commonly fortified with vitamins and minerals include breakfast cereals, infant formulae, meal replacements, sports drinks and slimming products [239]. Foods for vegans, foods such as soy products, are often fortified with vitamin B_{12}, and many infant foods are fortified with iron and vitamin D [239]. Minerals now commonly added to foods are copper, manganese, selenium, molybdenum and chromium [235]. Added calcium is available in bottled water, e.g., Danone Active, which is targeted, amongst others, at parents whose children refuse milk. Selenium enrichment is topical because of increasing localised dietary deficiencies associated with similar soil deficiencies in selenium. An example of a selenium-enriched food is fortified milk Parmalat Plus [240]. So-called ACE drinks are drinks enriched with vitamins A, C and E; this concept has recently been expanded to include folic acid, hence the emergence of FACE drinks [241]. As FACE drinks are targeted specifically at pregnant women, they are closely related to functional foods.

The kinds of beneficial bioactivity typical of many unprocessed foods, especially fruit and vegetables, may be replicated to some extent in man-made foods by the addition of the relevant active principles. This is the most common scenario for what are called functional foods. Such enhancements may also be possible in the future through genetic modification of biological tissues, as an alternative to food compounding and processing. Some of the nutrient-enriched foods mentioned above can be conceptualised as functional foods, as can foods for particular nutritional uses. It is in fact difficult to draw the limits of the functional food category. One particular definition of the class includes all foods designed to "target and favourably affect particular functions of the body" [242]. This might be related to physical or mental performance, or to protection from degenerative diseases, e.g., cancer, cardiovascular disease and osteoporosis. It might even be argued that for each age group, and also in relation to gender, there are foods that are particularly suitable in opposing health risks. For example, certain functional foods are targeted towards women undergoing menopause; others are targeted towards people participating in high-performance or endurance sports, and so on. The special nutritional needs of small children have already been highlighted. Functional food product development reflects a major shift in attitude, away from removing potentially harmful food components, e.g., fat, cholesterol and salt, and towards adding beneficial ones [243]. The Institute of Food Research list a range of commercial functional foods that are already widely available in the UK [242]. These include added-bran breakfast cereals and reduced-fat dairy products, added-fibre fruit juices, bread fortified with folic acid and spreads containing fish oils or olive oil. Other examples are dairy products containing bioactive cultures and drinks specifically formulated to provide a balanced replacement of fluids lost during exercise or to provide energy (sports drinks). Finally, the Institute include in their list carbohydrates that are slowly released in the body so that they supply energy over a prolonged period [242].

Foods provide the principal growth substrates for colonic bacteria, and pre- and probiotics are a class of functional foods targeted at intestinal health. Because of its resident bacterial flora, the colon is the most metabolically active organ in the human body [244]. Depending on its composition, the gut microflora plays a role either in disease development or in host wellbeing. One approach to gaining improved microflora management involves live microorganims as so-called probiotics (usually lactobacilli or bifidobacteria, i.e., lactic acid excretors) [244]. The other approach is the use of prebiotics, i.e., dietary components that have a specific fermentation directed towards certain populations of indigenous gut bacteria [244]. Probiotics must remain viable after ingestion to be beneficial. Fructooligosaccharides are obtained through partial hydrolysis of inulin. These oligosaccharides can stimulate growth of bifidobacteria in the colon to an extent that, after a short feeding period, they become predominant in the faeces [244]. Such materials are examples of prebiotics. Other prebiotics include raffinose, palatinose and stachyose and oligosaccharides that contain xylose, maltose and mannose [244]. Probiotics are thought to alleviate lactose malabsorption [244]. Widely available probiotics include Yakult and Danone's Actimel, which are fermented milks; there are also non-dairy options with biocultures, e.g., Pro Viva blackcurrant drink [245].

In addition to fermented dairy foods, margarines and spreads have recently been highlighted for their cholesterol-lowering benefits, the one having received the most attention being Benecol [243]. The active component responsible for lowering cholesterol is sitostanol ester [243]. Plant stanols are hydrogenation products of the respective plant sterols [246]. Plant sterols have a role in plants similar to that of cholesterol in mammals, i.e., in forming cell membrane structures [246]. The consumption of plant sterols and stanols lowers blood cholesterol levels by inhibiting the absorption of dietary and endogenously-produced cholesterol from the small intestine [246]. Nutribread Wholemeal Loaf is an example of a food enriched with n-3 unsaturated fatty acids [245]. Other products to which these PUFA have been added include Bertrams Exclusive Omega fruit juice drink, Columbus Healthier eggs, Freshlay Vita eggs and Vitaquell Omega-3 vegetable margarine [247]. The levels of n-3 unsaturated fatty acids in eggs can be raised through layer feeds which have been enriched with phytoplankton [248]. These eggs can then be used to produce pasta enriched with n-3 unsaturated fatty acids. Nutribread Brown Loaf for Women is designed for women during, or entering, menopause, or who experience pre-menstrual syndrome [245]. It contains both phytoestrogens and evening primrose oil [245]. Allied Bakeries' Burgen bread is labelled as "high in natural plant estrogens", reflecting its soy flour and linseed contents [245]. Energy drinks tend to be high in caffeine and frequently contain amino acids such as taurine, or botanicals such as ginseng or guarana [243]. Examples of these include Red Bull, Liptovan ACE, Liptovan B3, Purdey's Multivitamin energy drink and Solstis [247]. Sports drinks may be formulated to replace minerals, for rapid rehydration and for energy. Depending on the exact formulation, isotonic drinks are absorbed faster than water, and hypotonic drinks are absorbed fastest [249].

Vitamin and mineral supplements often contain other ingredients that are necessary in their manufacture, e.g., fillers and binders in tablets [65]. Tablets may also contain colours and/or flavourings, and may be covered with a protein coating. For example, many vitamin C tablets are coloured orange and made to taste sweet because of the association of the vitamin with oranges [65]. Supplements are available singly or as multivitamin, multimineral or multivitamin-and-mineral preparations. Vitamins A, C and E can also be combined with zinc, selenium, possibly iron, copper and manganese into antioxidant complexes which may, in addition, contain the amino acids glutathione and cysteine and antioxidant phytochemicals [65]. EFA supplements are available as concentrated oils, e.g., flaxseed or fish oil capsules for n-3 unsaturated fatty acids, and evening primrose or borage oil capsules for n-6 unsaturated fatty acids [65]. Bone mineral complexes to build healthy bones may contain calcium, magnesium, vitamin D, boron and a little zinc, vitamin C and silica [65]. Herbal remedies can be sold as food supplements only under the generic name of the herb [250]. Aloe vera is a cactus-like plant, and there are claims for various health-beneficial effects for this plant [250]. The gel used in supplements is extracted from its fleshy leaves [250]. Ginseng comes from the root of a family of plants found most commonly in South East Asia; it is claimed to improve feelings of well-being [250]. Chlorella and spirulina are fresh water algae, which are used as dietary supplements because of their great nutrient density [251]. It is currently estimated that over 10 million people world-wide take chlorella [251]. The outer layer of the chlorella cell wall consists of polymerised carotenoids, which are said to account for the detoxfying activity of the material, in particular, its ability to bind to heavy metals, pesticides and toxins such as polychlorinated biphenyls (PCB) [251]. Other herbal food supplements include guarana and royal jelly [250]; echinacea, which is thought to stimulate the immune system; garlic, which is traditionally used for its antiseptic action; and gingko biloba, which is thought to improve symptoms associated with dementia [252]. Herbs have a long tradition of use in healing, and herbal remedies are regulated in three ways: as licensed herbal medicines, unlicensed herbal medicines or as food supplements [252]. No direct health claims can be made for the latter, although claims can be made that suggest a health benefit. The type of extract and the dose used may vary with different products. Standardised extract means that a herb has been treated to extract the active ingredient, leading to greater batch-to-batch consistency [252].

6.6
Biological Contamination

Live organisms of various kinds can contaminate a food and affect its quality in various ways. The most important aspect of spoilage resulting from biological contamination is that of food safety, and in this context, microbial infections and intoxications currently show the highest profile. However there are, in addition, a number of parasites of public health significance. Finally, so-called pests, usually small animals, birds and insects, can gain access to foods. These too can act as disease carriers, or they may simply cause physical damage to a food, damage likely to set in motion a range of deteriorative reactions.

Almost all foods are potential substrates for unwanted microbial growth, which causes them to deteriorate and, ultimately, spoil. The composition of the spoilage microflora in each case depends on the nature of the food and on the storage environment. Some spoilage microorganisms found in, or on, a food merely bring about changes in the palatability of the food. Others lead to food poisoning if a contaminated food is consumed in a sufficient quantity. The three main types of microorganims found in food are bacteria, yeasts and moulds. All of these are single-cell organisms, which will replicate themselves as long as the environmental conditions are favourable. In addition, there are foodborne viruses; these need to take over the metabolic apparatus of a host cell in order be able to reproduce. There are certain key environmental aspects that influence growth and metabolic activity of bacteria, yeasts and moulds, i.e, presence of appropriate nutrient substrates, sufficient water activity (α_w), suitable temperature, pH and atmosphere, in particular, presence or absence of oxygen. Many of these organisms are ubiquitous in the environment. For them to be able to attack a food, they must first gain access to it. Intact plant tissues, e.g., fruit and vegetables, are protected by their structure, especially external features such as skin and peel. To some extent, this is also true for intact animal carcasses. Man-made foods use packaging, usually in combination with specific anti-microbial processing, to mimic this effect. Many foods have complex structures, which allow individual microenvironments to coexist. These microenvironments, and the interfaces between them, may harbour different types of microflora. This can lead to product instability. For example, feta cheese preserved in a vegetable oil can become unstable if moisture-rich vegetables, e.g., peppers are added to the product and allowed to settle at the surface, as this will allow mould or yeast growth to take place. Certain bacteria, e.g., bacilli and clostridia, are able to develop resistant spores under adverse conditions, which enable them survive these conditions. Bacilli are either aerobes, i.e., they require oxygen for growth, or facultative anaerobes; clostridia are anaerobes, which means that they become active only when oxygen is absent [162]. Mould growth on food is readily recognised by its fuzzy appearance, which is due to the formation of filaments. Some moulds produce toxic metabolites, i.e., mycotoxins. Unlike moulds, yeasts reproduce by budding or fission [162].

When the microorganisms involved in the spoilage of a food are pathogenic to humans, they constitute a hazard in public health terms. Bacterial food poisoning includes both illnesses caused by the ingestion of toxins elaborated by such organisms (intoxications) and illnesses resulting from infection of the host with the organism via their intestinal tract [162]. The two most important bacterial food intoxications are botulism, which is caused by toxin produced by *Clostridium botulinum*, and staphylococcal intoxication, which is due to toxin from *Staphylococcus aureus* [162]. Food infections can be divided into two types, those in which the food merely carries the organism, and those in which the food acts as a culture medium, allowing a pathogenic microorganism to multiply to an infectious level [162]. However, this division between intoxications and infections is not straightforward, as some of the pathogens growing in foods produce toxins there. The severity of risk presented by any pathogen depends on a combination of factors, namely, the pathogen itself, the susceptibility to the disease of the individual consumer, and on the dose of the exposure. Viruses neither multiply, nor do they produce toxins in foods. Instead, they are transmitted to the victim via foods from the intestines of fellow humans [253]. The two main foodborne virus infections in the UK are viral gastroenteritis, which is due to small, round-structured viruses of the Norwalk group (Norwalk-like viruses, NLV) and, less frequently, hepatitis A [253]. The most clearly implicated foods in the transmission of viruses are the bivalve moluscs, e.g., oysters, clams, cockles and mussels [253]. These concentrate viruses from sea water. In other outbreaks of viral gastroenteritis, prepared salads, buffet meals and sandwiches are often implicated, with contamination usually arising from food handlers [253]. Contamination can also be due to ineffectively sanitised fruit and vegetables [254].

In the UK, the extent of the public health problem posed by foodborne pathogens is increasing annually, with *Campylobacter*, *Salmonella* and *E. coli* the leading causes of foodborne bacterial infections [255]. Although entero-virulent *Escherichia coli* are also important, the problem associated with them is numerically small [256]. Bacteria responsible for common food poisoning, and the foods with which they are associated, are listed in Table 6.12 [257].

The following human pathogens may also be foodborne: *Aeromonas* species (shellfish), *Brucella abortus* (milk), *Coxiella burnetti* (milk), *Listeria monocytogenes* (milk, milk products, meat pâtés), *Shigella* species, *Vibrio cholerae* (shellfish), and *Yersinia enterocolitica* (pork, pork products) [258]. Pathogenic *E.coli* belong to a group of microorganisms commonly known as coliforms, not all of which are pathogenic. These coliforms have long performed a useful function in food analysis, i.e., as indicator organisms for the presence in a food of faecal contamination. This represents a warning that pathogens may be present in an affected food. Raw cows' milk may carry significant numbers of these coliforms, and it may also be contaminated with food poisoning pathogens [259]. *Bacillus cereus* is a spore former, whose spores are common contaminants of rice and other cereals. The spores vary in terms of their heat resistance, and they may be able to survive domestic cooking conditions. The optimal growth temperature for *B. cereus* is 30 °C, with a minimum at 10 °C and a maximum of 49 °C [162]. Extremely large numbers or the organims, i.e., 10^8 per g of food, need to be ingested

Table 6.12. Common food poisoning bacteria and the foods with which they are associated [257]

Food poisoning bacteria	Associated foods
Bacillus cereus	Rice, pasta, cereals
Campylobacter species	Poultry, cooked meats, milk and milk products
Clostridium botulinum	Preserved foods, canned, vacuum packed
Clostridium perfringens	Stews, rolled roasts, pies
Salmonella species	Poultry, eggs, meats and products made from them
Staphylococcus aureus	Cooked meats, poultry
Escherichia coli (VTEC)	Meat and meat products, milk and milk products
Vibrio parahaemolyticus	Shellfish

for symptoms to develop in humans [162]. Campylobacteriosis is a gastrointestinal infection caused by *Campylobacter jejuni* and is currently the commonest cause of food poisoning in the UK [260]. Campylobacters occur widely as part of the normal intestinal flora of many warm-blooded animals, including chickens and turkeys, and enter the human food chain during the slaughter of animals. Nearly half of raw chickens on sale in England and Wales (and almost three-quarters in Scotland and Northern Ireland) are contaminated with the organism [260]. *C. jejuni* is highly virulent, i.e., it causes infection at low numbers; it is also invasive [162]. It is inactivated when a food containing it is cooked and, in any case, does not grow readily in food [260]. Control measures for this organism should therefore focus on the prevention of both contamination and cross-contamination.

Foodborne botulism is an intoxication, which is associated with the consumption of a food in which *C. botulinum* has grown and produced toxin. The severity of any poisoning episode depends on the amount of toxin consumed. Botulinal toxins bind to the receptors on nerve endings and block the release of the neurotransmitter acetylcholine at the neuromuscular junction [261]. Death is usually due to asphyxia caused by muscular paralysis. Botulinal toxins are simple proteins and generally heat-labile. There are seven types of *C. botulinum*, based on the antigenic specificity of their toxin, namely, A-G. These differ in their tolerance to salt and α_w, in their minimum growth temperature, and in the heat resistance of their spores [261]. There are four groups (I-IV) of *C. botulinum* based on metabolic and serological similarities [261]. Spores of *C. botulinum* are widely distributed in the soil, in lakes and coastal waters; in addition, fruit and vegetables, and the intestinal tract of fish and animals are often in contact with these spores [261]. Non-proteolytic strains of *C. botulinum* do not necessarily render a food unpalatable. These are particularly dangerous in the sense that they do not give any warning during the consumption of a contaminated food. However, the risk of botulinal toxin production is present only in low-acid foods, under anaerobic conditions. Increasing nitrite, salt, or heat treatment, adding isoascorbate, polyphosphate or nitrate, or reducing storage temperature can significantly reduce toxin production due to *C. botulinum* [262]. In refrigerated, minimally heat processed foods, which are produced with the use of only mild, or sometimes no, preservatives, e.g., cook-chill and sous-vide (foods which are vacuum packed before pasteurization), there is concern that *C. botulinum* may be emerging as a significant pathogen [263]. However, a major outbreak of foodborne botulism in 1989 was associated with contaminated hazelnut yoghurt. This followed a reformulation of the product, in which the sugar in the hazelnut conserve had been replaced by aspartame [264]. Food poisoning outbreaks due to *Clostridium perfringens* are usually associated with fresh meat that has only been lightly cooked, and has subsequently been stored under poor refrigeration. This enables some clostridial spores to survive heating, as well as allowing any surviving spores to germinate and grow rapidly in the stored meat [265].

More than 1300 species of *Salmonella* are known. *Salmonella* infections associated with food may be caused by any of a large number of serovars (strains of different antigenic complements) [162]. Usually the infecting bacterium has grown in the food to attain high numbers, increasing the likelihood of infection, and of mass outbreaks, but infectivity varies among species [162]. Minimal temperatures for growth in foods range from 6.7 °C to 7.8 °C, and salmonellae grow well at room temperatures [162]. The intestinal tracts of humans and animals are, directly or indirectly, the source of the contamination of foods with salmonellae. People can be carriers without exhibiting any symptoms of salmonellosis themselves. Whilst the overall incidence of human salmonellosis has levelled out, illness caused by *S. typhimurium* DT 104 is gradually increasing [266]. The unique feature of this particular strain is that it is resistant to many commonly used antibiotics. The organism has been found in UK beef, pork, chicken and cereal and a supermarket's own brand of salami [266].

Food intoxication due to *Staphylococcus aureus* is caused by the enterotoxin which is formed in food during growth of certain strains of the organism. The toxin is called an enterotoxin because it causes gastroenteritis or inflammation of the lining of the intestinal tract [162]. The temperature range for both growth and toxin production is about 4 °C to 46 °C, depending on the food. Staphylococci are normally present on human skin, and the food poisoning strains generally come from human sources. Pasteurisation of a food destroys the organism, but not its toxin. However, about 75 % of staphylococcal food poisoning outbreaks occur because of the inadequate cooling of foods [162]. *Escherichia coli* constitute part of the normal human gut microflora, and most strains do not cause disease. Among the minority that do, there is a group called EPEC (enteropathogenic *E. coli*), and this includes *E. coli* O157:H7. These are distinguished by their ability to adhere to intestinal tissues in a particular manner [267]. EPEC may have acquired genetic material that turned a limited pathogen into a serious one, transferring upon it the ability to produce a verocytotoxin; transmission of bacterial genes from one strain to another via bacterial viruses (phages) is a well known phenomenon [267]. *E. coli* O157 is a highly virulent organism, with a low infectious dose [268]. Danger foods include undercooked burgers, and any food that has been contaminated with the organism, including apple juice [269]. Infections with *Vibrio parahaemolyticus* cause gastroenteritis and are particularly common in Japan [162]. The organism is a so-called halophile, requiring 1–3 % salt in its growing medium, and it grows over a temperature range of 10 °C to 44 °C [162]. It is a marine organism that can readily be killed by proper cooking of seafood.

Listeria monocytogenes and *Yersinia enterocolitica* are pathogenic bacteria that have the ability to grow at refrigeration temperatures, i.e., they are psychrotrophs. Psychrotrophs demand particular attention in view of the growing market for chilled foods and the tendency for such foods to spend increasing periods in storage and distribution. Both organisms are inactivated in food during cooking. Listeriosis is a severe disease, manifested in meningitis, abortion and septicaemia, and with a high case-fatality rate (20–30 %); however, host susceptibility plays a major role in this [270]. The foodborne transmission of listeriosis was first demonstrated in 1981 during an outbreak in Canada which was caused by coleslaw; in a second outbreak, pasteurised milk (post-pasteurisation contamination) was implicated [270]. Subsequently, pâtés and cheeses have been associated with the organism. However, many strains of *L. monocytogenes* are harmless to human health. Typing results of *L. monocytogenes* strains have shown that most strains accounting for human infection differ in certain respects from strains typically found in food [270]. *Listeria* are ubiquitous and may be found in soil, on vegetation, in water, in sewage and in the faeces of humans and animals [271]. With the exception of pregnant women, it is extremely unusual for healthy adults and children to become infected [271]. *Yersinia* may be transmitted via contaminated food, in particular, raw pork and pork products, via water, and through direct contact with contaminated persons or animals [272]. Symptoms include diarrhoea and abdominal pain.

Mycotoxins are metabolic products of moulds, which are secreted into a wide range of agricultural crops, both human foods and animal feeds. Moulds also grow on meat surfaces and in man-made foods. The main mycotoxins of interest here are aflatoxins, ochratoxin A (OA), fumonisins, patulin, moniliformin, sterigmatocystin, tricothecenes and zearalenone [273]. Aflatoxins, produced predominantly by *Aspergillus flavus* and *Aspergillus parasiticus*, are acutely, and chronically, toxic and are recognised as human carcinogens [273]. They are found in a wide range of foodstuffs including nuts, cereals, dried fruit and milk. Aflatoxins are relatively heat-stable, although roasting will partially degrade them – depending on type (B1, B2, G1, G2 and total) (K. Aidoo, 2000, personal communication). A range of fungal species from the genera *Aspergillus* and *Penicillium* are able to produce OA [273]. Contamination is most commonly associated with cereals, pulses and coffee. The toxin is associated with a chronic, progressive kidney disease [273]. Patulin is produced by a number of *Penicillium* and *Aspergillus* species. Of these, *P. expansum* is noted for its synthesis of patulin in decaying apples [273]. The presence of patulin in apple juice is generally attributed to the use of mouldy fruit, and it may also occur as a contaminant of products derived from such fruit. It is considered to be mutagenic [273]. Fumonisins are an increasingly important new group of mycotoxins and were first isolated in 1988 from *Fusarium moniliforme*, which is associated with contaminated maize [273].

Histamine belongs to a groups of chemicals referred to as biogenic amines. Histamine poisoning is a foodborne intoxication caused by foods that contain unusually high levels of histamine as a result of bacterial L-histidine decarboxylation [274]. Freshly caught sardines contain high levels of bacteria located mainly on the skin and in the gills; these invade and grow rapidly in sardine muscle. The high level of free histidine in sardines, and the susceptibility of its muscle to histamine and cadaverine formation, may explain its increasing implication in incidents of histamine poisoning [274]. Historically, histamine poisoning has been referred to as scombrotoxic fish poisoning (histamine, scombrotoxin) because of its association with the consumption of scombroid fish, e.g., tuna and mackerel. Basically, histamine production in fish is the result of temperature abuse in the handling of the fish after capture [275]. Another significant source of biogenic amines in foods is raw fermented sausages. In these, poor fermentation control can lead to the formation of amines such as tyramine, putrescine and cadaverine [276].

The main agents of parasitic infections transmitted by foods are *Anisakis* species (a nematode), *Endamoeba histolytica*, *Taenia saginata* (beef tapeworm),

Diphyllobothrium latum (fish tapeworm), *Taenia solium* (pork tapeworm) and *Trichinella spiralis* [162]. Human trichinosis is characterised by nausea, diarrhoea and related symptoms, and usually results from the consumption of raw or incompletely cooked pork containing the encysted larvae [162]. *Anisakis simplex*, common in the guts of fish, produces a protein that can cause anaphylactic shock, a potentially lethal allergy, even when the organism itself has been killed during food processing [277]. Both *Anisakis simplex* and cestodes (*Diphyllobothrium* species) are of public health signifiance in the USA [275]. A number of helminths (trematodes, nematodes, and cestodes) are present in finfish and shellfish; however, the majority of them present no risks to humans [275]. Parasited fresh herring already harbours 2.4–4% of parasites at the time of capture, with a mean of 0.25–0.33 nematodes in the flesh of each herring; higher values have been found in mackerel (5%) and saithe (10%) [278]. Such fish are therefore said to require processing for safety, i.e, adequate freezing, heating, salting, or pickling [278]. The most important waterborne and foodborne protozoan parasites potentially infecting humans are *Cryptosporidium parvum, Giardia intestinalis (lamblia), Sarcocystis* species and *Toxoplasma gondii* [279]. *C. parvum* is an enteric pathogen with a world-wide distribution, endemic in some areas and with infection rates highest in developing countries and in children [280]. In severely immunocompromised patients, e.g., AIDS sufferers, cryptosporidiosis may become chronic and serious, and sometimes fatal [280]. *C. parvum* cannot grow in food, but oocysts will survive in wet, or moist, foods. Raw milk, raw sausages and offal are the most likely foods to be contaminated; fruit and vegetables could be at risk if in contact with manure or contaminated water; cooked foods are not thought to be a risk [280]. Recently, a large outbreak of diarrhoeal illness in the USA was diagnosed as an infection with *Cyclospora cayetanensis,* a coccidian parasite believed to have been associated with fresh fruit [279].

Four major shellfish toxic syndromes share a common aetiology, i.e., the ingestion of toxins produced by marine microalgae [281]. Many algal species survive as cysts in the winter, but will form surface blooms under certain conditions in the summer [281]. Algal toxins can become concentrated in cockles, mussels, scallops, clams, oysters, crabs and lobsters, but do not taint them so that there is no warning about any contamination that may be present [281]. Each toxic syndrome is linked to specific algal species. Paralytic shellfish poisoning (PSP) is characterised by oral, then digital numbness or tingling, limb weakness, a floating sensation and respiratory distress [281]. The three known families of PSP toxins are all water-soluble and heat stable [281]. Diarrhetic shellfish poisoning (DSP) toxins have less severe effects, i.e., diarrhoea, nausea and vomiting [281]. The associated toxins are regularly detected in shellfish harvested in UK fishing waters [281]. Amnesic shellfish poisoning (ASP) causes both gastrointestinal and neurological symptoms, including a loss of short-term memory [281]. ASP occurs in scallop fisheries on the Scottish West coast. Unlike PSP and DSP, toxin levels remain high in the scallops after algal blooms subside [282]. Neurotoxic shellfish poisoning (NSP) includes numbness or tingling in the mouth, progressing to the extremities, and the hot to cold temperature reversal phenomenon, which affects the perception of hot and cold surfaces [281]. The algae responsible for this are known to produce Florida's famous red tides [281]. Ciguatera is a gastrointestinal and neurological disease following the consumption of certain tropical reef fish, e.g., snappers and barracuda, the original source of the toxin being dinoflagellate algae [275]. The toxins are not affected by cooking [281].

6.7
Chemical and Foreign Matter Contamination

Any contamination of foods with chemical substances or foreign matter may be either adventitious or deliberate. Environmental chemicals may contact foods through the air or via food contact materials, e.g., packaging. Various kinds of chemicals are applied to food deliberately, and these take on the official mantle of contaminants only if their presence exceeds permitted levels. This is the case with crop protection agents, veterinary medicines, food additives and the like. Occasionally, a food is contaminated maliciously with either chemicals or foreign bodies (tampering). Chemical contamination also takes place in a wider sense where food is adulterated to defraud consumers. Commission Regulation (EC) No. 194/97 sets limits for certain contaminants in foodstuffs and is enforced in the UK by the Contaminants in Foods Regulations 1997 [283].

Conventionally produced food crops are grown with the use of chemical fertilisers, and may, in addition, be treated with a range of pesticides, in particular, herbicides, fungicides and insecticides. Although nitrate is naturally present in vegetables, levels increase with the use of nitrogen fertilisers [3]. Green, leafy vegetables contain the highest concentration of nitrate and are the major source of nitrate in the diet. Nitrate can be metabolised in the stomach to form N-nitroso compounds such as N-nitrosamines and N-nitrosamides, which have been implicated in gastric cancers [3]. A proportion of ingested nitrate may be converted to nitrite, which may contribute to infantile methaemoglobinaemia, where the nitrite reacts with haemoglobin thereby impairing its oxygen-carrying capacity [94]. The European Commission has set maximum levels for nitrate in lettuce and spinach [283]. Pesticides are in common use all over the world. In the UK, some 300 active ingredients are currently on the market [284]. Also in the UK, once ministerial approval has been given for the use of a pesticide, it is the responsibility of the Pesticides Safety Directorate (PSD) to register it and control its advertisement, sale, supply storage and use [284]. It is because pesticides are highly bioactive that they pose potential risks to people and to the wider environment. The relevant regulations contain maximum residue levels (MRL) for over 6000 pesticide/commodity combinations, and cover the more important components of the UK diet [284]. There is an annual surveillance of 2000–3000 samples of food to establish whether MRL are being exceeded [284]. In a recent EU survey of pesticide residues on fruit, vegetables and cereals, 2% of UK produce exceeded MRL [285]. However, when local PSD surveillance data were analysed by Friends of the Earth, a green lobby group, it was revealed that nearly half of fruit and vegetables sold by UK retailers contained pesticides residues, albeit mostly at permitted levels [286]. Several crops have been highlighted as problem areas at different times, e.g., carrots in 1995. In that case, the findings led to the official advice that consumers should peel carrots for safe use

[287]. More recently, pears were shown to have been contaminated with chlormequat, a so-called growth regulator, which can improve the shape, size and yield of pears, but one that is not approved for fruiting pear trees in the UK [288]. Imported pears are supposed to comply with the UK's statutory MRL [288].

One of the features of the intensive rearing of animals is their high level of exposure to veterinary drugs [289]. For example, antibiotics may be used not only to treat existing infections, but also as a preventive measure. Where they are used prophylactically in this way, antibiotics are commonly referred to as growth promoters. Such routine use of antibiotics raises the spectre of the development of antibiotic-resistance in human pathogens, which may lead to difficulties in treating infections in humans [290]. Other animal medications and production chemicals that may cause food to become contaminated include wormers, udder medications and sanitisers for udders [289]. Another issue in animal-derived food production concerns hormones. Hormonal growth promoters include 17β oestradiol and its derivatives, testosterone, progesterone, trenbolone acetate, zeranol and melengesterole acetate [291]. As their use in meat production has been banned in the EU since 1988, countries which permit the use of growth hormones are required to guarantee that no animals and no meat coming from animals to which they have been administered will be exported to the EU [291]. Bovine somatotropin (BST) is a proteinaceous hormone naturally produced by all cows which, when injected into an animal, can increase the milk yield by minimising the rate of yield decline after peak lactation [292]. Modern biotechnology has resulted in the development of a recombinant BST (rBST) [292]. BST treatment of dairy cows has been routine in the USA since 1994, but not in the UK [292].

The most immediate source of environmental contaminants in foods lies in food contact materials, the main categories of which are processing plant, food-coating substances, and food packaging or wrapping materials. Contamination of a food with chemicals from processing plant may be due either to detergent or sanitiser residues, or to the inappropriate use of lubricants. Chemical migration of food coating or packaging chemicals into a food is a function of the compatibility of the two materials. Plastics are of special concern as they may contain low-molecular-weight additives that can contaminate foods under certain conditions. The control of chemical migration from food contact plastics is addressed in the Plastic Materials and Articles in Contact with Food Regulations 1998 [293]. However, paper and boards are possible sources of chemical contaminants too. Carton board for food packaging may have surface coatings for printing and general appearance. These coatings usually consist of a filler and a synthetic resin binder, e.g., styrene/butadiene or styrene/acrylate copolymers [294]. Where styrene/butadiene-type binders have been used, a number of volatile substances present as by-products from the polymerisation process have been identified; of these, 4-phenyl cyclohexene has a particularly strong odour, which is reminiscent of synthetic latex [294]. In the past, there have been a number of incidents of food taint contamination from paper/board packaging which have originated from chlorophenols; these have been identified as by-products of chlorine bleaching of wood pulp [294]. Cans are potential sources of contamination of a food with metals. Research on the levels of tin in canned tomatoes in 1998 revealed some products in which concentrations exceeded statutory limits, and these had to be withdrawn from sale [295]. The leaching of metals into foods is a function of the aggressiveness of the foods, e.g., their acidity [296].

Other chemical contaminants enter food from the environment generally, i.e., through the air and through water. Heavy metal contamination of foods can originate from traffic fumes or as industrial pollution. Marine and other aquatic food animals also pick up heavy metal contamination in significant quantities. Contamination of foods with lead, arsenic, cadmium, mercury and other metals is subject to regular governmental surveillance. Dioxins and polychlorinated biphenyls (PCB) are persistent environmental pollutants, traces of which are present in many food sources, particularly fatty foods, including milk and breast milk [297]. Chemical manufacturing plant and waste incinerators are the main sources of dioxins. Dioxin is a generic term for polychlorinated dibenzo-p-dioxins and dibenzofurans, with 2,3,7,8-tetrachlorodibenzo-p-dioxin (TCDD) of particular concern in terms of its level of toxicity [297]. Both dioxins and PCB are bio-accumulated as they proceed through the food chain, eventually forming significant concentrations in the fatty tissues of land animals and fish [297]. The main source of human exposure to these compounds therefore is through the consumption of fatty foods such as meat, fish, milk and dairy products. So-called oily fish, e.g., herring and salmon, are especially affected by build-up of dioxins and PCB [298]. In 1999, Belgian animal feed mills supplied contaminated products to customers throughout the EU farming industry, causing widespread contamination with dioxins and PCB [299]. Among the products affected were poultry, eggs, pigs and dairy products. The UK government also regularly surveys foods for the presence of radioactive nuclides [300].

A number of undesirable, potentially harmful substances may be formed in foods as process by-products. For example, polycyclic aromatic hydrocarbons (PAH) are produced in foods heated to very high temperatures [301]. Some of these, especially 3,4-benzopyrene and benzanthracene, are carcinogenic. Wood smoke contains varying amounts of PAH which, if the smoke is used to preserve or flavour a food, may be absorbed by that food [301]. The formation of PAH is minimised by removing as much fat as possible and by not charring meat. Similarly, the nitrites used in cured meat products can lead to the appearance of carcinogens in these foods [302]. This is because nitrites undergo nitrosation reactions with secondary amines in the meat, especially at high temperatures. The best known of the N-nitrosamines is N-nitrosopyrrolidine, its presence in fried bacon having been established in the early 1970s as frequently lying above 100 ppb [302]. Nitrosamine formation is particularly associated with the fattier cuts of bacon and with elevated levels of polyunsaturated fatty acids in pig fat [302]. Beer is another product that may contain nitrosoamines, this being linked to the malt kilning process [302]. Yet another carcinogen that may be present in food due to processing is 3-MCPD (3-monochloropropane-1,2-diol). It is formed during the manufacture of acid-hydrolysed vegetable protein (acid-HVP) [303]. It has been found in a number of soy sauce products for sale in the UK [303].

The food processing industry has been confronted by a new threat: increasing consumer lawsuits stemming from foreign objects found in food [304]. Small pieces of metal and glass can cause significant injury to a consumer. Metal shavings can be created as a result of wear in processing plant. Kellogg Company recalled toy cars that were packed inside some cereal boxes; tires had been found to detach from the cars, posing a choking risk for small children [305]. All manner of foreign matter can gain access to foods accidentally. For example, foreign matter associated with nuts includes fibres, hair and skins (of nuts), nails, wood splinters and sand, stones glass and broken shells [306].

Adulteration means to "render something poorer in quality by adding another substance, typically an inferior one" [307]. A concept closely related to adulteration is authenticity; this means that something is genuine and of undisputed origin [307]. Food fraud has a long history and generally, unless it is suspected to take place in the first place, remains undetected. This is because analytical methods are typically designed, and set up, to detect specific compounds. In 1990, 16 out of a total of 21 samples of orange juice taken during governmental food surveillance were found to have been extended illegally by companies who had added sugar, malic acid and/or extracts from the washing of orange pulp to the juice [308]. A survey of meat products, published in 1999, revealed that 15% of samples contained varying amounts of species of meat, which had not been declared on the food label [308]. Fraud raises serious public health concerns. In 1985, the Viennese health authorities impounded large quantities of white wine, which had been adulterated with diethylene glycol to increase its viscosity [309]. Diethylene glycol can cause kidney failure and death [309]. Adulteration of olive oil with cheaper oils is a temptation that can become hazardous, as in the case of the Spanish Toxic Syndrome in 1981, which was the result of olive oil having been adulterated with aniline-containing rapeseed oil [310]. Counterfeit spirits present more of an everyday hazard in that many will contain methylated spirits, which could cause blindness [311]. Added water in meat products is another issue. In 1997, 16 out of a total of 614 ham, bacon and gammon products samples from major UK food retailers were found to contain significant amounts of undeclared added water [312]. In a survey published in 2000, almost a third of frozen whole chickens exceeded the (7%) limit for added water [313].

One of the biggest stories concerning food contamination in recent years has been that of BSE. First emerging in UK cattle in 1986, the disease was officially linked, after ten years of uncertainty, with the human brain disorder vCJD [314]. It is difficult to classify the disease agent: is it biological or chemical in nature, is it generated in the body or introduced from the outside? The answer probably is a little of all of these. The predominant scientific hypothesis concerning the origins of BSE has been that the disease is caused by infectious rogue proteins called prions [315]. An infective prion is one whose shape has become distorted and when one of these reaches the brain cell membrane of a host, it causes normal prions to adopt a similarly distorted shape [315]. The appearance of BSE in UK cattle is generally linked with them having been fed commercial cattle feed concentrates, which contained meat and bone meal derived from sheep presumed to have been infected with scrapie which, like BSE and vCJD, is a prion disease.

Conclusion

As has been outlined in Part 1, the quality attributes of a food are initially identified, from among the multitude of possible product attributes, by using subjective filters. However, once the parameters of quality have been established (quality of design), quality of conformity becomes the focus of attention. These selected attributes must be set into a framework of verifiable specifications. Part 2 has provided an overview of a range of aspects of food science and technology as they relate to food quality. The initial focus has been on the native components of edible biological tissues, e.g., fruits, vegetables, meat, and so on. These have been shown to be able to adopt multiple functionality. For example, structural proteins provide textural qualities to a food. They also serve as dietary fuels and, as far as the indispensable amino acid (IAA) content of a protein is concerned, as nutrients. The structures in edible biological tissues tend to be complex and highly organised, and dependent on the specific, native food components present. However, in man-made, compounded foods, structure needs to be newly created. Both physical processing and the use of functional food ingredients, including additives, may play a part in this. Examples of the physical generation of structures include the homogenisation of emulsions and the extrusion of macromolecules into snack foods. In addition, some of the naturally protective mechanisms that exist in biological tissues have to be mimicked in man-made food structures. This includes preservative and antioxidant actions. Thus vitamins C and E may be present naturally in, or they may be added to, a food. Certain beneficial substances are also added to man-made foods in order to increase some aspect of their value to a consumer. This is the case with nutrient-fortified foods and functional foods. Biological structures, once removed from their life support systems, are unstable. This is one of the reasons why many foods need to be processed in order to increase their keeping quality. Food safety is paramount among food quality attributes. Whilst grossly contaminated food is unsafe for the public at large, in many instances, safety is a subjective issues, e.g., food allergies, and individual susceptibilities to *Listeria monocytogenes*. Food additives, functional ingredients and food contaminants are food components that are not present in the original biological tissue, but are added to a food either deliberately or accidentally. Larger-scale production demands increased precision in all aspects of process control, and therefore, purer ingredients with more closely defined functionality. A schematic change occurs when food additives exceed their legally permitted level in a particular food. In such a case, additives are, perceptionally, transformed into contaminants.

References

1. Bender AE, Bender DA. *A Dictionary of Food and Nutrition*. Oxford: Oxford University Press, 1995
2. Sadler MJ, Saltmarsh M (ed). *Functional Foods. The Consumer, the Products and the Evidence*. Cambridge: The Royal Society of Chemistry, 1998
3. Anon. *Research Requirements Document*. London: Food Standards Agency; Issue 5, July 2001
4. HMSO. *The Miscellaneous Food Additives Regulations 1995* (S.I. 1995 No. 3187). London: HMSO

5. Anon. Antibiotics in Food. Which? Magazine 1997; March: 18–20
6. Which? E numbers on eggs? Seriously... *Which? Inside Story* May 1990
7. Aguilera JM, Stanley DW. *Microstructural Principles of Food Processing and Engineering.* 2nd edition. Gaithersburg: Aspen Publishers, 1999
8. Larsson K. *Lipids. Molecular Organization, Physical Functions and Technical Applications.* Dundee: The Oily Press, 1994
9. Pitchford P. *Healing with Whole Foods: Oriental Traditions and Modern Nutrition.* Revised edition. Berkeley: North Atlantic Books, 1993
10. Tung AM, Paulson AT. Rheological Concepts for Probing Ingredient Interactions in Food Systems. In: Gaonkar AG (ed). *Ingredient Interactions. Effects on Food Quality.* New York: Marcel Dekker, pp 45–83, 1995
11. Roos YH. *Phase Transitions in Foods.* San Diego: Academic Press, 1995
12. Pomeranz Y. *Functional Properties of Food Components.* 2nd edition. San Diego: Academic Press, 1991
13. Hedley CL, Bogracheva TY, Lloyd JR, Wang TL. Manipulation of Starch Composition and Quality in Pea Seeds. In: Fenwick GR, Hedley C, Richards RL, Khokar S (ed). *Agri-Food Quality. An Interdisciplinary Approach.* Cambridge: The Royal Society of Chemistry, pp 138–148, 1996
14. Potter NN, Hotchkiss JH. *Food Science.* 5th edition. New York: Chapman & Hall, 1995
15. Penfield MP, Campbell AM. *Experimental Food Science.* 3rd edition. San Diego: Academic Press, 1990
16. Fox BA, Cameron AG. *Food Science, Nutrition & Health.* 6th edition. London: Edward Arnold, 1995
17. Martin-Cabrejas MA, Selvendran RR, Waldron KW. Ripening-related Changes in Cell Wall Chemistry of Spanish Pears. In: Fenwick GR, Hedley C, Richards RL, Khokar S (ed). *Agri-Food Quality. An Interdisciplinary Approach.* Cambridge: The Royal Society of Chemistry, pp 212–215, 1996
18. Edwards M. Change in cell structure. In: Beckett ST (ed). *Physico-Chemical Aspects of Food Processing.* London: Blackie Academic & Professional, pp 212–233, 1995
19. Van Marle N, Biekman E, Ebbelaar M, Recourt K, Yuksel D, van Dijk C. The Texture of Processed Potatoes; From Genes to Textures. In: Fenwick GR, Hedley C, Richards RL, Khokar S (ed). *Agri-Food Quality. An Interdisciplinary Approach.* Cambridge: The Royal Society of Chemistry, pp 181–183, 1996
20. Waldron KW, Parker ML, Parr A, Ng A. Chinese Water Chestnut: The Role of Cell Wall Phenolic Acids in the Thermal Stability of Texture. In: Fenwick GR, Hedley C, Richards RL, Khokar S (ed). *Agri-Food Quality. An Interdisciplinary Approach.* Cambridge: The Royal Society of Chemistry, pp 216–221, 1996
21. Hart R. Technology and food production. *Nutrition & Food Science* 1997; (2): 53–57
22. Patil SK. Application of Starches in Foods – Part 2. *Food Technology International Europe.* London: Sterling, pp 25–29, 1991
23. Lamberg I, Olsson H. Starch gelatinization temperatures within potato during blanching. *International Journal of Food Science and Technology* 1989; 24: 487–494
24. Dalgleish DG, Hunt JA. Protein-Protein Interactions in Food Materials. In: Gaonkar AG (ed). *Ingredient Interactions. Effects on Food Quality.* New York: Marcel Dekker, pp 199–233, 1995
25. Stryer L. *Biochemistry.* 2nd edition. New York: W.H. Freeman and Company, 1981
26. Streitwieser A, Heathcock CH. *Introduction to Organic Chemistry.* 3rd edition. New York: Macmillan Publishing Company, 1985
27. Taylor RJ. *The Chemistry of Proteins.* A Unilever Educational Booklet, 1972
28. MLC. *Overcoming the safety problem of aitch bone hanging.* Milton Keynes: Meat and Livestock Commision. Leaflet undated (ca. 1998)
29. Lawrie RA. *Lawrie's Meat Science.* 6th edition. Cambridge: Woodhead Publishing, 1998
30. Grigson J. *English Food.* London: Penguin Books, 1977
31. Lavéty J, Afolabi OA, Love RM. The connective tissues of fish. IX. Gaping in farmed species. *International Journal of Food Science and Technology* 1988; 23: 23–30

32. Hastings RJ. Comparison of the properties of gels derived from cod surimi and from unwashed and once-washed cod mince. *International Journal of Food Science and Technology* 1989; 24: 93–102
33. Shewry PR, Tatham AS, Greenfield J et al. Manipulation of Wheat Protein Quality. In: Fenwick GR, Hedley C, Richards RL, Khokar S (ed). *Agri-Food Quality. An Interdisciplinary Approach*. Cambridge: The Royal Society of Chemistry, pp 117–123, 1996
34. Shewry PR, Broadhead J, Fido R et al. Patterns of Hordein Synthesis and Deposition in Relation to Barley Malting Quality. In: Fenwick GR, Hedley C, Richards RL, Khokar S (ed). *Agri-Food Quality. An Interdisciplinary Approach*. Cambridge: The Royal Society of Chemistry, pp 124–126, 1996
35. Anon. An attractive alternative. *Food Manufacture*. May 1994: 25–26
36. Varnam AH, Sutherland JP. *Milk and Milk Products*. London: Chapman and Hall, 1994
37. Dalgleish DG. Structures and Properties of Adsorbed Layers in Emulsions Containing Milk Proteins. In: Dickinson E, Lorient D (ed). *Food Macromolecules and Colloids*. Cambridge: The Royal Society of Chemistry, pp 23–33, 1999
38. Shimizu M. Structure of Proteins Adsorbed at an Emulsified Oil Surface. In: Dickinson E, Lorient D (ed). *Food Macromolecules and Colloids*. Cambridge: The Royal Society of Chemistry, pp 34–42, 1999
39. Van Aken GA. A Phenomenological Model for the Dynamic Interfacial Behaviour of Adsorbed Protein Layers. In: Dickinson E, Lorient D (ed). *Food Macromolecules and Colloids*. Cambridge: The Royal Society of Chemistry, pp 43–49, 1999
40. Leaver J, Horne DS, Davidson CM, Brooksbank D. Influence of Charge on the Adsorption of Proteins to Surfaces. In: Dickinson E, Lorient D (ed). *Food Macromolecules and Colloids*. Cambridge: The Royal Society of Chemistry, pp 90–94, 1999
41. Renard D, Axelos MAV, Lefebvre J. Investigation of Sol-Gel Transitions of β-Lactoglobulin by Rheological and Small-angle Neutron Scattering Measurements. In: Dickinson E, Lorient D (ed). *Food Macromolecules and Colloids*. Cambridge: The Royal Society of Chemistry, pp 390–399, 1999
42. Howell NK, Lawrie RA. Functional aspects of blood plasma proteins. II. Gelling properties. *International Journal of Food Science and Technology* 1984; 19: 289–295
43. Zayas JF. *Functionality of Proteins in Food*. Berlin Heidelberg: Springer-Verlag, 1997
44. Le Meste M, Simatos D, Gervais P. Interaction of Water with Food Components. In: Gaonkar AG (ed). *Ingredient Interactions. Effects on Food Quality*. New York: Marcel Dekker, pp 85–129, 1995
45. Herrington TM, Vernier FC. Vapour pressure and water activity. 1–16 In: Beckett ST (ed). *Physico-Chemical Aspects of Food Processing*. London: Blackie Academic & Professional, pp 1–16, 1995
46. Fellows PJ. *Food Processing Technology. Principles and Practice*. New York: Ellis Horwood, 1988
47. Talbot G. Fat eutectics and crystallisation. In: Beckett ST (ed). *Physico-Chemical Aspects of Food Processing*. London: Blackie Academic & Professional, pp 142–166, 1995
48. Pond CM. *The Fats of Life*. Cambridge: Cambridge University Press, 1998
49. Johansson D, Bergenståhl B, Lundgren E. Sintering of Fat Crystal Networks in Oils. In: Dickinson E, Lorient D (ed). *Food Macromolecules and Colloids*. Cambridge: The Royal Society of Chemistry, pp 418–425, 1999
50. Brady JP, Buckley, Foley J. Factors influencing the adhesion-cohesion forces between butter layers. *Journal of Food Technology* 1978; 13: 469–475
51. Tabouret T. Technical note: Detection of fat migration in a confectionery product. *International Journal of Food Science and Technology* 1987; 22: 163–167
52. Beckett ST. Chocolate confectionery. In: Beckett ST (ed). *Physico-Chemical Aspects of Food Processing*. London: Blackie Academic & Professional, pp 347–367, 1995
53. Bergenståhl B. Emulsions. In: Beckett ST (ed). *Physico-Chemical Aspects of Food Processing*. London: Blackie Academic & Professional, pp 49–64, 1995

54. Wijnen ME, Prins A. Disproportionation in Aerosol Whipped Cream. In: Dickinson E, Lorient D (ed). *Food Macromolecules and Colloids*. Cambridge: The Royal Society of Chemistry, pp 309–311, 1999
55. Bee RD, Birkett RJ. Reflectance Studies on Ice-Cream Models. In: Dickinson E, Lorient D (ed). *Food Macromolecules and Colloids*. Cambridge: The Royal Society of Chemistry, pp 297–308, 1999
56. Department of Health. *Dietary Reference Values. A Guide*. London: HMSO, 1991
57. Benton D. *Food for Thought*. London: Penguin Books, 1996
58. Romano R. *Dining in the Raw*. New York: Kensington Books, 1992
59. Foster LH, Sumar S. Selenium in the environment, food and health. *Nutrition & Food Science* 1995; September/October (5):17–23
60. COMA. *Statements. Definition of Dietary Fibre for Nutrition Labelling Purposes*. London: NHS Executive, 1999
61. Clarke J. *Body Foods for Life*. London: Orion Publishing, 1999
62. IFST. Information statement. Dietary fibre. *Food Science & Technology* 2001; 15(3): 34–36
63. Gurr MI. *Role of fats in food and nutrition*. 2nd edition. New York: Elsevier Science, 1992
64. Gurr MI. Dietary Fatty Acids with Trans Unsaturation. *Nutrition Research Reviews* 1996; 9: 259–279
65. Holford P. *The Optimum Nutrition Bible*. London: Judy Piatkus (Publishers), 1997
66. BNF Information. *Vitamins*. British Nutrition Foundation, 1998. Retrieved on 6th July 2000 from http://www.nutrition.org.uk/Facts/energynut/vitamin.html
67. Finglas P. *Folic Acid – An Essential Ingredient in Making Healthy Babies*. Food Information Sheet 06.00. Norwich: Institute of Food Research, 2000
68. Institute of Medicine. *Dietary Reference Intakes for Vitamin C, Vitamin E, Selenium and Carotenoids*. Washington: National Academic Press, 2000
69. BNF Information. *Minerals*. British Nutrition Foundation, 1998. Retrieved on 6th July 2000 from http://www.nutrition.org.uk/Facts/energynut/mineral.html
70. Fairweather-Tait S. *Calcium and Bone Metabolism*. Food Information Sheet 06.00. Norwich: Institute of Food Research, 2000
71. Lesser M. *Nutrition and Vitamin Therapy*. Wellingborough: Thorsons, 1985
72. IFR. *Food Sources of Essential Minerals and Trace Elements*. Food Information Sheet 06.00. Norwich: Institute of Food Research, 2000
73. Fox T. *Selenium*. Food Information Sheet 06.00. Norwich: Institute of Food Research, 2000
74. Khetarpaul N, Chauhan BM. Effect of germination and pure culture fermentation on HCL-extractability of minerals of pearl millet (*Pennisetum typhoideum*). *International Journal of Food Science and Technology* 1989; 24: 327–331
75. Indumadhavi M, Agte V. Effect of fermentation on ionizable iron in cereals-pulse combinations. *International Journal of Food Science and Technology* 1992; 27: 221–228
76. Tomás-Barberán FA, Robins RJ. Introduction. In: Tomás-Barberán FA, Robins RJ (ed). *Phytochemistry of Fruit and Vegetables*. Oxford: Oxford University Press, pp 1–9, 1997
77. Costain L. *Super Nutrients Handbook*. London: Dorling Kindersley, 2001
78. Winkler JT. The Future of Functional Foods. In: Sadler MJ, Saltmarsh M (ed). *Functional Foods. The Consumer, the Products and the Evidence*. London: The Royal Society of Chemistry, 1998
79. Meletis CD, Bramwell B. Helping Patients to Age Well. *Alternative & Complementary Therapies* 2001: 27–31
80. Erasmus U. *Fats that Heal, Fats that Kill*. Burnaby BC Canada: Alive Books, 1993
81. Ahmad JI. Omega three fatty acids – the key to longevity. *Food Science and Technology Today* 1998; 12(3): 139–146
82. Clifford M. Are polyphenols good for you? *Food Science & Technology* 2001; 15: 24–28
83. Johnson I. Use of food ingredients to reduce degenerative diseases. In: Henry CJK, Heppell NJ (ed). *Nutritional Aspects of Food Processing and Ingredients*. Gaithersburg: An Aspen Publication, pp 136–165, 1998

84. Hertog MGL, van Poppel G, Veerhoeven D. Potentially anticarcinogenic secondary metabolites from fruit and vegetables. In: Tomás-Barberán FA, Robins RJ (ed). *Phytochemistry of Fruit and Vegetables*. Oxford: Oxford University Press, pp 313–329, 1997

85. Jackman RL, Smith JL. Anthocyanins and betalins. In: Hendry GAF, Houghton JD (ed). *Natural Food Colorants*. 2nd edition. London: Blackie Academic & Professional, 1996

86. Duthie GD, Duthie SJ, Kyle JAM. Plant polyphenols in cancer and heart disease: implications as nutritional antioxidants. *Nutrition Research Reviews*. 2000; 13: 79–106

87. Joubert E. Effect of batch extraction conditions on extraction of polyphenol from rooibos tea (*Aspalathus linearis*). *International Journal of Food Science and Technology* 1990; 25: 339–343

88. IFST. *Current Hot Topics. Phytoestrogens*. London: Institute of Food Science and Technology, 2001. Retrieved on 1st November 2001 from http://www.ifst.org/hottop34.htm

89. Britton G, Hornero-Méndez D. Carotenoids and colour in fruit and vegetables. In: Tomás-Barberán FA, Robins RJ (ed). *Phytochemistry of Fruit and Vegetables*. Oxford: Oxford University Press, pp 11–27, 1997

90. Hutchings JB. *Food Colour and Appearance*. London: Blackie Academic & Professional, 1994

91. Astley S. *Do Carotenoids Stimulate DNA Repair?* Science Brief. Norwich: Institute of Food Research, 2000. Retrieved on 11th July 2000 from http://ifrn.bbsrc.ac.uk/science/ScienceBriefs/carotenoids.html

92. Sambale C. New lycopene formulations. *Food Technology International* 2000: 12

93. Edens NK. Functional Foods. Representative Components of Functional Food Science. *Nutrition Today* 1999; 4: 152–154

94. D'Mello F. Toxic compounds from fruit and vegetables. In: Tomás-Barberán FA, Robins RJ (ed). *Phytochemistry of Fruit and Vegetables*. Oxford: Oxford University Press, pp 331–351, 1997

95. Sen CK, Roy S, Packer L. α-Lipoic Acid: Cell Regulatory Function and Potential Therapeutic Implications. In: Packer L, Hiramatsu M, Yoshikawa T (ed). *Antioxidant food supplements in human health*. San Diego: Academic Press, pp 111–119, 1999

96. Ministry of Agriculture, Fisheries and Food. *MAFF UK – Survey of Biologically Active Principles in Mint Products and Herbal Teas*. London: Ministry of Agriculture, Fisheries and Food. Food Surveillance Information Sheet Number 99, 1996

97. Food Science Australia. *Greening of potatoes*. Factsheet, February 1994

98. El Tinay AH, Bureng PL, Yas EAE. Hydrocyanic acid levels in fermented cassava. *Journal of Food Technology* 1984; 19: 197–202

99. Lloyd Parry R. Blowfish delicacy kills Japanese man. *The Independent*, 12th April 2001

100. BSI. *BS 5929:Part4:1986. British Standard Methods for Sensory analysis of foods. Part 4. Flavour profile methods*. British Standards Institution, 1986

101. Carluccio A. *In invitation to Italian Cooking*. London: Pavilion Books, 1991

102. Mosimann A. *Cooking with Mosimann*. UK: Papermac: UK, 1989

103. Jellinek G. *Sensorische Lebensmittelprüfung. Lehrbuch für die Praxis*. Pattensen: Verlag Doris & Peter Siegfried Pattensen, 1981

104. Stone H, Sidel JL. *Sensory Evaluation Practices*. 2nd edition. London: Academic Press, 1993

105. Hinreiner EH. Organoleptic evaluation by industry panels – the cutting bee. *Food Technology* 1956; 31(11): 62–67 (cited in 104)

106. Caul JF. The profile method of flavor analysis. *Advances in Food Research* 1957; 7: 1–40 (cited in 104)

107. Brandt MA, Skinner E, Coleman J. Texture profile method. *Journal of Food Science* 1963; 28: 404–410 (cited in 104)

108. Szczesniak AS. Classification of textural characteristics. *Journal of Food Science* 1963; 28: 385–389 (cited in 104)

109. Stone H, Sidel J, Oliver S, Woolsey A, Singleton RC. Sensory evaluation by quantitative descriptive analysis. *Food Technology*. 1974; 28(1): 24, 26, 28, 29, 32, 34 (cited in 104)

110. Johnson PB, Civille GV. A standard lexicon of WOF descriptors. *Journal of Sensory Studies* 1986; 1: 99–104 (cited in 104)

111. Muñoz AM. Development and application of texture reference scales. *Journal of Sensory Studies* 1986; 1: 55–83 (cited in 104)
112. ISO. *BS 5098:1992 (ISO 5492). Glossary of Terms relating to sensory analysis.* Geneva: International Standards Organization, 1992
113. MacDougall DB. Instrumental Assessment of the Appearance of Foods. In: Williams AA, Atkin RK. *Sensory Quality in Foods and Beverages.* Chichester, UK: Ellis Horwood, pp 121–139, 1983
114. The Tea Council. *A Glossary of Tea.* Retrieved on 8th November 1999 from http://www.teacouncil.co.uk/glossary/gloss4.htm
115. Halley RJ, Soffe RJ. *The Agricultural Notebook.* 18th edition. Oxford: Blackwell Scientific Publishers, 1998
116. Francis FJ. Quality as Influenced by Color. *Food Quality and Preference* 1995: 149–155
117. Gair A. *The Artist's Handbook.* Leicester: Abbeydale Press, 1998
118. Kent M, Porretta S. Food Colour: Measurement & Standardisation. *European Food & Drink Review* 1992: 53–59
119. McFarlane I. Colour Measurement. *European Food & Drink Review* 1990: 33–35
120. Schröder MJA. Aspects of sensory quality in milk and unfermented milk products. In: Williams AA, Atkin RK (ed). *Sensory Quality in Foods and Beverages.* Chichester, UK: Ellis Horwood, pp 401–411, 1983
121. Rahman FMM, Buckle KA. Pigment changes in capsicum cultivars during maturation and ripening. *Journal of Food Technology* 1980; 15: 241–249
122. Thomas P, Janave MT. Effect of temperature on chlorophyllase activity, cholorphyll degradation and carotenoids of Cavendish bananas during ripening. *International Journal of Food Science and Technology* 1992; 27: 57–63
123. Maccarone E, Maccarone A, Rapisarda P. Technical note: Colour stabilization of orange fruit juice by tannic aid. *International Journal of Food Science and Technology* 1987; 22: 159–162
124. Hendry GAF. Natural pigments in biology. In: Hendry GAF, Houghton JD (ed). *Natural Food Colorants.* London: Blackie Academic & Professional, pp 1–39, 1996
125. Petropakis HJ, Montgomery MW. Improvement of colour stability in pear juice concentrate by heat treatment. *Journal of Food Technology* 1984; 19: 91–95
126. Amiot MJ, Fleuriet A, Cheynier V, Nicolas J. Phenolic compounds and oxidative mechanisms in fruit and vegetables. In: Tomás-Barberán FA, Robins RJ (ed). *Phytochemistry of Fruit and Vegetables.* Oxford: Oxford University Press, pp 51–85, 1997
127. Houghton JD. Haems and bilins. In: Hendry GAF, Houghton JD (ed). *Natural Food Colorants.* London: Blackie Academic & Professional, pp 157–196, 1996
128. Blythman J. *The Food We Eat.* London: Michael Joseph, 1996
129. Warriss PD, Akers JM. The effect of sex, breed and initial carcass pH on the quality of cure in bacon. *Journal of Food Technology* 1980; 15: 629–636
130. Clifford MN. Astringency. In: Tomás-Barberán FA, Robins RJ (ed). *Phytochemistry of Fruit and Vegetables.* Oxford: Oxford University Press, pp 88–107, 1997
131. Bakker J. Astringency in wine. *BBSRC business* 1997 (October): 10–11
132. Brennan JG. Measurement of particular textural characteristics of some fruits and Vegetables. In: Williams AA, Atkin RK (ed). *Sensory Quality in Foods and Beverages.* Chichester, UK: Ellis Horwood, pp 173–187, 1983
133. Nute GR, Jones RCD, Dransfield E, Whelehan OP. Sensory characteristics of ham and their relationships with composition, visco-elasticity and strength. *International Journal of Food Science and Technology* 1987; 22: 461–476
134. Szczesniak AS. Textural Characterization of Temperature Sensitive Foods. *Journal of Texture Studies* 1975; 6: 139–156
135. Schenker S. The nutritional and physiological properties of chocolate. *British Nutrition Foundation Nutrition Bulletin* 2000; 25: 303–313
136. Wyeth L, Kilcast D. Sensory analysis technique and flavour release. *Food Technology International Europe* 1991: 239–242
137. Szczesniak AS. Consumer awareness of and attitudes to food texture. II. Children and teenagers. *Journal of Texture Studies* 1971; 3: 206–217

138. Rouseff R, Gmitter F, Grosser J. Citrus breeding and flavour. In: Piggott JR, Paterson A (ed). *Understanding Natural Flavors.* London: Blackie Academic & Professional, pp 113–127, 1994
139. Verhagen LC. Beer flavour. In: Piggott JR, Paterson A (ed). *Understanding Natural Flavors.* London: Blackie Academic & Professional, pp 211–227, 1994
140. Baigrie BD. Cocoa flavour. In: Piggott JR, Paterson A (ed). *Understanding Natural Flavors.* London: Blackie Academic & Professional, pp 268–282, 1994
141. Birt R. Foods for special medical purposes: where food meets medicine. *Food Science and Technology Today* 1998; 12(4): 201–203
142. Tunaley A, Thomson DMH, McEwan JA. Determination of equi-sweet concentrations of nine sweeteners using a relative rating technique. *International Journal of Food Science and Technology* 1987; 22: 627–635
143. Land DG. Savoury flavours – an overview. In: Piggott JR, Paterson A (ed). *Understanding Natural Flavors.* London: Blackie Academic & Professional, pp 298–306, 1994
144. Bakker J, Law BA. Cheese flavour. In: Piggott JR, Paterson A (ed). *Understanding Natural Flavors.* London: Blackie Academic & Professional, pp 283–297, 1994
145. Anon. Comprehensive Training Method. *Perfumer & Flavorist* 1989; 14: 30–44
146. Eriksson C. Cereal flavours. In: Piggott JR, Paterson A (ed). *Understanding Natural Flavors.* London: Blackie Academic & Professional, pp 128–139, 1994
147. Anon. *Tasting Whisky. The Tasting Wheel.* Retreived on 25th April 2000 from http://www.scotchwisky.com/english/tasting/how_to/enwords.htm
148. Tan SM, Piggott JR. Descriptive sensory analysis of soy sauce. In: Williams AA, Atkin RK (ed). *Sensory Quality in Foods and Beverages.* Chichester, UK: Ellis Horwood, pp 272–276, 1983
149. Noble AC, Williams AA, Langron SP. Descriptive analysis and quality of Bordeaux wines. In: Williams AA, Atkin RK (ed). *Sensory Quality in Foods and Beverages.* Chichester, UK: Ellis Horwood, pp 324–334, 1983
150. Hellemann U, Tuorila H, Salovaara H, Tarkkonen L. Sensory profiling and multidimensional scaling of selected Finnish rye breads. *International Journal of Food Science and Technology* 1987; 22: 693–700
151. Mottram D. Meat flavour. In: Piggott JR, Paterson A (ed). *Understanding Natural Flavors.* London: Blackie Academic & Professional, pp 140–163, 1994
152. Noble AC. Wine flavour. In: Piggott JR, Paterson A (ed). *Understanding Natural Flavors.* London: Blackie Academic & Professional, pp 228–242, 1994
153. Anon. FLAIR and the European Consumer. *FLAIR-FLOW EUROPE* 1994; F-FE 152/94
154. van Gemert LJ, Nijssen LN, de Bie ATHJ, Maarse H. Relationship between the capsaicinoid content and the Scoville Index in Capsicum oleoresins. In: Williams AA, Atkin RK (ed). *Sensory Quality in Foods and Beverages.* Chichester, UK: Ellis Horwood, pp 266–271, 1983
155. Victory SeedsTM. The Pepper Heat Scale. Retrieved on 13th July 2000 from http://www.victoryseeds.com/information/scoville.html
156. Mottram DS. Chemical tainting of foods. *International Journal of Food Science and Technology* 1998; 33: 19–29
157. Fogerty AC, Whitfield FB, Svoronos D, Ford GL. Effect of heat on the fatty acids and aldehydes of veal meat phospholipids. *International Journal of Food Science and Technology* 1989; 24:529–534
158. Pinthong R, Macrae R, Rothwell J. The development of a soya-based yoghurt. I. Acid production by lactic acid bacteria. *Journal of Food Technology* 1980; 15: 647–652
159. Whitfield FB. Microbiology of food taints. *International Journal of Food Science and Technology* 1998; 33: 31–51
160. IFST. *Shelf Life of Foods – Guidelines for its Determination and Prediction.* London: Institute of Food Science and Technology (UK), 1993
161. *The Food Labelling Regulations 1996.* (S.I. 1996 No. 1499). London: HMSO
162. Frazier WC, Westhoff DC. *Food Microbiology.* 4th edition. New York: McGraw-Hill, 1988
163. Thomas P, Joshi MR. Reduction of chilling injury in ripe Alphonso mango fruit in cold storage by temperature conditioning. *International Journal of Food Science and Technology* 1988; 23: 447–455

164. Olorunda AO, Tung MA, Kitson JA. Effect of post harvest factors on quality attributes of dehydrated banana products. *Journal of Food Technology* 1977; 12: 257–262
165. Emsley J. Birth of the smart tomato. *The Independent*, 21st March 1995
166. Schröder MJA. Effect of oxygen on the keeping quality of milk. I. Oxidized flavour development and oxygen uptake in milk in relation to oxygen availability. *Journal of Dairy Research* 1982; 49: 407–424
167. Saltveit ME. Physical and physiological changes in minimally processed fruits and vegetables. In: Tomás-Barberán FA, Robins RJ (ed). *Phytochemistry of Fruit and Vegetables*. Oxford: Oxford University Press, pp 205–220, 1997
168. Martinez MV, Whitaker JR. The biochemistry and control of enzymatic browning. *Trends in Food Science and Technology* 1995; 6(June): 195–200
169. Church N. MAP fish and crustaceans – sensory enhancement. *Food Science and Technology Today* 1998; 12: 73–83
170. Ke PJ, Ackman RG, Linke BA, Nash DM. Differential lipid oxidation in various parts of frozen mackerel. *Journal of Food Technology* 1977; 12: 37–47
171. Schröder MJA. Origins and levels of post pasteurization contamination in the dairy and their effects on keeping quality. *Journal of Dairy Research* 1984; 51: 59–67
172. Cerny G. Packstoffsterilisation beim aseptischen Abpacken. *Zeitschrift für Lebensmitteltechnik* 1990; (1 /2): 54–58
173. Otto I, Herrmann V. Doppelt informiert besser. *Lebensmitteltechnik* 1995; (9): 18–20
174. CSIRA Australia. *Active Packaging*. Division of Food Science and Technology Fact Sheet. April 1994
175. Anon. Modifying the Atmosphere in Food Packs. *FLAIR-FLOW EUROPE* 1992; F-FE 48/92
176. IFST. *Good Manufacturing Practice. A Guide to its Responsible Management*. 4th edition. London: Institute of Food Science and Technology (UK), 1998
177. Brocklehurst TF. Delicatessen salads and chilled prepared fruit and vegetable products. In: Man CMD, Jones AA (ed). *Shelf Life Evaluation of Foods*. London: Blackie Academic & Professional, pp 87–126, 1994
178. Department of Health. *Chilled and Frozen. Guidelines on Cook-Chill and Cook-Freeze Catering Systems*. London: HMSO, 1989
179. IFST. Position statement. The use of irradiation for food quality and safety. *Food Science and Technology Today* 1999; 13(1): 32–36
180. Symons H. Frozen foods. . In: Man CMD, Jones AA (ed). *Shelf Life Evaluation of Foods*. London: Blackie Academic & Professional, pp 296–316, 1994
181. *The Dairy Products (Hygiene) Regulations 1995*. (S.I. 1995 No. 1086), as amended in 1996. London: HMSO
182. Lenges J. Aseptische Lebensmittelherstellung – eine Übersicht. *Zeitschrift für Lebensmitteltechnik* 1989; (1 /2): 6–14
183. Gaze JE. Pasteurisation: criteria for process design. *Food Technology International Europe*. London: Sterling, pp 41–44, 1994
184. Jamieson L, Williamson P. The potential of electro-technologies for the processing of foods. *Food Science and Technology Today* 1999; 13(2): 97–101
185. Richardson P. Microwave technology – the opportunity for food processors. *Food Science and Technology Today* 1991: 5(3): 146–148
186. Bengtsson N, Ohlsson T. Industrielles Erhitzen mit Mikrowellen und seine Bedeutung für gefrier- und hitzebehandelte Lebensmittel. *Zeitschrift für Lebensmitteltechnik* 1990; (6): 392–399
187. IGD. *Beef. Benefiting from best practice*. Watford: Institute of Grocery Distribution, 1999
188. Chamberlain N, Collins TH, McDermott EE. Alpha-amylase and bread properties. *Journal of Food Technology* 1981; 16: 127–152
189. Davidson A. *The Oxford Companion to Food*. Oxford: Oxford University Press, 1999
190. Hom K. *Ken Hom's Asian Ingredients*. Ten Speed Press: Berkeley, 1996
191. Rhodes G. *New British Classics*. London: BBC Worldwide, 1999
192. Carluccio A. *An Invitation to Italian Cooking*. London: Pavilion Books, 1991

193. Beesley PM. Sugar functionality reviewed. *Food Technology International Europe*. London: Sterling, pp 87–89, 1995
194. Soebstad G. Surimi ... a novel product for the food industry. *Food Technology International Europe*. London: Sterling, pp 161–162, 1989
195. Penny C. Setting Standards in Gelling. *Food Ingredients & Processing International* November 1991: 19–22
196. Anon. *Pectin. General Description B4*. Brochure Copenhagen Pectin 10.92.1000
197. Gelatin Manufacturers of Europe. Gelatin replacers: the great pretenders? *Food Technology International Europe*. London: Sterling, pp 18–20, 2000
198. Croghan M, Mason W. 100 years of food starch innovation. *Food Science and Technology Today* 1998; 12 (1): 17–24
199. Van den Boomgaard Th, van Vliet T, van Hooydonk ACM. Physical stability in chocolate milk. *International Journal of Food Science and Technology* 1987; 22: 279–291
200. Helcke T, Froment V. Gelatin, the Natural Fat-Replacer in Low-Fat Spreads. *Food Technology International* 1992: 45–46
201. Fillery-Travis A, Clark D, Robins M. Emulsion stability – how oil and water mix. *Food Science and Technology Today* 1990; 4: 89–93
202. Flack E. Food emulsifiers –current status and recent trends. *Food Technology International Europe*. London: Sterling, pp 185–191, 1987
203. Flack E. Functional ingredients in frozen desserts. *Food Technology International Europe*. London: Sterling, pp 203–205, 1991
204. Bonekamp A. Reinheit durch Ultrafiltration. *Lebensmitteltechnik* 2000: 68–70
205. Casey JN. The use of phospholipids in instantising and release applications. *Food Technology International* 2000: 87–88
206. Blenford D. Acting at the interface. *Food Ingredients and Analysis International* 1995: 8–10
207. Winwood R. Viskosität nach Wunsch. *Lebensmitteltechnik* 2001: 46–47
208. Calorie Control Council. *Fat Replacers. Food Ingredients for Healthy Eating*. Atlanta, Georgia 1992
209. Bertoli C. Ersatz im Einsatz. *Lebensmitteltechnik* 1996: 39–40
210. MAFF. *Food Sense. About Food Additives*. London, Ministry of Agriculture, Fisheries and Food, 1997
211. McWeeny DJ, Shepherd MJ, Bates ML. Physical loss and chemical reactions of SO_2 in strawberry jam production. *Journal of Food Technology* 1980; 15: 613–617
212. Giannuzzi L, Rodriguez N, Zaritzky NE. Influence of packaging film permeability and residual sulphur dioxide on the quality of pre-peeled potatoes. *International Journal of Food Science and Technology* 1988; 23: 147–152
213. Wedzicha BL, Kaputo MT. Reaction of melanoidins with sulphur dioxide: stoichiometry of the reaction. *International Journal of Food Science and Technology* 1987; 22: 643–651
214. IFST. *Preservatives in Foods*. London: Institute of Food Science and Technology (UK), 1988
215. Anon. Prolonging Freshness. *Food Ingredients and Analysis International* 1995: 24–28
216. Timmermann F, Adams WF. Antioxidative Wirkung von Tocopherolen und Ascorbylpalmitat. *Zeitschrift für Lebensmitteltechnik* 1989; 40: 22–27
217. Buck DF, Edwards MK. Anti-oxidants to prolong shelf-life. *Food Technology International Europe*. London: Sterling, pp 29–33, 1997
218. *The Colours in Food Regulations 1995* (S.I. 1995 No. 3124). London: HMSO, 1995
219. *The Sweeteners in Food Regulations 1995* (S.I. 1995 No. 3123). London: HMSO, 1995
220. *The Flavourings in Food Regulations 1992* (S.I. 1992 No. 1971). London: HMSO, 1992
221. Stich E. Die Kraft der Farben. *Lebensmitteltechnik* 2001: 44–46
222. FDA. *Color Additives. FDA Backgrounder*. Washington: Food and Drug Administration, February 1990
223. Timberlake CF. Plant pigments for colouring food. *BNF Nutrition Bulletin* 1989; 14: 113–125
224. Choubert G, Blanc J-M, Courvalin C. Muscle carotenoid content and colour of farmed rainbow trout fed astaxanthin or canthaxanthin as affected by cooking and smoking procedures. *International Journal of Food Science and Technology* 1992; 27: 277–284

225. MAFF. *Foodsense factsheet No. 4 – The main types of food additives*. London: Joint Food Safety and Standards Group, 21 April 1999
226. Krüger C. Die richtige Balance finden. *Lebensmitteltechnik* 2001 (1–2): 47
227. Montijano H, Borrego F. Developments in high-intensity sweeteners in the European Union. *Food Technology International Europe*. London: Sterling, pp 91–95, 1998
228. Hardinge J. Flavourings: a Recipe for Regulation. *Chemistry and Industry* 1990: 694–698
229. Grundschober F. Regulation and safety evaluation of flavourings. *Food Technology International Europe*. London: Sterling, pp 52–55, 1998
230. Swaine RL, Swaine Jr. RL. Citrus Oils: Processing, Technology, and Applications. *Perfumer and Flavorist* 1988; 13: 1–20
231. Heath HB. An overall view of herbs and spices. *Food Technology International Europe*. London: Sterling, pp 255–260, 1989
232. Ramsay I. Flavouring of vegetarian products. *Food Technology International Europe*. London: sterling, pp 53–54, 1997
233. Bohrer B. Hefeextrakt in Bio-Qualität. *Lebensmitteltechnik* 2001 (5) :54
234. Owens JD, Mendoza LS. Enzymically hydrolysed and bacterially fermented fishery products. *Journal of Food Technology* 1985; 20: 273–293
235. IFST. *Addition of Micronutrients to Food*. Institute of Food Science and Technology (UK) 1997
236. *The Bread and Flour Regulations 1995* (S.I. 1995 No. 3202). London: HMSO
237. *The Spreadable Fats (Marketing Standards) Regulations 1995* (S.I. 1995 No. 3116). London: HMSO
238. Richardson DP. The addition of nutrients to foods. In: Henry CJK, Heppell NJ (ed). *Nutritional aspects of food processing and ingredients*. Maryland: Aspen, pp 1–23, 1998
239. BNF. *Foods with added nutrients*. BNF Information. British Nutrition Foundation, 1998
240. Waldner H, Buth I. Selenangereicherte Lebensmittel. *Lebensmitteltechnik* 2000 (9): 48–49
241. Anon. Functional Drinks. ACE ergänzt um Folsäure. *Lebensmitteltechnik* 2001 (10): 45
242. IFRN. *Functional Foods and Drinks*. Food Information Sheet 06.00. Norwich: Institute of Food Research, 2000
243. Hasler CM. A new look at an ancient concept. *Chemistry and Industry* 2 February 1998: 84–89
244. Gibson GR, Fuller R. The role of probiotics and prebiotics in the functional food concept. In: Sadler MJ, Saltmarsh M (ed). *Functional Foods. The Consumer, the Products and the Evidence*. Cambridge: The Royal Society of Chemistry, pp 3–14, 1998
245. Anon. Functional foods. *Health Which?* June 2000: 8–11
246. IFST. Current Hot Topics. Phytosterol esters (plant sterol and stanol esters). June 2000. Retrieved on 21st November 2001 from http://www.ifst.org/hottop29.htm
247. Bristow A. Functional Food. The Facts and the Fiction. *Which?* 2001; July: 34–36
248. Anon. Functional Food. Fitmacher und vieles mehr? *Lebensmitteltechnik* 1999; (12): 10
249. Anon. Drinks to make you a better sport? *Which?* 1990; November
250. Anon. Neither food nor medicine. *Which?* 1996; September: 24–25
251. Mason R. Chlorella and Spirulina. *Alternative & Complementary Therapies* 2001; June: 161–165
252. Anon. Nature's cures. *Which?* 2001; March: 30–33
253. Green J. Putting viruses on the food safety agenda. *Abstract. Food Science and Technology Today* 1997; 11(1): 38
254. ACMSF. *Expert report on foodborne viral infections published*. London: Department of Health, 1998/0493
255. Hinton J, Wells J. *Pathogenesis of infection caused by food-borne bacteria and the influence of environmental stress*. Norwich: Institute of Food Research. Retrieved on 11th July 2000 from http://www.ifrn.bbsrc.ac.uk/Safety/Pathogenesis.html
256. Rowe B. Food-borne disease – The changing face of food microbiology. *Abstract. Food Science and Technology Today* 1997; 11(1): 37–38
257. Anon. *Foodborne Illness – Table I: Common Microbiological Foodborne Diseases*. Retrieved on 18th January 2000 from http://www.scotland.gov.uk/food/fb-ill-tab1.html

258. Anon. *Foodborne Illness – Table II: Diseases which may be Foodborne.* Retrieved on 18th January 2000 form http://www.scotland.gov.uk/food/fb-ill-tab2.htm
259. ACMSF. *Meeting of the Advisory Committee on the Microbiological Safety of Food.* London: Department of Health, 1997/260
260. IFST. *Foodborne Campylobacteriosis – IFST position paper.* London, Institute of Food Science and Technology (UK), 1995
261. Rhodehamel EJ, Reddy NR, Pierson MD. Botulism: the causative agent and its control in foods. *Food Control* 1992: 3(3): 125–143
262. Roberts TA, Gibson AM, Robinson A. Factors controlling the growth of *Clostridium botulinum* types A and B in pasteurized, cured meats. *Journal of Food Technology* 1981; 16: 267–281
263. Peck M. *Understanding and predicting the behaviour of bacterial foodborne pathogens.* Norwich: Institute of Food Research, 2000
264. O'Mahony, Mitchell E, Gilbert RJ, Hutchinson DN, Begg NT, Rodhouse JC, Morris JE. An outbreak of foodborne botulism associated with contaminated hazelnut yoghurt. *Epidemiology and Infection* 1990; 104: 389–395
265. Roberts TA, Derrick CM. The effect of curing slats on the growth of *Clostridium perfringens (welchii)* in a laboratory medium. *Journal of Food Technology* 1978; 13: 349–353
266. IFST. Current Hot Topics. *Salmonella typhimurium* DT 104. November 1997. Retrieved on 13th November 2001 from http://www.ifst.org/hottop20.htm
267. Anon. Abstract. A certain combination of genes. *Food Science and Technology Today* 1999; 13(1):9
268. The Pennington Group. *Report on the circumstances leading to the 1996 outbreak of infection with E. coli O157 in Central Scotland, the implications for food safety and the lessons to be learned.* Edinburgh: The Stationery Office, April 1997
269. Lewis C. *Critical Controls for Juice Safety.* 1999, Food and Drug Administration. Retrieved on 17th July 2000 from http://www.fda.gov/fdac/features/1998/598_juic.html
270. Rocourt J. Risk factors for listeriosis. *Food Control* 1996; 7(4/5): 195–202
271. Skinner. *Listeria*: the state of the science, Rome 29–30 June 1995. Session IV: country and organizational postures on *Listeria monocytogenes* in food. *Listeria*: UK government's approach. *Food Control* 1996; 7(4/5): 245–247
272. PHLS. *Yersinia* Information. 2000. Public Health Laboratory Service. Retrieved on 2nd March 2000 from http://www.phls.co.uk/facts/yers-int.htm
273. MAFF Food Contaminants Division. Mycotoxins – current aspects of chemical safety. *Food Science and Technology Today* 1999; 13(1): 19–25
274. Ababouch L, Afilal ME, Benabdeljelil, Busta FF. Quantitative changes in bacteria, amino acids and biogenic amines in sardine (*Sardina pilchardus*) stored at ambient temperature (25–28°C) and in ice. *International Journal of Food Science and Technology* 1991; 26: 297–306
275. Ahmed FE. Review: Assessing and managing risk due to consumption of seafood contaminated with microorganisms, parasites, and natural toxins in the US. *International Journal of Food Science and Technology* 1992; 27: 243–260
276. Anon. Erfolgreiche Tagung in Ungarn. *Lebensmitteltechnik* 2001; (1–2): 14–16
277. Fairbairn R. Scientists find potentially deadly parasite in fish. *The Sunday Times* 16th March 1997
278. Huss HH. Fish parasites of public health significance. F-FE 125/94 Flair-Flow Europe 1994
279. IFST. *Development and Use of Microbiological Criteria for Foods.* Institute of Food Science and Technology (UK) 1999
280. IFST. *Cryptosporidium.* IFST Position Statement. *Food Science and Technology Today* 1997; 11(1): 46–48
281. Scoging A. Marine Biotoxins. *Food Science and Technology Today* 1995; 9(1): 39–43
282. Cairns C. Fears grow for scallop industry as poisons rise. *The Scotsman* 17th February 2000
283. *The Contaminants in Food Regulations 1997.* (S.I. 1997 No. 1499). London: HMSO

284. Anon. *The Work of the Pesticide Safety Directorate*. London: MAFF: Pesticide Safety Directorate, July 1995
285. Lichfield J. EU survey finds half of French food has residues of pesticides. *The Independent* 7th August 2001
286. Arthur C. M&S and Somerfield top pesticides finding. *The Independent* 17th August 2001
287. Anon. Carrot Tops. *Which? Magazine* 1995; November: 5
288. MAFF. *Chlormequat in pears*. 2000; MAFF News Release 123/00
289. Lampkin N. *Organic Farming*. Ipswich: Farming Press, 1992
290. Anon. Antibiotics in Food. *Which? Magazine* 1997; March: 18–20
291. MAFF. *MAFF stands firm on Commission Proposal for EU Beef Hormones Ban*. 2000; MAFF News Release 196/00
292. IFST. Bovine somatotropin (BST). IFST Position Statement. *Food Science and Technology Today* 1998; 12(3): 169–176
293. MAFF. *Testing Improved for Plastics coming in Contact with Food*. 1998; MAFF News Release 274/98
294. Tice P, Offen C. Taints and odours from packaging. *Food Technology International Europe* 1993: 214–216
295. MAFF. *Tin in Canned Tomatoes*. 1998; MAFF News Release 456/98
296. Hollaender J, Sedlmayr M. Wechselwirkungen zwischen Weissblechverpackungen und Füllgut. *Zeitschrift für Lebensmitteltechnik* 1989; 40(10): 606–614
297. IFST. Dioxins and PCBs in food. IFST Position Statement. *Food Science and Technology Today* 1998; 12(3): 177–179
298. Joint Food Safety and Standards Group. *MAFF UK – Dioxins and PCBs in UK and Imported Marine Fish* 1999; Food Surveillance Information Sheet Number 184
299. Joint Food Safety and Standards Group. *Dioxins Contamination of Belgian Animal Products – an Update*. 1999; Press Release FSA 15/99
300. MAFF FoodSense. *Understanding Radioactivity in Food*. London: MAFF, 1993
301. Silvester DJ. Determination of 3,4-benzopyrene and benzanthracene (PAH) in phenolic smoke concentrates. *Journal of Food Technology* 1980; 15: 413–420
302. Coultate TP. Contaminants from food processing. *Food Science and Technology Today* 1995; 9(1): 44–49
303. Joint Food Safety and Standards Group. *Manufacturers Told to Tighten Up on Chemical Contamination in Soy Sauce*. 1999; Press Release FSA 21/99
304. Dearden R. Automatic X-ray inspection for the food industry. *Food Science and Technology Today* 1996: 10(2): 87–90
305. CPSC. *CPSC, Kellogg Company Announce Recall of Toy Cars Inside Boxes of Cereal*. News from CPSC Release # 00-136. US Consumer Product Safety Commission, 2000. Retrieved on 17th July 2000 from http://www.cpsc.gov/cpspub/prerel/prhtml00/00136.html
306. Anon. Der Klang der Haselnüsse. *Lebensmitteltechnik* 1999 (9): 59–61
307. Pearsall J (ed). *The New Oxford Dictionary of English*. Oxford: Oxford University Press, 1998
308. MAFF. *Mixed Results in Food Surveys*. 1999; MAFF News Release 117/99
309. Patel T. Real juice, pure fraud? *New Scientist* 1994; May: 26–29
310. Kiritsakis AK. *Olive Oil. From the Tree to the Table*. 2nd edition. Food & Nutrition Press Inc, 1998
311. Anon. Blind drunk. *Which? Magazine* 1996; October: 4
312. Cooper G. Revealed: Cheats who sell watered down meat. *The Independent* 29th October 1997
313. Food Standards Agency. *Chicken Falls Foul of Added Water Test*. Press releases 2000/0048
314. Connor S. Portrait of a nation fed a diet of reassurances. *The Independent*, 27th October 2000
315. IFST. Bovine Spongiform Encephalopathy (BSE). IFST Position Statement. *Food Science and Technology Today* 1996; 10(4): 231–243

Part 3

Understanding the Food Consumer

Part 3 of the book draws together some of the earlier themes. The main purpose of this is to arrive at a more detailed understanding of individual food choices and food-related behaviours. In addition, this part expands on some of these themes, especially where it examines the role of physiological factors in eating and drinking. Some of the fundamental questions about food choice that will be asked are: What is eaten? In what amounts and proportions is it eaten? On what occasions is it eaten, how frequently, and why? People can survive without any food for some considerable length of time. In fact, the occasional fast would be of great benefit to most individuals. Yet the consumption of water is a daily necessity. Thirst is the motivational state of readiness to consume water and reflects a physiological need [1]. But does hunger reflect a physiological need? Clearly, ingestive behaviours generally, and the eating of food in particular, do not necessarily reflect either bodily deficiencies or other arousal states. People commonly eat and drink in anticipation of hunger and thirst, for convenience or social reasons, and so on. It is interesting that, depending on the context, a food can have either arousing (e.g., gourmet meals and unfamiliar foods) or calming effects (e.g., highly familiar and desired, so-called comfort foods).

Intake of food and drink is influenced by a complex system of interrelated factors, and has been studied from a variety of perspectives, including gastrointestinal physiology, neurology, endocrinology, metabolism and psychology [2]. There are two basic phenomena which are studied here, i.e., feeding behaviour (food choice) and food intake. The two different approaches are characteristic of two schools, that of appetite control and that of energy balance, respectively [3]. In the appetite field, phenomena such as the frequency and duration of eating events are measured in relation to both individual and food factors, i.e., food(s) chosen, portion sizes, and so on. Among researchers into energy balance, the phenomenon being studied is usually called either dietary intake or energy intake [3]. Whilst it is understood that eating and drinking are motivated by a combination of different factors, including physiological, pharmacological, social and psychological ones, the central question about food intake remains to be answered. This question concerns both the nature of the signal(s) controlling food intake, i.e., the amount of food consumed, and the nature of the sensors that detect the signal(s) [2]. Indeed, it is unclear that such a single stimulus exists at all. The maintenance of body weight relies on a simple mechanism, i.e., the balancing of body energy intake and energy expenditure. Yet, based on current childhood obesity levels, it is estimated that half the population of Europe will be obese within the next thirty years [4], and slimming diets are already an obsession for a large number of people [5]. Vicious circles are set up that start from a person worrying about their body weight, leading to simultaneously craving high-caloric foods and feeling that one should restrict caloric intake, which leads to more worrying, hence even more desire for high-caloric foods, and so on. Today it appears that when it comes to food intake, appetite and satiety, many people feel out of control. Among the many commentators on these issues, there are voices that attribute this state of affairs to a perceived deviation from the consumption of foods they believe humans evolved to eat, towards foods that are too denatured and too nutrient-dense [6]. Certainly, it seems easier to interpret satiety signals obtained from a fruit-based diet than signals from the typical modern Western diet. Innate food aversions, e.g., against irritants (e.g., chilli pepper) and bitter substances (e.g., coffee) may be reversed under the influence of cultural forces [7]. Food preferences are most readily predicted from knowledge of the cultural or ethnic group to which an individual belongs. This indicates that many food preferences are culturally acquired. Cultures develop food-related rules that constitute their cuisine, and that may have adaptive value, e.g., the processing of maize in Central America to enhance its nutritional value [7]. A more common purpose of shared cuisine is as a social marker (see Chapter 3). At the level of individual preferences, a food's palatability, i.e., a liking for the food, is the best predictor for it being chosen. Dislike of a food is an even better predictor for it being avoided.

Part 3 starts with Chapter 7, and a close look at the homeostatic and psychological bases of eating and drinking. Homeostasis refers to the attempts of the body to maintain a stable [8] and optimal [9] environment through self-regulatory processes. Through homeostasis, body levels of a range of parameters are maintained within a set narrow range. Caloric, body fluid, glucose and body temperature homeostasis exemplify this type of balance. However, whereas thirst, hunger and salt appetite fall within the homeostatic tradition, it appears that the ingestion of some very important food components does not [7]. Minimum levels of intake apply to nutrient requirements, i.e., for vitamins, minerals, EFA and IAA (see Chapter 4), and excesses are metabolised, excreted or stored, but without any satiety effects [7]. Nutrient deficiencies are, by definition, characterised by specific deficiency diseases, and the condition of malaise caused can motivate changes in food selection. In this way, through the mechanism of reinforcement, adequate nutrient supplies may be restored. Whatever their origin, both appetitive and satiety mechanisms are controlled by the nervous system. Chapter 7 therefore provides an overview of the nervous system, and in particular, of the senses, both autonomic and special, insofar as a role has been identified for them in food choice and feeding behaviours. The chapter concludes with a discussion of the role of visual, textural and flavour attributes in food liking and food choice. This builds on Chapter 5, where sensory attributes were introduced in a product-focused context.

Chapter 8 is concerned with various aspects of the food consumer as a participant in food markets. It starts with the general issue of the nature of segmentation within the food market, examining the role of consumer values and expectations in structuring the market. In making rational choices, food consumers need to carry out risk assessments. Food safety concerns specific types of risk, that have come to the fore in consumers' minds in recent times, largely as a result of a series of major food scares. However, costs, benefits and risks have to be weighed up in a more general sense whenever decisions need to be made, and the way alternative choices are framed has an important role in the outcome of such evaluations. The bottom line in every case is the risk to be disappointed by the consequences of a decision. The chapter also deals with consumer skills development and food-related learning generally. There are two major aspects of food-related learning. One, the acquisition of tastes through social learning, has already been mentioned. Consumers also learn to make good choices (see Chapter 3). However, how good these choices are, and how rational the individual consumer eventually becomes, depends to a large extent on the quality of the information about a particular purchase that is available to them. Food skills are transmitted through familial and other social group modelling, and learned through classical and operant conditioning (reward, reinforcement), through official education, marketing, entertainment (TV chefs, food journalists and cookery book writers) (see Chapter 3). Skills levels, or the level of sophistication in particular food choices, are also a determinant of market segmentation. In other words, less educated consumers can be persuaded more readily by food marketers and advertisers to make purchases that they would not have made if they had had a better understanding of the nature of the foods in question. In the last section of Chapter 8, the context of consumption, and the role of this in consumption value is investigated. This relates both to the physical context, e.g., home, food service or retail environment, and the emotional context, e.g., expectations and the consequent consonance or dissonance experienced. The examination of the consumption context covers issues such as convenience, performance (e.g., sports foods) and conspicuous consumption. All food preferences are context dependent, independently of whether they are innate or acquired. Clearly, a person with a strong liking for sugar and salt would put the sugar in the cake batter and the salt into the savoury broth, and not vice versa.

In recent years, there has been an explosion of interest in the way in which food influences what people think and feel [10]. Whilst some individual foods are blamed for making people ill, both mentally and physically, others are promoted as superfoods. The supposed effects of certain foods and eating patterns on brain function can range from minor mood changes to mental disorders. In general discourse, including journalistic writings as well as (TV and radio) food entertainment, individual foods are often picked out as particularly good or very bad, or described in terms with similar meanings. Clearly, these terms derive from individual evaluations and therefore do not describe a food in objective terms. Chapter 9 offers an analysis of some of these familiar concepts. It starts with a look at the perceptions of healthiness in foods and dietary patterns, including a discussion of so-called junk foods. The discussion refers back to Chapter 4, in particular, the characterisation of foods as sources of energy, nutrients and bioactive compounds. The next example of goodness in foods refers to their emotional value to the individual. The examination of this concept is wide-ranging, taking into account aspects such as comfort and mood enhancement, luxury and indulgence, nostalgic associations (traditional and ethnic) and social eating (especially celebration meals), novelty and variety, and stimulation (including aphrodisiacs). The next concept to be discussed is that of purity and authenticity in a food, with related terminology, e.g., real, natural, traditional and classic. The last issue deals with food ethics as a food value and includes aspects of religion and spirituality.

7 Homeostatic and Psychological Bases of Eating and Drinking

7.1
Introduction

Eating and drinking are examples of behaviours that illustrate the stimulus-response model of perception (Fig. 3.7). Communication processes rely, in the first instance, on the presence of sensors capable of receiving signals from the environment. In living organisms, the so-called autonomic senses respond to metabolic stimuli, whereas external stimuli are received by sensors located in specialised sense organs, such as the eye. In the case of the autonomic senses, the cognitive stages of the stimulus-response model are omitted. The autonomic and special senses together collect information to allow various systems within the organism to respond to stimuli in an appropriate, life-supporting manner. In interpreting sensory experiences involving the special senses, the brain refers to past events as well as contextual factors. In this case, the individual has at least some control over what is being experienced, and how the experience is interpreted. This is in contrast to the autonomic senses, which cannot normally be influenced through thinking power. Mechanisms such as body temperature regulation, water and electrolyte balance and the maintenance of blood glucose levels are typical of those involving the autonomic senses. Stimulation of the special senses and, in some cases, of the autonomic senses may be subject to classical conditioning, which is also known as Pavlovian conditioning [8]. Classical conditioning plays a pervasive role in the regulation of eating behaviours. Conditioned stimuli paired with individual foods, or with defined eating patterns, have the ability to elicit motivational states, in particular, appetite and satiety. Appetite may be elicited through a glance at one's watch, which may tell one that lunchtime is approaching, with all the associated imaginings and anticipations. Similarly, satiety may be elicited by eating one's dessert at the end of a multi-course meal. This provides the signal that a meal, which has also been a "proper" meal, has been suitably concluded. Food aversions are also commonly acquired via mechanisms related to classical conditioning.

There are many factors that influence eating and drinking, including what is consumed, and in what quantities and combinations. The motivational framework of ingestive behaviours is complex, and individual components difficult to disentangle. As indicated above, physiological factors play a role, and so do psychological and social factors, e.g., the palatability of the food and the social setting of the eating event. However, all human behaviours, whatever their nature and purpose, are controlled by the nervous system. The structure of the nervous system will be briefly outlined in the following. There are two divisions of this, i.e, the central nervous system (CNS), located inside the skull and spinal column, and the peripheral nervous systems (PNS), which lies outside them [9]. The PNS is split further into a somatic and an autonomic division. The somatic nervous system (SNS) consists of spinal and cranial nerves, which convey sensory information to and from the sense organs and skeletal muscles [9]. The autonomic nervous system (ANS) governs the regulation of smooth and cardiac muscles and of the glands, working continuously and in an involuntary, reflexive manner, to assure normal internal functioning [11]. The ANS is divided into three parts, i.e., the sympathetic, parasympathetic and enteric nervous systems. The sympathetic division of the ANS controls functions that accompany arousal and expenditure of energy [9]. It is called into action when "fight or flight" responses are appropriate. It causes blood pressure to increase, glucose to be released into the bloodstream and digestion to slow down [11]. On the other hand, the parasympathetic division supports activities that are involved with increases in the body's supply of stored energy [9]. During the so-called "rest and digest" response, the ANS works to conserve energy. Blood pressure decreases and digestive processes accelerate, including the production of saliva and gastric juices [11]. Finally, the enteric system innervates the viscera, e.g., the gastrointestinal tract, the pancreas and the gall bladder [11]. An overview of the structural organisation of the human nervous system is given in Fig. 7.1.

The brain, rather than simply processing information, is a self-activating organ that issues commands to somatic, autonomic and endocrine action systems [1]. Although these may be modified by various signals, it is the brain that produces the behaviour. Signals are integrated into neural programs for actions that are either innate or acquired [1]. It used to be thought that there were brain centres providing specific controls of eating and drinking, but this is no longer the case. Instead, the involvement of non-specific arousal mechanisms in motivated

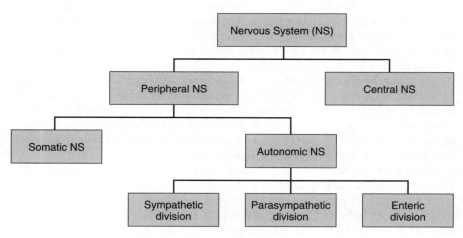

Fig. 7.1. Organisation of the human nervous system (based on [11])

ingestive behaviours is now attracting the interest of researchers [12]. Brain mechanisms operate in the arousal and satiation of feeding behaviour, being involved in the autonomic and endocrine control of the internal organs and the gastrointestinal system in particular [13]. The control of hunger and food intake involves external sensory input (special senses), internal sensory stimulation (gastric contractions and gastric distension), internal environmental factors (insulin, nutrients), and modification by learning and experience [13].

Muscles, nerves and synapses are the key anatomical units required for behavioural functioning [14]. Muscle action is used when placing food in the mouth, and when chewing and swallowing it. Nerves send electrical impulses around the body as informational signals. Cranial nerves are particularly important in eating because they carry taste and smell signals to the muscles of the jaw [14]. The term synapse refers to gaps between nerves and muscles, or between adjacent nerve cells. Arrival of a nerve impulse at a synapse causes chemical substances, called neurotransmitters, to be secreted into it. This ensures that neural signals do not terminate at the end of nerve cells. In addition to neurotransmitters, neuromodulators and hormones transmit information between cells in the body. All of these mechanisms require the presence both of cells that release the relevant chemicals and of specialised protein molecules (receptors) that detect them [9]. Both neurotransmitters and -modulators are released by the so-called terminal button of nerve cells; whereas the former are detected by receptors located a short distance away, the latter, which are typically peptides, tend to travel farther and are dispersed more widely [9]. Most hormones are produced in cells located in the endocrine glands, but others are produced by specialised cells located in various organs, e.g., the stomach, the intestines, the kidneys and the brain [9]. They are secreted into the extra-cellular fluid around capillaries and are then transported to the rest of the body through the bloodstream [9]. Whilst hormones transmit messages from a secreting gland to a target tissue, pheromones carry messages form one animal to another. In mammalian species, most pheromones are detected by means of olfaction [9].

The CNS contains numerous neurotransmitters, some of which can be modulated by pharmacologically active agents. For example, amphetamine interferes with the noradrenalin transmitter system, naltrexone blocks the endogenous opioid system, and fenfluramine prevents re-uptake of serotonin into synapses [14]. The neurotransmitter acetylcholine (ACh) is chiefly responsible for muscular movement. In fact, the toxin produced by the anaerobic microorganism *Clostridium botulinum* (see Chapter 6) acts as an ACh antagonist [9]. Epinephrine (adrenalin), norepinephrine (NE, noradrenalin) and dopamine are neurotransmitters which share their chemical classification as catecholamines. They cover a wide area of the brain and are involved in increasing or decreasing brain activity in specific regions [9]. The neurotransmitter dopamine is involved in movement, attention and learning [9]. The precursor for the biosynthesis of both NE and dopamine is tyrosine, an IAA (see Chapter 4). Both NE and ACh are present in the ANS [9]. The hormone epinephrine is produced within the adrenal gland, which is located just above the kidney; it also serves as a neurotransmitter in the brain, but is less important than NE [9]. The behavioural effects of serotonin, an indolamine neurotransmitter, are complex. Serotonin plays a role in the regulation of mood, eating, sleep, arousal and pain [9]. The precursor for serotonin is the amino acid tryptophan [9]. All the above neurotransmitters are synthesised within neurons. Some neurons secrete simple amino acids as neurotransmitters, with at least eight involved in the mammalian CNS, three of them occurring commonly, i.e, glutamate, gamma-butyric acid (GABA) and glycine [9]. The neurons of the CNS also release a large variety of peptides [9]. Although most of these peptides appear to act as neuromodulators, some act as neurotransmitters [9]. Endogenous opioids are especially important here. Several different neural systems are activated when opiate receptors are stimulated; one type produces analgesia, another inhibits defensive responses and a third is involved in reinforcement (behavioural reward) [9]. Several peptide hormones are found in the brain, where they serve as neurotransmitters and neuromodulators [9].

The receptors for neurotransmitters, neuromodulators and hormones are specialised proteins that bind with certain molecules. The sensory receptors in the special senses differ fundamentally from these in being specialised nerve cells adapted to detect specific physical phenomena. At least five special senses are recognised: sight, hearing, smell, taste and somatosensation (touch) [9]. The somatosenses provide information about what is happening on the surface of the body and inside it [9]. They detect changes in pressure, temperature, vibration and limb position, and events that damage tissue (i.e., that produce pain). One might think that included amongst these should be the mechanoreceptors associated with the gastrointestinal tract, which can produce a sensation of (excessive) fullness. In fact, the human gut is the largest contact organ with the external environment, its mucous membrane extending to approximately 300 m^2 [15]. However, Grill and Kaplan, in distinguishing between exteroceptors and interoceptors, include amongst the latter both gastric mechanoceptors and intestinal chemoceptors [16]. Carlson refers to the organic senses, i.e., those senses that arise from receptors in and around the internal organs, providing both unpleasant sensations, such as stomach aches and gallbladder attacks, and pleasant ones, such as those provided by a warm drink on a cold winter day [9]. The sense of kinaesthesis provides information about body position and movement and arises from receptors in joints, tendons and muscles [9]. An overview of the human special and autonomic senses is given in Fig. 7.2.

Homeostasis refers to self-regulation of vital bodily systems such as thermoregulation and energy balance. The physiological value of eating and drinking ultimately derives from vital systems being kept in balance, and from vital processes resulting in feelings of wellness. Here, the regulation of fluid balance through the ingestion of salt and water is much better understood than mechanisms underlying food selection and intake, and nutrition and energy balance [13]. In the case of hunger for calories, it is still not clear what it is exactly that is being regulated [13]. The cellular need for metabolic fuels is continuous, and fuels are therefore supplied to the cells automatically [12]. They are derived from the excess calories consumed during past eating episodes, and are obtained either from the intestines or from endogenous depots. Water is slowly lost from the body by respiration, evaporation and urinary excretion. However, the consequences of its loss are buffered by the osmotic movement of water from large cellular stores [12]. Lost water can thus be replaced by episodic drinking behaviour.

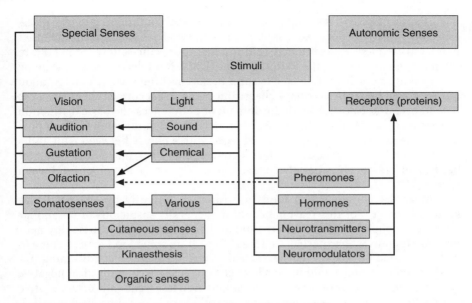

Fig. 7.2. The Human Special and Autonomic Senses (based on [9])

Stricker identifies as the key variable in food intake control the alteration be-tween the two phases of caloric homeostasis, in which metabolic fuels are sup-plied either from recently consumed food that still is in the alimentary canal or from accumulated triglyceride, glycogen and protein stores [12]. According to this perspective, the shift from satiety to hunger is not associated with some metabolic deficit. Rather, shift occurs when animals enter the postabsorptive state and begin to mobilise calories from stores. Food intake has been most of-ten related to energy balance and caloric needs, but Stellar emphasises that it may also be driven by a more general arousal mechanism [13]. In this particular view, hunger develops in response to metabolic signals caused by a decrease in calories coming from the intestine and liver, as well as from inputs from the spe-cial senses. The motivation for food intake is then controlled by oropharyngal, gastric and hepatic signals [13]. However, with regard to gastric signals, there is also a view that these govern intake only when the food being ingested is famil-iar. This suggests that signals mediated by the stomach have to be calibrated by a process of learning, and that there is a calibration specific to each food [17].

Motivated behaviour generally is a response to the state of the internal bod-ily environment, including mental states which, in turn, responds to internal and external stimulation. Eating and drinking are rewarding behaviours, and satia-tion may be understood as a reduction of a motive state [13]. Three general mo-tives for food acceptance or rejection may be identified: sensory-affective moti-vation, anticipated consequences of ingestion (e.g., harm) and ideational (based on knowledge of the origin or nature of a particular food) [7]. In general, expo-sure tends to increase liking and may even provide an important explanation for it [7]. Exposure provides the opportunity for classical conditioning to take place, especially if early exposures take place in positive social settings. On the other hand, too-frequent exposure may lead to sensory specific satiety (SSS) [18]. In the phenomenon of SSS, the liking for a particular food decreases as it is being eaten. Liking, or affect, is reflected in some preferences but not all. A dieter might like ice cream better than cottage cheese but prefer to eat cottage cheese because it contains fewer calories [7]. In the domain of preferences, the basic distinction is between foods liked for their sensory properties and foods eaten for their positive consequences [7]. A hungry child faced with a choice between fruit and a manufactured snack product, such as a confectionery bar, may prefer the latter because of previously experienced levels of satiety for each type of food when hungry [19].

Culture determines which foods are eaten customarily by any particular group of people. For this reason, people are generally suspicious of unfamiliar foods, i.e., people's attitude towards unfamiliar foods is characterised by neophobia. Humans who experience nausea after eating some food tend to come to dislike the food (an acquired distaste). This is due to Pavlovian conditioning. According to Rozin and Schulkin, this happens much less frequently after negative gastrointestinal events other than nausea, and even less frequently after negative nongastrointestinal events such as skin eruptions or respiratory problems [7]. According to these authors, after negative events other than nausea, the associated food tends to become classified as dangerous. However, recent research in the UK questioned any unique role for nausea in the (recalled) acquisition of sensory aversion to a particular food [20]. Foods are rejected for four main reasons, i.e, distaste, perceived danger, perceived inappropriateness and disgust [7]. Distastes refer primarily to dislikes of sensory properties, e.g., beer, spinach. Dangers may be perceived in potential contamination or the presence of allergens or carcinogens. Inappropriates are rejected primarily because of a negative response to the idea or nature of the item, e.g., the presence of sand or grass. Finally, disgusts include rejection on ideational grounds, but with a strong distaste component as well, e.g., presence of parasites, insects and faeces in foods. Whilst neophobic traits are present naturally in people, partly as a protective mechanism against poisoning, neophobia is balanced by neophilia. This balance depends on the exact nature of the unfamiliar foods, how much and what is known about it, and the personality and present state of mind of the individual considering its consumption.

As well as intervening in the appetite-satiety balance, eating and drinking impact mood and mental performance; conversely, mood affects food related behaviours. For example, a common research finding is that, compared with late morning, people's cognitive performance is reduced in the early afternoon. For some tasks, the post-lunch dip occurs whether lunch has been eaten or not [21]. Lunch has least effect on performance efficiency under conditions of high arousal [21]. In contrast, research has produced little evidence in support of the popular belief that missing breakfast adversely affects cognitive performance [21]. The composition of a food or meal, i.e. the proportions of the different types of dietary fuel present, has been linked to mood. Large meals with a high proportion of fatty foods are well known to induce drowsiness. However, in terms of effects on mood, the role of carbohydrates is of greater interest. It is widely thought that the consumption of a meal high in carbohydrates increases the ratio of the plasma concentration of tryptophan, the precursor of serotonin, leading to enhanced mood [21]. However, the evidence for a stress-relieving effect of carbohydrate-rich meals, and the opposite effects of protein-rich meals, is mixed [21]. In addition, eating disorders, such as obesity, are typically linked to emotions arising from psychosocial issues. In overeating, basically mood-enhancing foods may quickly aronse negative feelings, as guilt routinely follows overindulgence.

Food cravings are the strong desire for a particular food [10]. Chocolate is recognised as one of the most commonly craved foods, which is reflected in the popular concept of the chocoholic [10]. Food cravings are usually highly specific. If people crave chocolate, another sweet food will not satisfy the need; if potato crisps are craved, another salty food will not be satisfying [10]. Is chocoholism a physical response or a psychological one? Chocolate contains caffeine, but so do cola drinks, and these are not the subject of cravings [10]. Chocolate also contains phenylethylamine, which is also produced naturally by the body where it is associated with sex [10]. Most importantly, chocolate has highly desirable sensory properties (sensuous). There is evidence both for and against homeostatic control mechanisms in relation to meal size. Evidence for the maintenance of a constant energy intake in meals in human subjects comes from studies in which the energy density of single-item or multi-item meals was altered [14]. Other studies show individuals consuming constant weights of food regardless of energy content. However, appropriate responses to energy dilution may require multiple exposures in order to procude change [14]. In general, meals terminate because of combinations of a variety of factors, some physiological, some psychological (becoming bored), some social, some practical (the plate is empty) [14]. Due to SSS, the perceived attractiveness of a food diminishes continually whilst it is being eaten. Therefore, the greater the variety of palatable food items available in a meal, the more food is likely to be consumed overall. If these foods are predominantly refined, energy-rich types, and if no control mechanisms are present, intake may lead to overeating. Control mechanisms include social factors, e.g., who is present at a meal, and personal commitment devices, e.g., portion controls and other heuristics. There is a combined role in the regulation of ingestive behaviours for the brain, the special senses and the autonomic senses. Memory, sometimes in concert with mood, and physiological signals can elicit appetite without the intervention of the special senses. On the other hand, the sight and/or smells of a food can elicit appetite through conditioning. Olfactory, orosensory, postingestive and postabsorptive signals influence both appetite and satiety. They are mediated by mood, including the effects of the social context, i.e., the presence of others at a meal.

7.2
Physiological Controls of Eating and Drinking

The term physiology describes the functioning of living organisms in relation to both their internal and external environments. One of its aspects is metabolism, i.e., the inter-conversion of chemicals. In the human body, several physiological systems work together to achieve balance and effective overall body functioning. Thus, circulation of blood in the cardiovascular system allows oxygen, nutrients, hormones and waste products to be delivered to their destinations. Oxygen is supplied to the blood by the respiratory system, which also serves to remove carbon dioxide waste. The endocrine system is characterised by special glands, which produce hormones and secrete them into the blood as well as other body fluids. The digestive system (stomach, oesophagus, salivary glands, liver, gall bladder, pancreas, intestines) stores and digests food, delivers water, fuels and nutrients around the body, and eliminates waste. The urinary system (bladder, urethra, kidney) eliminates waste products and maintains water balance and chemical balance. The ANS controls salivation, gastric and intestinal motility, the secretion of digestive juices, and increased blood flow to the gastrointestinal system [9]. Basal metabolic rate, the speed of utilisation of energy, varies between individuals, e.g., due to differences in muscle mass and thyroid gland activity [22].

Physiological self-regulatory processes keep the internal characteristics of an organism constant in the face of external variability [9], as exemplified by body weight, temperature and fluid homeostasis. In common with all regulatory systems, homeostasis operates within a framework of stimuli, sensors and responses. Identification of a stimulus and a receptor are interrelated problems. Characterisation of the stimulus requires that its effect on a specialised receptor be demonstrated, and identification of the receptor demands that it be activated, or altered, by an adequate stimulus [23]. Unlike the special senses, which are fixed, the autonomic senses are mobile. Similarly, metabolic stimuli, such as the levels of oxygen, glucose and hormones, circulate in the blood, where they provide continuous feedback on the status of the relevant system. Except for oxygen, the activities of eating and drinking supply all the inputs into this internal world. How do people then manage to consume the correct amounts and types of foods and beverages at each eating or drinking event? What are the physiological factors that motivate people to engage in ingestive behaviours in the first place, and how are motivations translated into actual food choices? Due to the delay between ingestive behaviours and the replenishment of depleted stores, such behaviours are controlled by satiety mechanisms as well as by detectors that monitor the system variables [9]. There are joint roles for physiology and behaviour in homeostasis. For example, thirst during dehydration is accompanied by renal water conservation, and salt appetite during sodium deficiency is accompanied by sodium conservation [12]. Within such pairs of responses, physiology and the behaviour complement each other, in the sense that either response can be increased when the other is constrained; thus, urine is more concentrated when drinking water is withheld [12].

The regulation of body fluids involves the organism's digestive, respiratory, circulatory and excretory systems [9]. The volume regulation of both cells and blood is interrelated. Thirst, defined as a tendency to seek water and to ingest it, causes drinking. This behaviour continues until satiety mechanisms, which monitor water present in the digestive tract, stop further ingestion of water, in anticipation of the replenishment that will occur later [9]. Sodium appetite is a strong motivation of animals to seek, obtain and consume salty tasting fluids and foods [24]. Because cell membranes are impermeable to sodium, water is osmotically retained within extracellular spaces. This enables the vascular system to expand and, given the importance of blood circulation to life, the maintenance of adequate supplies of sodium in the extracellular fluid is essential [24]. Excesses of water and salt are excreted via the kidneys. The immature kidney does not have the capacity to maintain isotonicity on its own, and cannot defend against overhydration [25]. In fact, the consequences of feeding salty foods to babies can be fatal.

Because loss of water may be either intracellular or intravascular, thirst may be described as osmometric or volumetric, respectively. However, normal dehydration is accompanied by both these phenomena. Volumetric thirst relates to the volume of blood plasma; osmometric thirst occurs when the solute concentration around the cells increases, drawing water from the cells, which consequently shrink (cellular dehydration) [9]. Receptors in the digestive system signal to the brain that water has been received and is making its progress along the way to absorption. Having water placed directly into the stomach appears to be less thirst quenching than drinking it, although the satiety produced by receptors in the mouth or throat lasts only a short while [9]. Regulation of water intake is an example of integration of physiology and behaviour to promote homeostasis, where water deficiency simultaneously stimulates a conserving hormone and elicits intake [13]. There are a variety of physiological controls to preserve body sodium. Prominent among them is the adrenocortical hormone aldosterone, which promotes the conservation of sodium in the kidneys [24]. The peptide hormone angiotensin can elicit salt appetite when applied directly to the brain, and it is thought that it exerts some of its effect through synergistic action with aldosterone [13]. Receptors in the mouth do not appear to play a role in satiety associated with salt intake, satiation occurring only when the salt solution is permitted to accumulate in the stomach [9]. Sodium receptors in the liver may contribute to the satiation of a salt appetite [9]. Salt appetite is innate [25], and the fact that the human tongue contains receptors that specifically detect the presence of sodium chloride may be due to its vital role in the maintenance of body fluid homeostasis [9]. However, episodes of salt appetite are less common than episodes of thirst, as it is easy for healthy people to take in sufficient sodium through their diet [9]. In fact, ingestive behaviour cannot generally be understood simply as responses to physiological requirements [25].

There are two types of storage reservoir for metabolic fuels, one short-term and the other long-term. The short-term stores consist of glycogen and are located in the liver and the muscles; glycogen is synthesised in the liver from glucose, mediated by the pancreatic peptide hormone insulin [9]. Glucagon, like insulin a peptide hormone, has the opposite effect, i.e., it stimulates the conversion of glycogen back into glucose when the receptors in the pancreas and in the brain detect a fall in glucose [9]. The stores in the liver are primarily reserved for the CNS. Small, short-term carbohydrate reservoirs in the muscles serve as a fuel for quick bursts of muscular work, e.g., a sprint run. The long-term reservoir of metabolic fuels consists of fatty, adipose tissue, which is located under the skin and in the abdominal cavity [9]. The fat reservoir keeps the body working when the digestive tract is empty. Once the carbohydrate reservoir is depleted, fat cells start converting triglycerides into fuels that the cells can use and releasing these fuels into the blood stream. Fatty acids can be directly metabolised by all body cells except the brain, which requires glucose. The liver takes up glycerol and converts it to glucose. During prolonged fasting the liver converts some of the fatty acids present in the blood into ketones, which supplements the brain's diet [9]. Glycogen has limited storage potential, whilst fat has plenty, as adipose cells can expand enormously.

The liver and the brain are generally considered to be the sense organs for detecting metabolic events that control food intake, although the specific receptor cells have not been identified [23]. However, it may be stated that the detectors in the brain monitor the nutrients available on their side of the blood-brain barrier, and the detectors in the liver monitor the nutrients available to the rest of the body [9]. Because the brain can only use glucose, its detectors are sensitive only to glucoprivation (depriving cells of glucose), and because the rest of the body can use both glucose and fatty acids, the detectors in the liver are sensitive, in addition, to lipoprivation (depriving cells of lipids) [9]. The brain contains detectors that monitor the availability of glucose, its only fuel, inside the blood-brain barrier, and the liver contains detectors that monitor the availability of glucose and fatty acids outside the blood-brain barrier [9]. The detectors in the liver appear to be cells that are sensitive to their own internal rate of metabolism, not to the presence of particular nutrients outside them [9]. Circulating insulin is detected by receptors in the brain, but the CNS role of insulin, unlike its peripheral role, is not to facilitate glucose uptake or utilisation [26]. Because both the secretion and plasma levels of insulin are closely correlated with adiposity in individuals, it is thought that insulin acting at the brain plays a pivotal role in regulating fatness [26]. Insulin acting within the brain appears to cause a coordinated and generalised catabolic response, which includes consuming less food and losing weight [26].

The physiological signals that cause a meal or an eating episode to begin differ from those that cause it to end [9]. A fall in blood glucose level (hypoglycaemia) is a potent stimulus for hunger. Eating may start because the supply of nutrients and fuels has fallen below a certain level, but people do not stop eating because the level of these substances has been restored to normal [9]. Instead, the amount eaten is inversely related to the amount of nutrients left over from the previous meal [9]. If the hunger signal is moderate, a moderate satiety signal will terminate the meal, but if the hunger signal is strong, only a strong satiety signal will stop it [9]. Short-term satiety signals come from the immediate consequences of eating. The special senses, stomach and duodenum, and the liver can all provide signals to the brain that food has been ingested and is progressing on its way towards absorption [9]. Long-term satiety signals arise in the adipose tissue and control the intake of calories by modulating the sensitivity of brain mechanisms involved in hunger [9]. The stomach plays an important role in satiety, one that is based not simply on the volume of food present but also on the presence of specific receptors sensitive to nutrients and metabolic fuels [9]. It has been shown that the rate of gastric emptying is controlled by inhibitory effects on gastric emptying by calories in the intestine [17].

Many endogenous compounds are purported to be able to reduce meal size, cholecystokinin (CCK) being perhaps the most widely studied, with both central and peripheral receptor sites being proposed for it [26]. This peptide hormone is secreted by intestinal cells. As it is more effective at reducing meal size when levels of endogenous insulin are relatively elevated, one way insulin might act in the brain is to make animals more sensitive to some peripherally acting satiety compounds [26]. The last stage of satiety appears to occur in the liver, where it is no longer anticipatory, as the liver is the first organ that learns that food is finally being absorbed [9]. Leptin, the long-term satiety hormone secreted by well-stocked adipose tissue, desensitises the brain to hunger signals [9] and may be involved in a CCK-sensitising mechanism [14].

The putative roles of angiotensin in thirst, CCK in satiety and aldosterone in salt appetite are not unambiguous. In each case the hormone has a marked effect on ingestive behaviour: angiotensin is a potent stimulus of aldosterone secretion during hypovolemia, CCK inhibits gastric motility and emptying, and aldosterone promotes renal sodium conservation [12]. However, present controversies concern the significance of those effects; the fact that exogenous hormone can affect behaviour does not demand that it does so normally [12]. Associative learning mechanisms are involved in the development of autonomic responses in feeding and drinking systems. One example is the conditioning of insulin release to meal-associated and anticipatory stimuli. Conditioned hypoglycaemia was produced in rats after they had been injected with a placebo rather than the insulin which they usually received [26]. Similarly, the sweet taste of saccharin has been shown to be capable of producing the effect [26].

Stricker's scheme of the physiological mechanisms controlling food intake proposes excitatory and inhibitory systems in the brain, although their anatomic locations are not specified [13]. Hunger develops when the excitatory system is aroused by metabolic signals caused by the decrease in calories coming from the intestine and the liver and, as eating commences, from input over special sensory pathways (taste and smell etc.). Activity of catecholamines arouses the organism generally and allows the feeding response to take place. The resulting calories from the intestine and liver, along with insulin release, contribute to satiation, inhibiting the metabolic arousal signals. This is thought to occur, in part, because insulin favours the uptake into the brain of tryptophan, the precursor of serotonin, which may play an important role in the central inhibitory system. The central inhibitory (serotonergic) and excitatory (catecholaminergic) systems act reciprocally on each other. In fact, hunger may not reflect a biological need for food, as thirst does for water, because metabolic fuels are continuously available to the organism [13]. And there may be no stimulus for hunger, rather hunger sensations may be due to a reduction of the inhibitory stimuli of satiety [13].

Serotonin may play a specific role in a feedback mechanism that regulates appetite in relation to carbohydrate intake, as it has been found that tryptophan entering the brain, and the associated synthesis of serotonin, are increased in carbohydrate-rich meals as compared to protein-rich meals [10]. There has been increasing interest in the relative satiating power of protein, fat and carbohydrate, and the consensus that has emerged has been that protein produces the most potent control of food intake [3]. Similarly, it is thought that carbohydrate balance, the relationship between intake and expenditure, like protein balance, is tightly controlled, whereas fat balance is poorly controlled [3]. Thus, a change in food habits favouring lipid intake has an immediate impact on energy intake and energy balance [3]. This is despite a common belief that eating fat stimulates sensations of fullness. In fact, research supports this, in that it has shown fat to generate potent pre-absorptive satiety signals in the intestine. These signals are thought to be mediated, at least in part, by a CCK mechanism [3]. However, they are not capable of preventing over-consumption of fat. This is perhaps not surprising because fats provide strongly positive oral stimulation, as well as positive mental effects. Milk is quieting to newborns, and this effect is mediated by both opioid and CKK-related pathways, with one of the trigger substances being fat [25]. In rats, endogenous CCK that is normally released by milk perfusing the gut helps calm infants during maternal absence after a meal [25]. Milk's calming influences are already available to the newborn (rat) during the first suckling episode. Milk may calm and reduce pain postabsorptively through the exopioid β-casomorphine, which is hydrolysed from milk casein [25]. An important characteristic of the opioid-mediated system is its persistence beyond termination of the proximate initiating stimulus [25].

Comparable conclusions regarding sweet taste have been reached for human infants, for whom sucrose, or aspartame, exerted a powerful calming, energy conserving effect [25]. Sucrose calms human newborns 1–3 days of age, and the effects extend beyond the period of sucrose administration. Similarly, sucrose reduces pain in 1 to 2-day-old infants as judged by crying during blood collection from the heel for PKU screening or during circumcision, but lactose does not. Sucrose also markedly reduces heart rate, sustains a low heart rate and increases the incidence of hand-to-mouth behaviour [25]. The latter aspect of sucrose has been interpreted as evidence of the engaging of a feeding system [25]. Mediation for all these sucrose effects is oral rather than postingestive, because they are achieved in humans within 15 s of administration [25]. Cholesterol performs essential functions, e.g., in cell membranes, as a precursor in hormone synthesis and for bile acids [10]. Low blood cholesterol levels may increase both depressive and aggressive responses because, when brain cholesterol decreases, serotonin sites also decrease [10]. Vitamin B_6 also plays an important role in the production of the brain chemicals and neurotransmitters serotonin, dopamine and noradrenalin [10].

There is evidence that the consumption of low-fat foods decreases the risks of over-consumption, due to lower energy density and because carbohydrates may operate as a more rapid physiological cue for satiety [27]. In addition, high-carbohydrate, low-fat foods exert a greater osmotic load than high-fat ones, leading to more drinking which, in turn, may assist in preventing excess food intake [27]. One of the commonest mistakes people make when low in energy is to eat something very sweet for a boost [28]. Snacks rich in easily absorbed sugar cause the pancreas to produce too much insulin, leading to a rapid fall in the sugar level, which makes people feel even worse. Whilst a hypoglycaemic response of the kind experienced in diabetes is uncommon in normal individuals eating a normal diet, many respond to feeling tired and a little low by eating a biscuit or a chocolate bar [10]. However, increasing blood glucose levels do not make people feel more energetic, although they can prevent them feeling tired when under pressure [10]. The insulin response to cooked potatoes and bread is similar to that of sugar, although it is delayed; in contrast, rice and corn cause a smaller reaction [10]. Alcohol facilitates the secretion of insulin in response to the presence of glucose in the bloodstream, thereby causing rapidly falling glucose values [10]. Sitting calmly during eating allows the body to concentrate its blood, and therefore oxygen, supply on the digestive system [28]. By contrast, circulating stress hormones in the bloodstream divert blood to the limbs in support of the fight-or-flight response.

7.3
The Role of the Special Senses in Eating and Drinking

Food acceptance has been defined by one particular author as a perceptual/evaluative construct, a feeling, emotion or mood with a defining pleasant or unpleasant character [29]. Under normal circumstances, a food is only eaten if its sensory properties are acceptable to the person eating it. Sensations, together with the perceptions deriving from them, are key components of the so-called stimulus-response model, which is depicted in Fig. 3.7. The process of the sensory evaluation of a food or beverage begins with the transduction of physico-chemical product attributes, via specialised receptor organs, into neural events in the PNS. The neural processing of sensory stimuli gives rise to the basic sensory dimensions of quality attributes (e.g., salty, cold, red), their magnitude (weak, strong) and their duration [29]. These dimensions represent the first stage of sensory perception, but do not say anything about the nature of the object, nor about its acceptability. Environmental sensory receptors are specialised neurons that detect a variety of physical events. Five so-called special senses, or sensory modalities, are generally recognised, i.e., vision or sight, audition or hearing, gustation or taste, olfaction or smell, and the somatosenses (Fig. 7.2). However, stimuli reaching the sense organs are not processed automatically. Instead, as part of the mechanism of attention, people can take some active role in their perceptions by singling out meaningful stimuli [30]. This is particularly true of active behaviours, such as consumption. Attention also influences the way information is processed. Stimuli that are being attended to become clearer and more vivid, whilst those that are not being attended to are excluded from a particular experience [30]. Sensory stimuli are registered only if a minimum threshold of stimulation has been reached. This is the detection threshold. The recognition threshold usually lies above this absolute threshold. In fact, the sensory system is more adept at sensing differences between stimuli than at registering absolute intensities of stimuli, with the just-noticeable difference (JND) between two stimuli playing a particularly important role in practice (see Chapter 3).

Although the duration of a sensation is directly related to the duration of stimulation, physiological factors also play a role. In the physiological phenomenon called adaptation, repeated stimulation of the receptor organ leads to reduced neural activity, i.e., one becomes increasingly less aware or the stimulus

[29]. Oral-olfactory adaptation, during an eating episode may well be the first step in the process that retards ingestion rate and eventually leads to the termination of eating [25]. This is related to the phenomenon of sensory-specific satiety (SSS) discussed earlier. Persistence is the opposite of adaptation, and this may occur in response to eating bitter, sour or pungent foods [29]. Such persistence may be due either to continued neural excitation in the absence of the stimulus or to the presence of residual stimulus near the receptor surfaces [29]. Few foods are homogeneous mixtures, and aqueous and fatty constituents in particular act as solvents for different flavour active compounds. When food is chewed, sensory stimuli are released from the food matrix at different rates. The persistence of individual stimuli, and the order in which stimuli are revealed can contribute significantly to the enjoyment of a food or drink, e.g., chocolate, coffee or wine. In chewing gum, the food that is chewed for a longer period of time than any other, it is particularly important that the gum matrix releases any flavourings evenly and over a prolonged period, with the overall flavour impression maintained throughout the chew. Time-intensity studies are concerned with the intensity of stimuli, their timing and duration. Typically, several or all of the special senses will participate in a behaviour such as eating.

The human eye is designed to detect visible light, i.e., electromagnetic radiation of wavelengths ranging from 380 to 760 nm [9]. For a person to be able to see, an image must be focused on their retina, the inner lining of the eye, which accommodates the photoreceptors (rods and cones) [9]. Although black-and-white vision is adequate for most purposes, colour vision facilitates perceptual organisation [30]. For example, it confers the ability to see objects against a varied background and to distinguish ripe from unripe fruit [9]. Colours vary along three perceptual dimensions, i.e., hue, brightness and saturation [9]. Hue is defined by the dominant wavelength of the light, brightness relates to light intensity, and saturation to the relative purity of the light. Individual foods are seldom viewed in isolation, whether it be in the (super)market or on the plate. Instead, they may form part of a complex scene which may even be illuminated by lamps with enhancing colours. A visually attractive display increases the desirability of a food and stimulates appetite prior to its consumption. Sound stimuli are produced by molecules of air which are in motion, vibrating at a frequency of approximately 30 to 20,000 times per second [9]. Sound varies along three dimensions, i.e., pitch (high or low frequency), loudness and timbre (complexity/overtones) [9]. Sound has various roles in the sensory perception of foods. At the point of sale, it generates positive associations, e.g., French music being played in the vicinity of a French cheese, or wine, display. During cooking and eating, it provides or strengthens textural cues, as in the sizzling of a steak or the crunchiness of an apple. Both the sound of the sizzling steak and the crunch of the apple may be associated with liquids being released from cells, and therefore, juicy texture on eating.

Cutaneous sensations are based on the stimulation of receptors in the skin, i.e., tactile sensations caused by mechanical displacement of the skin, temperature sensations caused by heating or cooling, and pain caused by stimuli that are potentially damaging to the skin [30]. The cutaneous senses are served by the somatosensory system. This system also includes proprioception, the sense of the position of the limbs, and kinaesthesis, the sense of limb movement [30]. Skin consists of subcutaneous tissue, the dermis, and an outer layer, the epidermis, and contains various receptors scattered throughout it [9]. To identify an object by touch, kinetic sensations from muscles and joints is needed together with cutaneous information [9]. This allows an object to be scaled and to be judged as hard, soft or slippery [9]. Mechanoreceptors are free-nerve endings that respond to intense pressure, which might be caused by something striking, stretching, or pinching the skin [9]. A second type of free nerve ending appears to respond to extremes of heat, to acids, and to the presence of capsaicin, the active ingredient in chilli peppers [9]. Another type of nocireceptive fibre contains receptors that are sensitive to adenosine triphosphate (ATP), which is released when the blood supply to a region of the body is disrupted or when a muscle is damaged [9]. Information about the internal organs is provided by receptors in the outer layers of the gastrointestinal system and other internal organs, and in the linings of muscles and the abdominal and thoracic cavities [9]. Many of these tissues are sensitive to stretch and do not report sensations when cut, burned or crushed. In an instance of adaptation, a moderate, constant stimulus applied to the skin fails to produce any sensation after it has been present for a while [9].

Gustation (taste) and olfaction (smell) respond to chemical stimulation, which is why they may be referred to as the chemical senses. These are most fully engaged during eating and drinking because the flavour of a food or beverage is a combination of taste, smell and, depending on cuisine, chemosensory irritation [31]. The chemical senses are considered as gatekeepers, because the stimuli they respond to are on the verge of being assimilated into the body [30]. In this respect, their task is twofold, i.e., to identify harmful substances, that should be rejected, and to identify beneficial substances, that should be consumed [30]. This gatekeeper function is reinforced by the large affective, or emotional, component of flavour perception [30]. Since the chemical senses are constantly exposed to irritants such as bacteria and dirt, their receptors undergo continuous renewal (neurogenesis), usually over a period of 5 to 7 weeks [30]. The sense of taste creates perceptions in response to liquid molecules that contact receptors in the mouth. The sense of smell creates perceptions in response to gaseous molecules that contact receptors in the nose. Preferences for food flavours are either innate or learned. For example, the liking for sweet taste is present at birth, whereas preferences for bitter tasting foods and beverages are thought to be acquired later in life [31]. Cultural transmission of flavour preferences probably begins before birth as the late-term foetus has functional chemosensory systems and the odour of amniotic fluid is altered by what the mother eats and drinks [31]. There is a view that the only specific ingestive appetites that are not acquired at all are for water and salt [25]. However, the taste of milk, fat and sucrose is rewarding to infants, and this is probably opioid-mediated [25]. A mother's milk also contains sensory cues reflecting the mother's diet and these cues influence dietary preference at weaning [32]. Because the flavour of human milk is altered by what the mother eats, the chemosensory world of the breast-fed infant, unlike that of its formula-fed contemporary, is forever changing [31]. Age-related loss of chemosensory sensitivity, especially olfactory sensitivity is common [31].

For a substance to be able to be tasted, it must first be dissolved in the saliva, where it will then be able to stimulate the relevant oral receptors, in particular, those on the tongue [9]. There are at least four qualities of taste (there may be additional ones, see Chapter 5): bitter, sour, sweet, and salty. The tongue is covered with a variety of papillae that give it its bumpy appearance. Although filiform papillae are the most numerous, they do not contain any taste buds; however, fungiform, foliate and circumvallate papillae do [33]. Fungiform papillae are distributed most densely at the tip and on the edges of the tongue, foliate papillae on the rear edges of the tongue, and circumvallate papillae on the rear of the tongue. There are also taste buds (without papillae) on the roof of the mouth [33]. Taste buds consist of clusters of taste receptor cells, which communicate with the surface of the papilla in which they are embedded via a taste pore, a small opening. Essentially, sweetness, bitterness, saltiness and sourness can all be perceived on all loci where there are taste receptors, and although there are variations in taste threshold around the perimeter of the tongue, these are small [33].

In order to taste salty, a substance must be ionised. Although the best stimulus for saltiness receptors is sodium chloride, a variety of salts containing metallic cations (Na^+, K^+ and Li^+) with a small anion (Cl^-, Br^-, SO_4^{2-} or NO_3^-) also taste salty [9]. The source of the saltiness is the sodium ion and not the chloride ion [34]. Sourness receptors appear to respond to the hydrogen ions present in acidic solutions [34]. Salty and sour substances produce relatively little individual variation in taste, because the chemical stimulus for both – a tiny positively charged ion – is relatively simple in structure [34]. Bitter and sweet substances are more difficult to characterise. These are usually large and complex organic molecules that undergo permutations and have varying effects [34]. The typical stimulus for bitterness are plant alkaloids such as quinine; for sweetness it is sugars such as glucose or fructose [9]. There are wider variations in what people refer to as sweet or bitter perceptions than in what they call sour or salty. Unlike sour and salty substances, different bitter substances act upon different receptor mechanisms, which may explain why some people are sensitive to some bitter foods but not to others [34]. Work on genetic variation in taste has identified so-called supertasters of the bitter substance PROP (6-n-propylthiouracil), individuals who are unusually sensitive to bitters and sweets as well as the burn from capsaicin in chilli peppers [33]. Supertasters appear to have more taste buds than other people, and since taste buds have trigeminal neurons (mediating pain) associated with them, there is an association between perception of taste and irritation. Supertasters of PROP are a subgroup among tasters, with some individuals incapable of sensing bitterness in the substance (nontasters). Some other sources of bitterness are perceived to be more intense by tasters of PROP, e.g., the bitter tastes of saccharin, of sodium chloride, of sodium benzoate and potassium benzoate, as well as of Cheddar and Swiss cheeses [33]. Saccharin, sucrose and neohesperidin dihydrochlcone (a sweetener made from citrus fruit peels) are also perceived as sweeter by PROP tasters than by nontasters [33].

The stimulus for odour consists of odorous volatile substances having molecular weights in the region of around 15 to 300 [9]. Almost all of them are organic, lipid-soluble chemicals [9]. There are innumerable odours, ranging from simple to very complex. Odours with positive emotional connotations are also referred to as aromas or fragrances. Fifty million olfactory receptor neurons reside within two patches of mucous membrane, the olfactory mucosa, located on the roof of the nasal cavity [9]. They can be accessed either directly, through the nose, or via the oral cavity by the retronasal route. The retronasal route provides for a degree of heating and separation of individual odourants. The olfactory mucosa also contains free trigeminal nerve endings [9]. These presumably mediate sensations of pain that can be produced by sniffing some irritating chemicals, e.g., ammonia [9]. The mucosa lies just below the olfactory bulb, an outcropping of the brain [30]. In humans there appear to be between 500 and 1000 different receptors, each of them being sensitive to different odourants [9]. However, any particular odourant binds to more than one receptor, creating different patterns of activity. Although people can discriminate among many thousands of different odours, there is a lack of a serviceable odour vocabulary [9]. The ability to recognise and name specific odours can be learned however. To achieve this it is important that subjects are provided with correct names, or labels, for the odours in question at the beginning of training [30].

The nature of the solvent systems within a food is important in how individual stimuli are perceived. For example, trigeminal irritants, e.g., capsaicin in chilli peppers, tend to be lipophilic [35]. This means that they are more readily soluble in lipids than in water. Many food and drink products have acceptance criteria that are not, strictly, sensory attributes, but are associated with them, e.g., attributes such as refreshing and thirst-quenching in beverages. Thirst-quenching attributes of drinks are associated most with sourness and least whilst sweetness and viscosity [36]. Surprisingly, carbonation of a drink appears not to be associated with it being thirst-quenching [36]. As basic sense data are conveyed through the nervous system, numerous intra- and cross-modal sensory interactions modulate the information flowing along different channels, organising it into object-relevant characteristics, drawing input from learning and memory [29]. It is here that taste and odour combine to form a recognisable flavour, and that sounds emitted during biting combine with kinaesthetic information to result, for example, in the perception of the crispness of a food [29]. As these integrated percepts evolve and as object recognition emerges, information from learning and memory creates a framework of contextual information and expectations that serves to modulate the percepts [29]. At this stage, hedonic experience is also evoked [29]. Like the percept it accompanies, this hedonic component is subject to a variety of factors unrelated to the stimulus itself, including previous experience, context, culture, expectations, and physiological status [29].

Appearance is an essential cue in the identification of flavour, e.g., sight of a pineapple or an apple. Where the original form has been lost through processing, e.g., puréeing, colours assist in the identification of a flavour. Where a flavoured product is colourless, colour may need to be added to allow identification to occur. There is no question that the appearance of food influences its desirability, and research indicates that people's ability to identify tastes and smells depends to some degree on colour [30]. When colour matches flavour (red for cherry flavour, yellow for lemon flavour), people are better able to identify a flavour than when it does not. In particular, the presence of an extraneous colour hampers the correct identification of a flavour. These phenomena appear to be related to the ease with which odour information may be retrieved from memory [30]. Odour-taste enhancement is a phenomenon whereby a sweet taste is judged as tasting sweeter when mixed with a tasteless odourant that is associated with sweet flavour (cherry, caramel). Neither training nor exposure influences taste enhancement, indicating that confusion between odours and tastes may be more cognitively impenetrable than may have been thought in the past [37]. Perceived acidity is strongly suppressed in beverages by sweetness, but sweetness is only weakly suppressed by acidity [38]. The saltiness of pâté is enhanced by having the salt sprinkled on the surface of a food rather than incorporated into the food [39]. The diner's perception of a food or dish, e.g., a piece of roast beef, derives from both intra- and intermodular effects, which act either concurrently or in sequence. The golden brown external, and pinkish internal, appearance; the crunchy, then chewy texture on the outside and the melting texture of the centre; these combine with complex and developing flavours to constitute the overall appeal of the food.

The totality of sensory experience can serve to establish a context, or set of perceptual and hedonic expectations, that alter the emotional or behavioural response to any single element of that overall experience [29]. Total meal liking is not a simple, unweighted linear function of the components of a meal [40]. There is evidence for a peak of liking for the main course [40]. The finding that dessert is the second most hedonically predictive component in a multi-course meal is compatible with the idea that the end of any sequence contributes disproportionately to the memory of that sequence [40]. If a meal does not satisfy the expectations of course components and sequence, satiety is incomplete. This can mean that thoughts about the next meal will arise more quickly. No matter how objectively "full" someone is after eating a starter and main course, they will usually find space for dessert, some cheese, more wine and some chocolate to go with the coffee. Satiety therefore requires sensory satisfaction as well as physiological stimuli. Food-related gustatory and olfactory stimuli have the ability to elicit salivary, gastric and pancreatic secretions, which act to facilitate the perception, consumption and absorption of foods and nutrients [29]. For example, release of insulin can be conditioned to meal-associated and anticipatory stimuli [26]. More generally, associations and conditioning influence the motivational states of appetite and thirst. The more desired a food, and the more attractive the presentation of that food, the more likely it is to elicit appetite. The greater the deprivation in respect of a desired food, the greater the potential appetite for it [8]. As the likelihood of a desired stimulus occurring increases, so does the appetite for it.

7.4
The Visual Appeal of Foods and Beverages

Food has always been the object of artistic study. Many still lifes incorporating fruit, vegetables and other foods, part of the legacy of the greatest painters in human history, testify to this. Natural structures typically incorporate symmetrical, and otherwise harmonious, elements, which appeal to people. Fruit and vegetables also exhibit a wide variety of colours. Visual examination of objects such as melons, lemons and onions therefore tends to elicit positive responses in people. Of course, the visual attractiveness of fruit and vegetables may derive, at least in part, from their association with the wider sensual pleasures of eating. Whether positive responses to pattern, form and colour are innate or acquired by the artist and the consumer is a matter for debate. However, the artist's tradition in relation to food continues today, both as such and in food photography and the related discipline of food styling. With the ever-increasing remoteness between the sites of food production and consumption, modern consumers depend on product appearance and image as quality cues. This includes the appearance of the packaging in which food, in particular, manufactured food, is presented for sale. The consumption value inherent in food displays, both at the point of sale and during an eating event, is not solely attached to the food itself. It also encompasses the food provider, whose skills, intentions and attitudes find physical expression in the way in which a food is presented. Appearance attributes in a food are particularly important in allowing consumers to categorise a product as familiar or unfamiliar, normal or abnormal, and desirable or undesirable. However, to a large extent, what is thought to be an appetising display of food is context-dependent, with cultural and fashion elements playing a major part. Similarly, the type of consumption value required is important, e.g., will the eating event be subject to emotional, fun or spiritual overtones. Sophistication is not appropriate in every context. In the wrong context, it will appear as either over- or under-elaboration. The sophistication of food presentation may be conceptualised on a scale running from chaotic, to rustic or informal, then to simple and delicate, from there to elaborate, and ending in over-decorated. This particular food presentation scale is anchored in negative terms at both ends, suggesting lack of care and skill on the one hand, and pretension and fussiness on the other.

The visual perception of a food by a consumer is an integral part of the overall impression generated during the purchasing or consumption situation. It is a commonplace that people buy and eat food with their eyes, i.e., appearance is a key factor in food acceptance. There are many aspects of appearance which are important in this, most notably, colour. However, size, shape, surface structure, pattern, uniformity, proportions, balance, gloss, clarity, translucency and turbidity may be similarly important. This depends on the food or beverage under consideration. It is thought that size and shape are taken for granted by consumers, not meriting much thought [29]. However, this too has to be seen in context. In fact, size frequently forms the basis of product differentiation. Confectionery bars have, at different times, been offered for sale as either standard, maxi, or mini versions, as well as praline-type versions of themselves. Shape too can be turned into a signalling tool, in particular, the shape of food packaging. The traditional fluted Coca Cola bottle is probably the most familiar example of this. However, in most food choice situations, shape is perhaps most important in terms of its cueing that a food is normal. Whilst little attention may be paid by a shopper to the shape of a food if that shape appears normal, an abnormal shape is likely to result in instant rejection. In the first instance, appearance provides cues about the edibility of a food, i.e., whether it is the sort of food one wants and whether it is fit to eat, e.g., free from mould and other defects. Faded colour, whether in the food or beverage itself, or in its packaging, is a cue of abnormality and induces rejection. Much of the rationale for the use of food colours for achieving batch-to-batch conformity, is based on the argument that consumers will reject a food if its appearance points towards failures in manufacturing control (see Chapter 6). Pigmentation may indicate whether a fruit it is ripe and ready to eat, e.g., the yellow-brownish colour of ripe bananas.

The assessment of the appearance of a food precedes its ingestion. However, appearance has predictive value in terms of its orally perceived sensory properties. For example, the creamy texture of a dessert can often be perceived visually as well as orally. The careful display and arrangement of foods serves to communicate certain positive attitudes on the part of the provider, e.g., thoughtfulness, humour, etc., thereby enhancing their overall appeal. Brand identity performs a similar function. When food is offered to children, attractive presentation is particularly important. A meal is an occasion in a child's day, not just a necessity, and it therefore benefits from an element of fun [41]. For example, rather than offering a plain jacket potato and poached egg to a child, the food may be assembled to represent a "potato man" (body: jacket potato; face: poached egg; eyes: peas; hair: shredded carrot) [42]. If a portion of a standard dish, e.g., shepherd's pie, is ladled out with little thought, the food may be negatively associated with school meals. On the other hand, the same dish looks stylish if presented in individual ramekins [42]. Ambience is closely related to presentation, and clutter is distracting in relation to both. Harmony at the dining table, and the harmonious presentation of food, increase enjoyment and improve digestion. The sight of an appetising display of foods influences mood in a variety of ways. It can be either calming and relaxing, or exciting. Anticipation and expectations are generated, and these are especially important during high involvement eating occasions, such as festive meals and expensive restaurant dinners. Both are eating events invested with hightened emotions. Food that looks as though it has been fussed over may be off-putting [43]. This is especially true of home cooking. An element of theatricality may however be expected in certain types of restaurant cooking. Appearance tends to be more important in starters and in desserts than in main dishes. It is the special role of starters to stimulate appetite. Food that looks appetising kicks the autonomic sensory system into action, switching it into rest-and-digest mode. Presented with an appetising-looking starter, the expectant diner will savour this food visually. Their imagination having been stimulated, they will then be able to relax into confident anticipation of all of that is yet to come. By looking closely at a food before eating it, the spectrum of sensations to be experienced will be able to be anticipated. However, the appearance of a starter does more than raise expectations about itself; it also raises expectations about subsequent courses and about the eating experience overall. The dessert, on the other hand, signals the end of a meal and, as a final parting shot, provides an experience of total indulgence, involving all the senses and offering strong stimuli to all of them.

Components of visual appeal are instantly assessed by the consumer and individual stimuli integrated in the process of perception. For example, a dish of bacon and egg will first be recognised as such. It will then be pre-tasted through memory recall. This will lead to an anticipation of an instant saltiness, followed by the revelation of more subtle flavours, the sound and texture of the bacon snapping and the mouth-coating effect of the egg. Form and colour both catch the eye and convey information via visual cues, e.g., via colour-flavour associations. For example, warm colours such as red and orange can raise expectations of sweetness in a food [44]. In one particular study, vanilla pudding, coloured to look like chocolate pudding was perceived as tasting of chocolate [45]. However, colour appears to have little effect on the perception of the saltiness of a food [44]. Colour contributes to the phenomenon of SSS, i.e., compared with monochrome presentation, attractive and varied colouring can increase food intake [44]. Products with inappropriate colours are likely to be rejected out of hand.

Translucency and haze represent points along a scale, which is anchored in the food attributes of transparency and opacity [46]. Any attributes on this scale can be either desirable or undesirable, depending on the food in question. At worst, certain cloudy products may be judged to be faulty. For some products, both clear and cloudy versions exist, e.g., apple juice. Consumers perceive clear apple juices as more thirst-quenching and more suitable for drinking with a meal than they do cloudy juices [47]. Cloudy juices, on the other hand, are perceived as more nourishing, representative of whole fruit [47]. Clarity is an issue in solid foods as well, with jellies an obvious example. Raw salmon flesh is translucent and pink, but turns opaque and lighter upon cooking [46]. Translucency is also reduced in fish that has been frozen. There is a slight difference in the appearance of smoked salmon manufactured from raw or frozen fish. The haze in very thinly sliced smoked salmon may, or may not, be noticed by a consumer and may, or may not, affect the overall acceptability of the product. The appearance attribute "gloss" is also associated with fish, i.e. in terms of freshness, and it also signals freshness in fruit and vegetable. The association between gloss and freshness is especially strong in the case of whole fish. Educated shoppers know that fresh fish have bright eyes, with black pupils and transparent corneas, and that the skin is shiny due to a coat of natural clear slime [48]. Lack of gloss in a positive sense is associated with dry foods, as well as with hung meat, although a soft sheen should remain present on the latter. Differences in gloss enable the diner to distinguish between components of mixtures such as fat floating on hot gravy [46]. The custom of coating freshly cooked vegetables with butter serves to preserve their gloss while the meal is being eaten [46]. Gloss makes a food more visible because of the increase in reflected light. Egg or milk washes are commonly used to finish baked and other goods to improve their visual appeal. The texture terms juiciness and succulence are both associated with gloss.

Patterns promote object recognition both in basic produce and in man-made foods. This includes colour patterns, e.g., in different varieties of apples. Very precise and regular patterns may give the impression that a product, e.g., biscuits, has been machine-shaped and therefore, mass produced [46]. Greater randomness tends to be associated with home- or handmade foods, although the regularity of items of patisserie may be attributed to the high skills of a master craftsman. The shape of a food may be used to present it in a way that highlights its origin and nature, e.g., by serving whole baby vegetables instead of large ones chopped into pieces. On the other hand, the origin and nature of a product may be disguised in the way it is presented for eating. Modern consumers, due to their lack of involvement in food production, typically prefer to gloss over the link between meat and a living, thinking and feeling animal [49]. This general attitude finds an extreme expression in the popularity of products such as fish fingers and chicken nuggets, where any anatomical connection with the raw material has been lost. The sight of intact fish, especially on the dining table, is off-putting to some consumers, but is attractive to others. Buying fish fillets not only removes the association with the animal, but is also more convenient. On the other hand, the freshness cues available from the whole fish will be lost. Buying whole fish and preparing it at home may be inconvenient because of the smell and waste problems associated with whole fish. Fishmongers should therefore be willing and able to skin, scale, fillet and bone fish according to requirement and request [48].

Shape can be used as a proxy for variety. Pasta is a simple food, consisting of no more than wheat flour mixed with water, with egg pasta including egg in the recipe as well. Yet in Italy, more than 300 different shapes of pasta have been invented, some short, some long, some smooth and some ridged [50]. Some varieties are coloured through the incorporation of coloured foods, e.g., spinach (green), beetroot (purple), tomatoes (red or pink), wild mushrooms (brown) and cuttlefish ink (black). This considerable range of basically similar foods is needed, ostensibly, to accompany a similarly wide range of sauces [50]. Whilst such a claim may seem somewhat exaggerated, in fact, because sensory signals are integrated between modalities, variety of shape, even if it does not impact flavour directly, is still experienced as variety. Besides, pasta shapes have different regional associations, and this introduces another variable, i.e., emotions held about a product. Sausages, balls and patties (burgers) are popular shapes for the presentation of chopped meats. These same shapes have been adopted in the development of many vegetarian foods, causing certain commentators to question, somewhat unfairly, why vegetarians still insist on eating sausages, haggis etc. In other words, why don't they just eat vegetables that still look like vegetables? Simple shapes reduce mess and have a role in portion control and this, rather than their early association with meat products, may be one of the reasons for their universal appeal.

Massed displays of uniformly shaped and coloured fruit and vegetables in peak condition are highly effective in drawing a shopper's attention to a product. In this scenario, quantity reinforces both quality, in the sense of conformity, and desirability, as mass displays signal actual demand for a product. Carefully arranged smaller displays are similarly attractive, although perhaps for different reasons, e.g. a platter of oysters, all graded according to size and shape. On the other hand, an individual plate, or a serving plate, may contain an assortment of items, perhaps a selection of different shellfish of different sizes. In this case, the consumption value displayed is that of variety, although the display is still harmonious. Other examples include assortments of cold meats as a starter, a medley of several fish as a main dish, or one of themed (e.g., chocolate) sweets as a dessert. Presentation tends to be a key element in traditional recipes, and may include garnishes and other decorations, as well as the correct crockery and drinking glasses for serving the item. A harmonious, patterned effect is achieved by laying out fans of potato slices on a hot pot, or of fruit on a tart. Japanese food, like Japanese culture generally, is designed to create a mood of simplicity and calm [51]. The underlying principles are those of Zen Buddhism. Foods are served in small portions, whose appearance is disciplined, said almost to be austere, and its arrangement on the plate is always three-dimensional [51]. Good Japanese restaurants are rated on three components, and in equal measure; they are the cooking, the presentation of the food and the general ambience [51]. In Chinese cooking, the size and shape of the individual ingredients of a dish, in particular, a stir-fried one, should harmonise with one another [52]. This is why food-cutting techniques are especially important in Chinese cuisine. Breads and pastry doughs for pies, and for encasing meat and fish, may be scored or shaped for decorative purposes. Bread may be presented on a scale from rustically uneven to highly decorative, as in breads used as part of religious rituals. In some Cretan villages, women still specialise in the intricate decoration of celebration breads [53]. In the UK, fancy pies may be similarly decorated with pieces of dough shaped into leaves, trellises and other forms. In the dish commonly known as "loup en croûte", a whole sea bass is encased in a pastry crust which is itself decorated to represent that very fish. Pastry has also been made into cages and used to present shellfish such as langustines, paralleling the caramel cages and spun sugar sculptures often used to frame elaborate desserts. All these techniques, and the skills with which they are executed, serve to communicate to the diner the mood, attitudes, concerns and skills of the cook or chef who is providing the food.

Whilst shape is the key visual attribute for the identification of categories of foods, in some cases, colour can assume a similarly important role. This is obvious where the original structure of a food has been destroyed as a result of processing. It also applies to certain fruit and vegetables, where the colour needs to be seen for the shopper to be able to distinguish between similar forms, oranges and grapefruit, and bananas and plantains. Modern foods typically cannot be sampled at the point of sale, so that food packaging has to represent the food and its sensory attributes. The representation of the product on the packaging, and the colour used, are important in communicating the nature of the product to the consumer. The use of added colour in food, to make it look more appetising, or for purely aesthetic reasons, or to give it a symbolic character, has been a feature of many cuisines world-wide at least since classical times [54]. Colour affects the mood, thus contributing to the pleasure of eating. People who suddenly lose their colour vision may find food difficult to look at whilst eating it [30]. Being presented with a piece of white fish, served on a white plate with a white sauce, is unimpressive [55]. Colour in foods and beverages is important because people expect foods to have a characteristic appearance. This is often used as a justification for adding colours to foods [56]. One of the most important variables in the cross-modal effects of colour on taste and odour is the appropriateness of the colour to a food and/or its flavour constituents [29]. In general, appropriate colours will increase the perceived intensity of colour-associated flavours [29]. Inappropriate colours, on the other hand, will increase thresholds and decrease flavour discrimination [29].

In earlier human cultures, many foods were chosen for consumption solely on the basis of their colour. The young men of ancient Greece were urged to eat red food, and drink red liquids to become more sanguine, cheerful and confident [29]. Bulgarians too delight in the colour red, which is considered healthy and invigorating [54]. Red-cooked rice is known as a happy food in Japan because the colour red is considered lucky and joyous [57]. The ancient Persians, who strongly believed in the satisfaction of the eye as well as of the palate, found little to gratify them about the fare of the ancient Greeks [58]. Indeed, they thought that the Greeks remained hungry much of the time because of the dreariness of their food. However, it appears that colour preferences in foods are the result of experience, culture and conditioning [29]. There is a peculiar regional preference for brown eggs among the inhabitants of the north-eastern USA, a colour preference that does not exist elsewhere in the country [29]. The colour brown has been associated with so-called wholefoods, e.g., wholemeal breads and brown rice, and the type of healthy eating associated with them. For a time, these brown foods also stood for a slight freakishness of sandal-wearing, austere individuals, who had deliberately dropped out of the so-called consumer society. This is one example of a minority taste which, over a period, has become mainstream.

The use of synthetic colours in staple food products, in particular, canned fruit, vegetables and meat, requires a degree of discretion [46]. This seems to indicate that consumers do not think of products of this nature as artificially coloured. On the other hand, strong, vivid, phantasy colours are widely acceptable in made-up products, especially sugar-based foods. Many people perceive ethical issues in the colouring of foods. The argument is that colour does not simply stand for itself, but usually represents some other kind of value. For example, a high-fat sausage coloured red might suggest a high lean meat content, and orange-coloured juice drinks a high fruit content. The deep yellow of egg yolk and the intense pinkness of salmon might be associated with healthy animals eating healthy, natural diets, rather than with animals being fed diets supplemented with canthaxanthin (see Chapter 6). Consumers will not be automatically aware of this practice because the presence of the synthetic pigment is not declared on food labels. There remains a strong argument that the use of colours in food formulations often misleads consumers about the true nature of the food. Flavoured waters, which have become popular on the UK market are not coloured. Instead, colour cues associated with each flavour can be found on each bottle, e.g., a large lime-coloured tick. Although this new approach may lead to mis-purchasing by confusing the product on the supermarket shelf with unflavoured water, the mistake would not be repeated, and the cost of learning would be small. It also seems obvious that bottled water, with its image of purity, could not possibly contain food colours.

Visual appeal is particularly important in low-odour, cold foods, because they lack the aroma that may stimulate the appetite for hot foods. While the eye lingers on a dish that is about to be consumed, in the shopping situation, visual cues usually need to be more immediate and impactful, especially for new product introductions. On the supermarket shelf, each product must compete against its neighbours, and it must do this in the course of a very short time interval. In fact, a large number of first purchases are impulse buys. Here the orderliness of a display signals the professionalism of the organisation. Poor presentation serves as a warning, and reduces the likelihood of a product being selected. For example, care taken with the pack design for sausages strongly influenced expected eating quality and safety [59]. In general, consumers are less aware of, and concerned about, a food's appearance than they are about its flavour and texture. Among a wide selection of common foods, the ten foods with the highest appearance value for consumers were brightly coloured fruit and vegetables, whilst the ten lowest appearance values were associated primarily with liquids [60]. Appearance may not be consciously identified by consumers as a major sensory attribute. Yet, in practice, any deviation from what is expected is likely to lead to the rejection of a food. This is best illustrated using the example of organically produced fruit and vegetable, which are often rejected by the wider food buying public on account of minor cosmetic defects, even though they may be thought to offer other, far more important, types of consumer value.

7.5
The Textural Appeal of Foods and Beverages

Texture is the sensory manifestation of the dimensions, structure and other physical properties of a food or beverage. The role of texture as a driver of the sensory acceptance of a food is a particularly interesting one. The liking, or disliking, of some textural attribute in a given food is influenced by individual, physiological factors on the one hand, and by psychological ones on the other. Social learning plays a major part in the textural preferences of the individual members of different cultures. Texture perception can involve several of the special senses, i.e., vision, hearing, touch and kinaesthesis. A food may be rejected on account of its texture even before it is taken into the mouth, e.g., because of its visual texture or after manual touching. Texture perception normally involves a food being taken into the body, although the food is not necessarily ingested at this stage. Most aspects of texture are perceived before a food is ingested, and texture thus serves as a valuable screening, and risk assessment, device when food is consumed. The presence of an abnormal texture in a food alerts the individual to the possibility of other, more serious abnormalities. Texture has an important role in the ease with which a food may be manipulated in the mouth, including swallowing it. Visual texture refers to appearance cues about texture, e.g., the viscosity of soups. A soup which is visibly viscous may signal to an observer not only how it will feel in the mouth, but also, how nutritious the soup is likely to be, and how filling after consumption. Any physical contact between a consumer and a food or beverage takes place by touching, either directly or indirectly, via some eating implement. Manual touching is followed by chewing, sucking or sipping a food, resulting in an overall perception of its physical structure [29]. Texture may often be assessed by manipulating a food with the hands. The ease of cutting, or pulling apart, a piece of meat will predict its tenderness in the mouth on chewing, and the ease with which it will be swallowed. In this manner, a parent is able to make texture assessments on behalf of a young child. Spreadability is another aspect of texture, although this has little predictive value in terms of the mouthfeel of a fat or spread. Texture is also assessed via the sense of hearing. For example, the snap of a bar of chocolate is an important aspect of the pleasure of consuming it, as is the characteristic sound associated with eating crisps and other snack products. Sound is closely related to texture perception, as characteristics such as crunchiness can, and must, be heard.

Basic fruit, vegetables, fish and meat animals, and other foods not modified by food technology invariably have complex structures, which originate in their anatomy. Some portions of such foods will display naturally attractive textures, in the sense of them being easy to manipulate in the mouth, whilst others are less suitable for human consumption. Seedless grapes and tomatoes may be eaten whole, and all aspects of their texture may be thought of as desirable, including the outer, bite-resistant skin, which is burst by the teeth, and the soft, juicy interior, which is subsequently revealed. Inedible structures include stones in fruit and bones in meat and fish, and these may have to be manipulated out of the mouth whilst a food is being eaten. Man-made foods present the opportunity for textures to be engineered in accordance with consumers' desires. Such foods, especially solid ones, rarely mimic naturally occurring textures convincingly, and clearly, texturally undesirable structural features will not be built into them. It is easy to produce biscuits with homogeneous texture, and to standardise their texture in terms of hardness, stickiness and other key textural attributes. If desired, biscuit textures can be made more complex by introducing shelled nuts, either whole or chopped, sultanas, or chocolate chips. French fries (potatoes) have a hard, crunchy skin that needs to be pierced by the teeth before the soft interior is revealed. This texture perception sequence is not unlike that for eating a grape, and both can be savoured in a similar way.

Oral motility, which involves the working together of the tongue, cheeks, palate and teeth, plays an important role in food texture acceptability. The physical skills required to break down food in the mouth and transport it to the throat for swallowing must be learned. At birth, only liquids can be consumed, and oral texture perception is therefore subject to considerable postnatal developmental processes [29]. From infancy through to adolescence, a child undergoes a developmental pattern of preferences for food textures that begins with soft, smooth and uni-dimensional textures and progresses to firm, rough and complex ones [29]. In young children, attitudes to texture are therefore shaped mainly by physiological factors, with cultural and social influences playing a secondary role [61]. The throat is a very sensitive part of the body, especially in children. A bolus that contains particles too large to swallow comfortably, or one that goes out of control while sliding down the throat, may invoke a gag reflex [61]. For this reason, textural characteristics that render a food difficult to control in the mouth are likely to cause it to be rejected [61]. Fish with many small bones, e.g., herring, exemplify the problem. On the other hand, the texture of cod, in which soft, juicy flakes of flesh come away easily from the cooked fish, is an example of a food for which oral control is easy. Ageing, and the associated loss of dentition, or wearing of dentures, may result in further changes in texture preferences, avoiding hard and sticky foods. Because dentures cover part of the palate and the gum, masticatory efficiency, and sensitivity to textural attributes, is sharply reduced in denture wearers [61].

Crispiness and crunchiness appear to be universally liked textures, whilst textures that give a sense of lack of control in the mouth are commonly disliked, e.g., soggy, watery, lumpy, sticky, slimy, crumbly and tough [62]. However, socially and culturally learned associations have a significant influence on consumer evaluations of texture. People's awareness of food texture is heightened either when expectations are being violated or when non-food associations are being triggered [62]. Many adults in Britain are disgusted by the thought of semolina pudding, something they were forced to eat at school. At school, the appearance and sliminess of the dish caused it to be compared to frogspawn. Other non-food associations relevant to slimy textures include mucous and vomit. Stringy, mushy and lumpy textures are difficult to manipulate in the mouth and may lead to children rejecting certain cooked vegetables. Among British 8-to-10-year old children, textural properties are strongly associated with aversions to certain vegetable preparations [63]. Fear of choking or gagging during mastication could be the basis for such aversions. In the UK market for yoghurt and other dairy-based desserts, products targeted at young children are invariably smooth in texture. However, neither children nor adults appreciate the presence of unidentifiable surprise objects in their food. Texture awareness plays an important role in food preparation techniques e.g., how a vegetable is cooked, soft boiled or stir fried. For adults to enjoy new and contrasting textures, a leisurely eating environment is normally indicated. Food can be left to linger in the mouth, and chewed more deliberately than usual. In such a setting, novel dishes and textures may be explored, every bite being savoured and information obtained, where necessary, from the host or from experienced fellow diners, about ingredients and preparation techniques.

The meals of babies and infants are taken, inevitably, in a social context, allowing the relevant food culture to be transmitted to the child. For specific dietary items, textural preferences vary in different parts of the world. For example, while almost all East Asians eat rice as a staple food, preferences differ between countries and regions. Long-grain varieties are preferred in China, more glutinous short-grain varieties in Japan and Korea, light, fluffy varieties in Bali in Indonesia, and small balls of glutinous rice in Northern Thailand [51]. Among Western cooking traditions, Italians prefer short-grain rice, to be turned into creamy risotto, whilst Spaniards prefer a longer drier grain for paella. According to Madhur Jaffrey, in the culinary art of the Far East, Western taboos concerning textures do not hold, as the soft, the smooth, the crunchy and the slithery are presented in many permutations [51]. A greater variety of textures may be experienced in oriental, than in traditional western, cuisines, and textural contrast is specifically valued in a meal. Balut represents an extreme of this idea. Balut it is a delicacy particularly associated with the Philippines, a boiled fertilised duck's egg savoured for the variety of textures within [54]. The perfect balut, to the Filipino, is 17 days old, when the chick is not mature enough to show feathers, claws or beak. Balut is also popularly believed to be an aphrodisiac [54].

The role of texture in food acceptance depends very much on the product [29]. Texture is more evident in some foods than in others, with foods that are either bland or mildly flavoured, or crunchy or crispy, eliciting the most texture awareness responses from consumers [60]. With some exceptions, e.g., carrots, such foods also tend to be neutrally coloured. Strong texture is associated with alertness and arousal, because chewing requires effort. Stimulating foods are likely to have distinctive and varied structures as well as flavours. Comfort foods tend to represent low-effort eating, which is why they are often soft, or have other melting-in-the mouth textures, e.g., meat pies. The crispiness of chips represents a man-made, rather than a natural, texture. Chips are popular comfort foods, which require little chewing effort, despite their crispness. The temperature of the mouth is important in the perception of temperature-sensitive products, e.g., butter, ice cream and chocolate as it affects the melting rate, viscosity of the melt and mouthcoating effect [61]. The mouthcoating effect of fat is long lasting, which makes it particularly satisfying. When presenting a multi-course meal, the repetition of textures in a sequence of dishes is to be avoided, and variety to be encouraged. For example, a host would not serve a creamy soup, followed by a creamy sauce, followed by a creamy dessert.

7.6
The Flavour Appeal of Foods and Beverages

Flavour perception is due to a combination of taste stimuli, sensed in the mouth, and odour stimuli, sensed in the nose. Foods are chemically complex materials (see Chapter 4), and each individual food contains a wide range of flavour-active, or potentially flavour-active, substances. Taste and odour stimuli therefore do not occur in isolation in a food or beverage. Any combination of stimuli is interpreted by the brain as a unique, distinctive, new flavour, e.g., the flavour of lamb and the flavour of garlic. In the hands of a skilled cook, the flavour of lamb and the flavour of garlic can be caused to subtly merge, resulting in new and different, and again distinctive, flavours. If, on the other hand, the garlic is applied to the lamb in a heavy-handed manner, it will dominate the finished dish. In this manner, recipes can be built up to yield increasing depth of flavour. Despite their chemical complexity, such flavours are perceived as harmonious, even simple. The process of cooking also causes flavours to alter significantly. In traditional regional dishes and in so-called classic cuisines, this process may be the outcome of many decades of experimentation and refinement, involving local ingredients used in varying proportions and employing different food preparation techniques, with many, largely anonymous cooks sharing in the process. This allows the affinities between different ingredients to be explored to the full. Such affinities may be based either on flavour contrast, e.g., lemon juice and olive oil, or consonance, e.g., vanilla and cream. In all cultures, some of these classic, perfected dishes become the stuff of myth and nostalgia for individuals who learned to appreciate their flavours at their mother's knee. By contrast, the dilettante cook will throw together different flavours for the mere purpose of making something different. Although experimentation with novelty may not result immediately in the sort of perfection that can be sustained over several decades or longer, it can provide impulses for change. For example nouvelle cuisine, which became popular in the UK in the 1980s, was considered by many commentators to have been an extreme departure from classic western cuisines. Yet, its lasting legacy has been that it provided impulses to rethink the heaviness of the earlier cuisines, with their liberal use of fat and flour. This heaviness had become inappropriate to more modern ways of thinking. In this way, top chefs may act as arbiters and re-definers of taste, as restaurant innovation filters down the food chain, eventually emerging in the shape of retailers' ready meals for a wider market. Consumption value associated with novelty initially lies in the intensive stimulation of the senses. There is no emotional value attached to such foods at that stage, as emotional value relies on memories of previous experiences of a food, in particular, the context(s) of earlier consumption.

The flavour of a material, together with its appearance, provides important cues for its identification. If the flavour is acceptable, the green light for swallowing the food is given. Flavour recognition includes the identity and the condition of a food, as well as learned cues about appropriate consumption contexts. Odours represent particularly persuasive signals in reinforcing the appeal of appetising foods and in warning against the consumption of spoiled foods. Alone among stimuli, odours cannot be ignored or avoided, because they enter the body with the air that is breathed into the lungs. Smells generally can be very memorable, and tied up with powerful emotions, in particular, feelings of nostalgia. The stimulating smell of breakfast cooking when one gets up in the morning is an everyday, food related example of this phenomenon. Examples of evocative smells associated with places include the environmental smell of a distillery located in a beautiful part of the country, and the smell of small specialist food shops, e.g., coffee or cheese shops. It is noteworthy that in modern supermarkets, individual produce sections no longer smell of themselves, e.g., the fruit and vegetable section and the cheese section. This new blandness may be appreciated by certain sections of the market. It may be lamented by others, i.e., those for whom the smell of the product forms part of the satisfaction of the overall consumption experience in relation to food.

Individual flavour-active chemicals can be perceived as either pleasant (e.g., vanillin) or unpleasant (e.g., ammonia), and so can compound and complex flavours. However, such perceptions also depend on the intensity of the stimulus and on the individual consumer. For example, the slightly ammoniacal-fishy odour given off by skate wings, and which can be strongly nauseating, will be actively admired by skate wing aficionados. There are many such examples of simultaneous food preferences and dislikes, especially in the area of fermented foods. Here differences in perception are explained largely by whether fermentation odours are interpreted as spoilage or as proper maturation. Food odours are commonly associated with specific primary taste. For example, vanilla, butterscotch and certain fruit aromas are usually associated with sweetness, other fruit flavours (lemon) with acidity and meaty aromas with salty, or savoury, tastes. This is largely due to cultural conditioning as clearly, fruit often forms part of savoury dishes, e.g., meat dishes. As discussed earlier in the chapter, salt appetite represents an innate motivation, which comes strongly into play when cell and blood salt levels are diminished. Over and above this physiological appetite, many cultures have acquired a liking for strongly salted foods, as they became accustomed to eating meats, fish and other foods preserved in salt. A liking for sweet foods may, or may not, be due to genetic causes, but is usually present at the time of birth. Positive responses to sweet foods generally survive into adulthood, and sweetness plays a major role in the attractiveness of many foods. People's rejection of bitter-tasting plants is often interpreted as a bilateral, adaptive mechanism, which benefits both the plant and the individual. On the one hand, it allows people to identify and avoid plants rich in toxic metabolites, e.g., bitter alkaloids; on the other hand, toxin production is deployed by plants for their own survival. However, bitter taste aversion is amenable to cultural modification, and in fact, many so-called adult foods and beverages display a degree of bitterness. One of the reasons for such learned preferences may be the positive, pharmaco-active, post-ingestive consequences that are due to some of these compounds, e.g., caffeine in coffee and quinine in tonic water. In addition, many medicines are bitter, and as medicines cure diseases, it is learned that bitter tasting substances can, after all, be good for you. The use of sour foods and food ingredients is common to many styles of cooking, e.g., the use of lemons and of fermented foods. Whilst mild sour notes may be required to balance the rich flavour of certain dishes, many people enjoy the occasional over-stimulation from food providing near-painful sensations, e.g., vinegar pickles.

Given the toolbox of flavourings available to food technologists today, acceptable flavour is often easier to create in man-made foods than acceptable texture. In addition, the use of flavourings allows the flavour of a food to be reproduced consistently in successive batches of a food. Consumers tend to experience disconfirmed expectations in the sensory characteristics of a food as highly negative, to the extent that they may avoid the product in the future. It is much more difficult for the food grower than it is for the food technologist to optimise, and then standardise, the flavour of products, e.g., the flavour of a fruit. In fact, unreliable flavour prediction is a major barrier preventing desired levels of fruit consumption in the UK [64]. In contrast, man-made, branded foods such as snack foods are highly predictable in terms of all of their sensory attributes, including flavour. Flavour standardisation is also less of a problem in foods and beverages that can be blended before sale, e.g., coffee and fruit juices. In this situation, variation between batches can be smoothed out. When compared to the roles of the appearance and texture of foods, flavour is a particularly important sensory attribute in beverages and, not surprisingly, in condiments [60]. In a home-made, or home-style, fish cake, if flakes of fish are clearly visible, the flavour intensity of the fish will be less important than in a mass-produced product, in which the pieces of fish will be smaller, and more even in terms of size distribution. In the home-made product, visual cues reinforce the impression of fishiness.

The sweetness of a (sweet) food increases with the sugar concentration. Its pleasantness first increases, then reaches a maximum, and subsequently decreases [29]. The same is true of the perception of saltiness, but not of that of bitterness and sourness [29]. Such intensity-pleasantness relationships are highly dependent on the model food used, i.e., whether a particular stimulus is perceived as appropriate in a given context. This supports the view that individual liking or disliking of flavours is largely the result of learned associations. Learning generally takes place as a result of exposure to information. Flavour preferences too are learned by exposure, especially if consumption is positively reinforced by social factors, such as presence of respected group members at a meal containing unfamiliar foods. This is the case when children first learn about the flavour preferences of the their own culture where, with repeated exposure, many new foods that children initially reject are accepted [65]. It continues to be the case when adults acquire the tastes of an alien culture. Conditioned flavour preferences may also develop based on post-ingestive consequences of fat intake, e.g., as a result of flavour becoming associated with the sensation of satiety [66].

Traditional cuisines from around the world have selected, refined and combined the best of many influences over the centuries. The Chinese approach to food is said to be most apparent in its emphasis on the freshness of its ingredients and the balance of its tastes. Despite regional variations, there is a sense of unity due, in part, to a common use of the trinity of Chinese seasoning – ginger, spring onions and garlic, as well as the use of soy sauce [67]. Flavourings are however used in moderation. Five tastes are balanced through the choice of appropriate ingredients, i.e., bitter (e.g., bitter melon), sweet, sour, salty and hot (e.g., chilli). The Chinese differentiate themselves from their south-east Asian neighbours – who, they allege, bury the flavours of food – and from their Japanese and central Asian neighbours – whose food, they say, is bland and tasteless [67]. From a more positive perspective, Thai cuisine is said to be wild, bursting with contrasting hot, sweet, sour and salty dishes [51]. It is believed to have taken the best ideas from Malays, Chinese and Indians, and added its own aromatic preferences [51]. A particularly popular and effective Thai salad incorporates shredded chicken, prawns, grapes, oranges, fried garlic, fried shallots and roasted peanuts, with a dressing of lime juice, sugar, salt and green chillies [51]. In Ayurveda, a 5000-year-old Indian healing system, considerable emphasis is placed on the composition of the diet, which is thought to affect the mind as well as the body [68]. The term Ayurvedic cooking evokes the aroma of delicate vegetable curries, hot chapatis and light basmati rice, but includes a variety of other sorts of vegetarian dishes, as long as they are fresh, easy to digest, and not fermented [68]. Six tastes are recognised, i.e., sweet, sour, salty, bitter, astringent and pungent, and each of these has a different effect on the body [68]. Ayurvedic foods are thus associated with relaxation, stimulation, warmth or soothing [68]. Sweet taste is perceived as cool, moist and heavy in its action in the body, and as pleasing, softening and relaxing. Sour taste is perceived as hot, wet and heavy in the body, increasing salivation and appetite. Salty taste is thought to be heating, moist and heavy. Bitter and astringent tastes are considered cool, light and dry. Finally, pungency is perceived to be hot, light and dry and, in small amounts, to stimulate salivation. In shōjin cooking, there are six basic tastes that must be balanced in individual dishes, i.e., bitter, sour, sweet, hot, salty and delicate; three so-called virtues also need to be incorporated, i.e., lightness and softness, cleanliness and freshness, and precision and care [69].

Whilst the basic food ingredients are, more or less, the same all over the world, culinary culture makes the finished products different. Very different flavours can be obtained from the same combinations of ingredients, e.g., onions, celery and peppers and garlic, by varying the amount of each, by adding a portion at the start of cooking and another at the end, and by the temperatures and times of cooking [70]. In order to maintain appetite, flavours, e.g., in the form of herbs and spices, should not be repeated in the sequence of a meal, but should change frequently, ideally with every bite [70]. However, if too much variation is used, the sensory experience may turn into too much of a challenge, an assault course for the senses. Flavours must mingle rather than individual ones standing out as discordant, large contrasts. However, expectations based on cultural differences can play an important role in this. For example in Germany, "Currywurst" a grilled sausage drenched in tomato ketchup and sprinkled with curry powder has long been a favourite snack. But when a US journalist was persuaded to try a Parisian chef's speciality, a "mille-feuille of langoustines with curry", he was indignant at being served a French dish with a curry powder applied cosmetically [71]. He complains that nobody had understood that curry is a whole technique of cooking that you have to understand. Aromas perceptible in a dish can be fleeting and elusive. A hint of kidney can add depth of flavour and mystery to a meat pie, but if expected flavours are too elusive, this will detract from the enjoyment of a dish. Food should therefore be well seasoned, the flavour neither overpowering nor elusive. In the Niçois speciality "estocaficada", the aggressive flavour of stockfish (unsalted wind-dried cod) is said to transmogrify itself, when combined with potatoes, cream, eggs, garlic and parsley, into something delicate and gently aromatic [72].

Low flavour impact in a food may be either desirable or undesirable, depending on the context. This is reflected in the terminology that may be applied to such foods, which ranges from insipid and bland to plain and subtle. It may be desirable that food ingredients should taste of themselves in a way that is unpretentious and eschewing perceived gimmickry. However, bland staples, such as rice and noodles require some sort of flavouring, and soup may be used as a relish to moisten bread. Certain neutrally flavoured foods can, in the consumer's, mind be associated with a wide range of dishes, eggs being a prime example. When eggs are used as a food in their own right, rather than as a food ingredient, tomatoes, salt and herbs transform them into a savoury dish (omelette), whilst a combination of sugar, milk and flour turns them into a dessert (pancake). Seasonings and condiments serve to adjust the blandness of a food. Overseasoning (ketchup or mustard with everything) is likely to swamp the subtle flavours of a dish but, on the other hand, it helps the neophobic, unadventurous consumer to accept unfamiliar dishes. A special enjoyment is experienced when eating spicy food – an elating feeling [73]. Tolerance varies from person to person, but once a taste for spicy foods has developed, they may be positively craved [73]. Highly spiced dishes need to be balanced with blander ones, e.g., rice, pasta or potatoes. This allows the individual to vary flavour concentration in each bite or mouthful, and in a shared meal, to adjust to individual tastes. Vegetables too can provide the bland foil for strongly seasoned food. High-impact flavouring is expected in products such as confectionery and relishes. Relishes, such as pickles and chutneys are designed to be mixed in with the blander components of a dish by the individual diner, whilst in confectionery, intense aroma is required to balance intense sweetness. This type of product exists naturally in many tropical fruit, which can have very strong and insistent flavours. Surprisingly, the sweet and sour flavours of different fruit and berries, and the richness of nuts and chocolate, match the taste of chilli [73]. In Mexico, street vendors sell fresh fruit powdered with ground chilli [73]. The hot flavour of ground chilli in desserts makes for a refreshing ending to a meal. When eating grapefruit, the intensive sour flavour may be suppressed by a sprinkling of sugar (or of salt).

Tastes in foods change over time, with different cuisines and styles of cooking coming to the fore according to fashion. However, people tend to retain a nostalgic attachment to childhood favourites. Similarly, food aversions acquired in childhood tend to be deeply ingrained. In Britain, this commonly includes milk puddings based on sago and similar starches (because of their association of frogspawn) as well as the smell of boiling cabbage, both associated with school. Every type of food will be appetising if well prepared and served in the appropriate context. Modern living means that foods served in public frequently need to be as widely acceptable and inoffensive as possible. There is a constant mixing of cultures in public places, the inoffensive motive being particularly relevant where people cannot get away from one another, as is the case on aeroplanes. A general blunting of appetites for high-odour foods has been noted, even in France, where traditionalists fear that traditional pungent French cheeses are under threat [74]. Similarly, because fat is being bred out of pigs and cattle, preferences for meat species tastes are being lost. Short growing seasons for produce in countries such as Britain have traditionally provided seasonal variety, with the fragile flavours of strawberries, asparagus and apples and the subtle flavour of wild salmon savoured at the allotted times. This is being undermined to some extent with the counter-seasonal availability of produce, providing further evidence for a blunting of people's tastes. However at the same time, evidence is emerging of greater sophistication and discrimination between products. British tea drinkers now desire a refreshing clean aftertaste instead of the traditional stewy bitterness that could be accommodated whilst tea was being drunk heavily sweetened and with milk [75].

8 The Food Consumer in Society

8.1
Introduction

Strictly speaking, the consumer is defined as the ultimate user of products, the final link in supply chains. In practice, consumption is often taken to include individual purchasing behaviours. However, the individual who buys or prepares a food may not do this only for themselves, but also on behalf of other household members, including companion animals. Cat food is extruded into the shape of little fishes, and cat meat tins decorated with cats' paw prints, not to try to be appetising to the domestic cat, but to appeal to his or her owner. Consumption links together individual consumers, or market segments, and producers and their offerings. Whilst today a food product continues to represent its producer, the mass production and mass provisioning of foods, and the complex supply chains associated with these, have led to increasing alienation within food supply chains. Various initiatives have been instigated in response to this, in particular, quality assurance schemes.

Food-related consumer behaviour, like all consumer behaviour, must be examined in the context of environmental factors and their underlying trends. Among environmental factors influencing consumption patterns, orientation towards specific social reference groups is particularly important. Western secular societies are experiencing rapid social change, with a trend towards individualism and away from group cohesion. Increasing multiculturalism and the widespread rejection of religion have weakened the role of food taboos and opened the door for experimentation with unfamiliar foods. In such a climate, food preferences tend to be subject to fashion. This can be clearly seen in the ethnic food market, where, at any one time, a particular national or regional cuisine tends to be promoted and, eventually, favoured by fashionable diners. As a consequence, there is a continuous need for consumers to learn about food, both about new and unfamiliar foods and about their physiological effects on them personally. Changes in family and household structures, as well as other lifestyle changes, e.g., in terms of work and leisure patterns and practices, and changes in disposable income, play an important role in food choice. The current chapter provides a closer look at some current consumer trends in food. Actual consumer choice depends on the channels of distribution through which a product is obtained, as well as the way in which innovation is encouraged to diffuse into larger markets. The UK food industry is constantly changing, with new product development, in particular in areas such as ready meals, playing a major role in its success. UK consumers, whilst they may feel nostalgic about traditional foods, do not want to return to eating such foods on a regular basis [76]. With the opening up of global markets, change is only likely to accelerate. As a consequence, a thorough understanding of differences in consumer needs and attitudes between different countries is becoming increasingly important [76].

Consumption occurs in response to demand, either existing or latent, and whilst consumers themselves obviously benefit from offerings with genuine value to them, at the same time the economy is energised [77]. Food producers and retailers need to constantly reassure themselves about the satisfaction of target consumers with a product. Users can be asked directly about product performance, or they can be observed as they use the product. The latter approach is particularly useful for consumer research involving children. Active dialogue between the producers and users of a product allows the latter to assume productive roles in the product lifecycle, as they actively explore and exploit the attributes of the product in terms of their contribution to its overall value to them. No product can be all things to all people, and advertising does not work unless there is credible communication with the target segment of the consumer market. Today even the simplest of foods and food ingredients, e.g., eggs and salt, are subject to considerable segmentation. Market segmentation is the process of so featuring a product that it will have a particular appeal to some identifiable part of the total market [77]. Market segments, by definition, differ from one another, and consumer analysis determines exactly how they differ. Segments influence each other, and changes in the behaviour of one can affect the market overall.

Markets can be segmented on the basis of economic, geographic, demographic, lifestage, psychographic and behavioural factors, amongst others [77]. Food discounters supply the cost-sensitive segment of the market, as do the value lines of the main food retailers. Food preferences, because they are culturally learned, are closely related to geographical factors. In fact, where somebody comes from is one of the best predictors of what foods they will like. Population size, structure and distribution may be linked to data on purchasing power or wealth in order to predict market demand [77]. The lifestage a consumer finds themselves in is a key factor in their tastes and preferences, in foods as in other areas of consumption. This may be linked to age, particularly in childhood. However, this is less true in later life. For example, at any given age, single status implies very different consumption habits from married-with-children status. Age alone therefore is not a good discriminator for lifestyle, i.e., the activities, interests and opinions of consumers. Psychographics and social change analysis in marketing is concerned with tracking consumers' attitudes and values. Social value groups fall within three major classifications referring to members' attitudes, i.e., sustenance-driven, outer-directed and inner-directed groups [78]. Members of sustenance-driven groups are motivated by the need for security and may, or may not, be socially disadvantaged [78]. Members of outer-directed groups are conspicuous consumers, motivated by the search for esteem and status, their criteria for success being external to themselves [78]. Members of inner-directed group are motivated by self-actualisation, and include social resisters, experimentalists and self-explorers [78]. Behavioural variables include the types of benefit derived from a product, the extent of use and the usage situation [77]. Benefit derived from a product includes values such as body fatness, beauty and health, providing the basis for calorie reduced and functional foods.

The degree of involvement of the individual in a consumption or purchase event influences the extent of problem solving during the decision process [77]. Involvement in food may find expression in an interest in food preparation and cooking, both for oneself and for others. To the extent that involvement is present, the consumer acts with deliberation to minimise the risks, and maximise the benefits, gained from purchase and use [77]. Highly involved consumers are more likely to be aware of differences between competing products than consumers with low involvement in a particular product category. Purchasing and consumption behaviours, when reflecting a high level of involvement, present opportunities for consumers to hone their risk assessment, and risk management, skills. Product knowledge and involvement in its purchase or consumption are related to each other in that knowledge enhances involvement, whilst involvement leads to greater knowledge. Learning about foods is part of the socialisation process, which takes place within the family and other social institutions. However, the industrialisation of food production and processing has led to a marked information deficit for consumers. Yet, consumers continue do learn about foods, and there are many mechanisms through which this can take place, including school education, food labels on individual products, food writing and food-related broadcasting, as well as public announcements on behalf of the government, in support of food policy. Consumers also learn by trying out new products for themselves, either in a restaurant setting or in the domestic environment. In this, consumers weigh up risks and benefits prior to the purchase or consumption of a product and, at least as far as the sensory performance of the food is concerned, learn from either confirmed or disconfirmed expectations. Domestic cooks learn by trying out recipes and having their family and friends evaluate the results. However, there are many areas that consumers do not have direct experience of. The food chain is therefore, to a large degree, a trust chain, and what consumers want in terms of information about food is simple messages, from a trustworthy source [76].

The chapter begins with an overview of the factors that play a role in the segmentation of the food market in the UK. Much of this can be extrapolated to the situation in countries which are fundamentally similar to the UK. From the issue of what is eaten, the discussion moves on to the question of how consumers develop food preferences. Specific attention is focused on the role of cost-benefit and risk assessment in food choices. The chapter concludes with a discussion of the role of the food purchasing and food consumption contexts in food acceptance.

8.2
Defining Food Consumers and Market Segments

Demographic and lifestyle analyses provide explanations for food consumption trends in modern society, e.g., the demand for safe foods, which are also convenient, healthy and gratifying. Food consumption reflects wider social trends, e.g., in household structures and in work and leisure patterns. Demographic data are particularly important for food retailers to be able to tailor their offer to the catchment area of a particular store. Whilst bulk buys and deals would be attractive to price sensitive households, a catchment in which affluent, professional households predominate needs to target its marketing towards different types of consumer value, including variety and attention to detail. The nature of the UK population, currently estimated at approximately 60 million [79], has been changing significantly over recent decades, and this has had a major impact on consumer markets. As in other EU countries, the population is ageing to the extent that by the year 2016 in the UK, the number of people aged 65 and over is expected to exceed those under 16 years of age [79]. However, the image of the pensioner is changing, and the desire to take full advantage of one's later years is likely to act as a stimulus for the development of foods associated with wellness, including functional foods. At the same time, increased opportunity for travel may lead to demand from this lifestage group for authentic foreign and ethnic foods. In the UK, one person in 15 belongs to an ethnic minority group, and such groups tend to have a younger age structure than the average [79].

The participation of women in the workforce has increased substantially during the 20th century, and there has been a strong upward trend in the share of higher socio-economic occupations, especially among men [79]. Increased economic independence for both sexes has opened up a wider variety of lifestyle choices, with the result that in 2000, almost three in ten households in Great Britain comprised a person living alone, more than two and a half times the proportion in 1961 [79]. Both the young and the old increasingly live on their own. Smaller households require smaller pack, or portion, sizes of foods than larger households. Living alone reduces the need for carefully planned household food provisioning and articulates with the so-called 24-h, impulse culture, where snacking throughout the day increasingly replaces formal meals at set times. Fast food consumption is on the increase, but is most popular with younger consumers [79]. The percentage of dependent children living in lone mother families more than tripled between 1972 and 2000, to almost one in five [79]. As the proportion of life-long marriages continues to decrease, individuals are more likely to form a series of partnership. In 1998–9, step families accounted for about 6% of all families with dependent children in Great Britain [79]. The increased choice and variability in family relationships has implications for consumption, as each major life change will require not only the acquisition of homes and household goods, but in addition, the re-negotiation of domestic habits, including those related to food consumption. The decline of the traditional family has been accompanied by a decline of the traditional family meal, which used to be prepared in the home from basic ingredients. Increased participation of women in the workforce has reduced the time that can be devoted to complex food preparation activities.

UK consumers feel that a new, time-pressured lifestyle has emerged in their country in recent years [76]. Under these circumstances, most consider the buying, preparing and cooking of food to be a chore, although it is generally acknowledged that these activities have become easier in recent years, due to the improved convenience of the retailer offer [76]. However, convenience foods are considered to be less beneficial in nutritional terms than fresh, whole foods; and taking the easy option may therefore give rise to feelings of guilt associated with the purchase of pre-packed convenience foods [76]. The desire for convenience in food preparation is matched by a preoccupation with body image and weight, especially among those in professional occupations. This is related to the interest of these consumers in fashionable clothing, which has increased greatly in the UK in recent years. Whilst sedentary lifestyles have played some part in the increasing prevalence of obesity, there is recent evidence that people are beginning to eat more healthily at home, e.g., increases of over a third in the amount of fresh fruit eaten since the mid-1980 s [79]. Along similar lines, there is also increased consumption of low-fat milks and spreads, at the expense of wholemilk and butter [79]. On the other hand, consumption of fresh potatoes has shown a downward trend and consumption of processed potato products, mainly chips and crisps, an upward trend [79]. The decline in the consumption of potatoes may also be accounted for by the increased popularity of rice and noodles as carbohydrate-rich staples to accompany ethnic foods. An important indicator of people's lifestyles is the amount of time that they spend on certain activities. The most common leisure-time activity outside the home among adults in Great Britain in 1995/6 was visiting a public house [79]. Watching television is the favourite home-based activity for both men and women, whilst participation in home-based activities generally varies by age and gender, e.g., needle work, gardening and do-it-yourself [79]. In 1995, British consumers took 59 million holidays for four nights or more, an increase of 43% on the number taken in 1971 [79].

Household expenditure has increased in real terms in most years over the past half-century, including expenditure on food [79]. However, expenditure on food as a proportion of total expenditure has declined progressively over recent decades. Increased disposable income has fuelled the demand for value-added foods, primarily foods with elements of food service or convenience built into them, and preferably without loss of sensory acceptability. The traditional chilled ready meal in the UK is based on the cook-chill system, which means that the consumer gets a reheated meal. However, current innovation focuses on meal kits, e.g., Marks and Spencer's kits of raw fish and vegetables that can be steamed in the microwave oven. Fresh fruit salads and mixed-leaf salad kits with a separate portion of dressing have been available from a number of UK food retailers for some time. Important aspects of convenience are keeping quality, easy-to-use packaging and portion- or calorie controlled meals. Foreign travel and other cross-cultural influences have caused significant numbers of UK consumers to become increasingly adventurous and knowledgeable about foreign foods, as well as previously unfamiliar food ingredients. Food related learning and exploration tends to be encouraged by trips abroad, by television programmes and by restaurant visits. Innovation typically diffuses from trendy restaurants down to the supermarket ready meal section, and finally to authentic ingredients being marketed by the supermarkets for use by the home cook. Sophistication of food knowledge, and alignment with the values of certain celebrity chefs, forms part of the consumer's self-definition. They do not actually need to be able to cook the dishes in question, but they need to be able to discuss them, and the celebrity chef's latest book on the book shelf will signal their food values to any passing visitor. As consumers' food preparation skills continue to decline, attitudes to the merits of traditional cooking styles may change, as may the definition of traditional cooking [76].

There are two pre-requisites for genuine consumer choice. Firstly, consumers need to be able to influence what is being placed on the market for them to choose from. Secondly, they have to have access to the available choices. Market mechanisms generally act in a way that satisfies demand. However, global sourcing by the large food retailers may prevent access to locally produced foods. In institutional contexts, i.e., school meals, choice will also be restricted. As a result, packed lunches have gained popularity among pupils and their parents, and the food industry has responded by providing lunchbox foods. There are varies kinds of barrier that limit access to the foods a consumer might wish to choose, e.g., educational, financial, geographical and physical. People living in deprived neighbourhoods are likely to have limited shopping access. In the case of the elderly poor, all three factors may act together, whilst the young poor may be disadvantaged by poor food knowledge and food preparation skills. In addition to commercial factors, regulatory activity too may restrict access to foods for which there is demand, although the question has to be asked whether this is reasonable. The fundamental rationale for the regulation of food is consumer protection. However, if a food regulatory and supervisory system is motivated more by the fear of litigation than the devising of suitable risk management to provide a wide range of foods safely, consumer choice becomes threatened [80]. Issues surrounding food risk are discussed further later in the chapter. Some commentators, [e.g., 81], warn that access, or freedom of choice, should not be equated with breadth of choice and suggest that there is little value in multiplying alternatives for its own sake, i.e., if the alternatives do not possess significantly different, desired attributes. The counter argument to this is that the market will eventually sort out this type of situation. In the UK, for many years, yoghurt meant flavoured dessert in various permutations. Today, the market is segmented in a more meaningful way. Plain yoghurt is available everywhere, made from various cultures, representing different textures, and made from different grades of milk, including organic milk. In this example, the sophistication of the offer has grown in line with the knowledge and sophistication of the consumer.

Where people live has a major impact on food habits. Out-of-town, one-stop supermarkets encourage the weekly shop, but are not readily accessible to all consumers. Domestic waste collection takes place on a weekly, and sometimes fortnightly, basis. This discourages the purchase of foods associated with smelly waste, e.g., whole fish, by urban consumers. Lingering cooking smells may also be undesirable in small living quarters. One of the key trends in food consumption is snacking, including the consumption of fast foods away from home. Snack foods are commonly understood to refer to manufactured foods specifically designed for the purpose of snacking, e.g., chocolate and cereal bars, and bagged savoury snacks, such as crisps. However, snacking may refer more generally to the amount of food consumed during an eating episode. For example, one or two beefburgers accompanied by vegetables on a plate is likely to be perceived as a meal, whereas a burger in a bun might be perceived as a snack. Eating out has grown as a leisure activity in recent years [82]. At the same time, the boundaries between eating out and eating in are becoming blurred, as retailers offer a large range of what are, in effect, take-away meals. A typical UK household undertakes one major shopping trip per week [82], which, clearly, has important implications for the shelf life requirements for modern foods. Recently, the strongest growth in sales has occurred in the fruit and vegetable market, followed by fish. Meat (product) sales also rose, despite a longterm decline in the sales of carcass meat [82]. The weakest sector were dairy products and sugar, although the cheese market is beginning to polarise, with budget-priced cheese blocks at one end of the market, and a growing range of premium and speciality cheeses at the other [82].

Organic food is perceived by most consumers as fresh and natural, and as the solution for anything that might be wrong with conventional food production systems [76]. Organic food consumption in the UK has increased phenomenally in recent years [83]. As organic farmers tend to get involved more directly with consumers than do farmers who operate conventional systems, the popularity of these foods may be due in part to consumers' loss of trust and confidence in conventional food production, triggered by a series of recent food scares, most notably, BSE. The issue has become politicised, attracting emotionally tinted comment both from the chairman of the Food Standards Agency ("organic foods do not offer value for money") [84] and from the Select Committee on Agriculture ("some of its (organic farming) apostles still proselytise with an almost religious fervour") [83]. These examples show that at government level, a true appreciation of what influences consumer motivation remains lamentably absent. However, the food marketing industry can, at times, be seen to be similarly inept. The organic strategy of the food retailer Iceland misfired because it did not understand its own customers, and could not attract the more sophisticated customers the strategy was designed to attract [85]. In the wake of Iceland's failure Tesco, the UK's top food retailer, has recently committed itself to mass marketing organic food [86]. Issues important to adults, such as organic food, GM food and the future of farming have not registered with 16- and 17-year olds [87].

Since organic foods are relatively expensive, it follows that the organics market is characterised by consumers with relatively high involvement in their food purchasing. Traditionally, motives for participating in this market have focused on food safety (fewer environmental contaminants) and ethics (environmental pollution, animal welfare). However, organic foods have become increasingly fashionable, as they have shed their erstwhile beards-and-sandals image. The offer no longer consists solely of agricultural and horticultural produce, supplemented by a few simple manufactured foods. Instead, it now encompasses a very full range of processed foods, including flour confectionery and chocolates. Today, organic foods can therefore be used as a lifestyle statement, and as a vehicle for conspicuous consumption by the wealthy [88]. It is not only organic foods that are highly fashionable at present, but an interest in foods generally. Fashion has a useful function in that, at the same time as encouraging change, it sets boundaries for individual experimentation. The increasingly popular farmers' markets serve to bring farmers closer to consumers. Television shows feature a number of chefs, with different personalities and cooking styles, and each having his or her own admirers. In fact, watching these celebrity chefs perform causes some hobby cooks to suffer anxiety when they are faced with having to entertain dinner guests at home [89]. Slightly less anxious individuals will take advantage of top-of-the-range ready meals, supplied by all the major supermarkets, to impress their guests [90]. Newspapers and television are viewed as the main sources of information about serious food issues; however, media commentary is not necessarily taken at face value [87].

Low involvement with food leads to limited information processing when making food choices, and limited evaluative activity. The readiness of consumers with low involvement to complain about a food is also low. Following an unsatisfactory consumption experience, they would rather switch between products or brands. Low involvement typifies the majority of food consumers. The fact that consumers generally do not invest time in improving their understanding of food production makes it difficult to correct any misunderstanding [91]. Few teenagers, particularly boys, have much interest or involvement in food, tending not to think about food in a wider context than just eating it [87]. When teenagers are feeding themselves, the time and ease of cooking and cleaning up are the most important factors in food choice [87]. UK consumers on the whole feel that they can get the products they want, although single person households tend to feel disadvantaged by the relatively high cost of single portion products, and the irrelevance to them of promotional offers requiring them to buy a large amount of a product [87]. Food production issues are rarely considered at the point of sale. Because they assume that the details of animal welfare standards might be unpleasant, consumers prefer to rely on the retailer or brand owner to ensure that appropriate standards are met [81]. Consumers are becoming increasingly divorced from food production, and the feeling that the reality of food production might be even worse than expected further reduces consumers' desire to become better informed [91]. Such attitudes create difficulties for the promotion of UK meat, i.e, whether to emphasise high welfare standards, or whether to back off because of the risk of reducing demand by reminding consumers of the link between animals and meat [92]. UK farmers' expressed attitudes towards their ultimate customers, fuelled by the urban-vs.-rural debate on issues such as the legitimacy of blood sports, can undermine partnership approaches. This may be illustrated by the farmer interviewed on the BBC "Farming Today" programme (12/2/02), who referred to people who express concerns about animal welfare as "bunny lovers". Consumers can be divided into three main groups based on their interest in improving their understanding of food processing [91]. Abdicators (48%) do not want to find out more, spectators (42%) feel they should be interested and investigators (11%) have actively sought information. However, when presented with information in a readily understandable format, consumers generally show considerable interest [91].

The food provisioning of a household depends on a variety of factors, including the presence of kitchen appliances. For example, a student who does not have access to a refrigerator, may forego highly perishable foods, and may even lose the habit of eating meat for good. A microwave oven is required to take full advantage of the convenience of ready meals and meal kits. Another important aspect of provisioning is the identity of the gatekeeper(s) for food purchasing. Whilst an unskilled individual is likely to shop for goods by reference to a recipe or shopping list, a skilled cook will be able to respond to the offer on the day. Gatekeepers control the food that will be available for a household to eat. The presence of children has a major influence on what foods are, and are not, consumed. For examples, a perceived need to feed children well may turn parents away from fast convenience foods and towards organic foods [93]. Children of between 5 and 15 years of age have a considerable effect on parental purchasing decisions and overall consumer trends, with food and drink one of the main market sectors [94]. Food shopping with children is difficult. In particular, offering character-branded food at higher prices than conventional brands makes a shopping trip much harder when mothers shop with their children [87]. A current example of character licensing is Disney's Monsters Inc., which is being used to market McDonald's Happy Meals, amongst others [95].

8.3
Benefits, Risks and Costs in Food Choices

All genuine choices contain some element of risk. No matter how carefully a meal is planned – in terms of foods, context and participants – the actual outcome remains uncertain until the end. Risks perceived in the context of the purchasing and consumption of consumer goods cover a wide area. In general terms, risk relates to the product not performing as promised, or as expected. This implies financial risks, i.e., paying over the odds for a product. If food is substantially less enjoyable than expected, the implications may be more far-reaching, e.g., the ruination of a dinner party and possibly, loss of face by the host. Perceived risk in purchasing may increase as a function of certain characteristics of the product itself, e.g., price, length of time the product needs to be retained, switching costs, additional products needed to be consumed with the product; it also depends on the characteristics of the consumer [96]. Purchase of a whole salmon for a party, and first-time bulk purchasing of some prime olive oil as a larder item both exemplify, in food terms, relatively major risks, whereas the purchase of a single-portion pack of a new snack product or sweet represents a relatively minor risk. However, in the context of food consumption, the discourse about risk most commonly focuses on food safety issues, due of the potential seriousness of unsafe foods. The prospective consumer's uncertainty about the suitability of a food is particularly pronounced during its first-time purchase. Strategies commonly employed by individuals to reduce this risk include loyalty to specific retailers and brands, and product endorsements by trusted celebrities and other personalities in the public eye. The consumer's involvement in a product category plays an important role in risk perception and in the motivation to actively manage risk.

In common language usage, the terms hazard and risk tend to be used interchangeably. In fact, the dictionary offers at least two alternative definitions of a hazard, i.e, as a chance or probability (neutral frame) and as a danger or risk (negative frame) [97]. Detriment is a further term often used interchangeably with the term risk [98]. Interestingly, where uncertain choices are concerned, the language appears to offer a far greater number of terms associated with possible negative outcomes than with positive outcomes (rewards). This perhaps reflects the fact that in such choices, people tend to be more concerned about incurring a possible loss than they are excited about the chance of receiving some gain. The value function (Fig. 3.6) developed by Kahnemann and Tversky provides the theoretical basis for this phenomenon. Unlike common usage, for the purposes of food safety management, hazard and risk are carefully defined, together with a range of terms relating to each of them [99]. Thus hazard refers to the nature of some adverse effect – which can be biological (pathogenic), chemical or physical – whilst risk refers to a function of the probability of an adverse effect and the severity of that effect [99]. The Royal Society Study Group on Risk Assessment in 1983 defined a hazard as an intrinsic situation that in particular circumstances could lead to harm, and a risk as the probability that a particular adverse event occurs during a stated period of time or results from a particular challenge [100]. Nevertheless, the question whether there is such a thing as an absolute, objectively specifiable, risk remains keenly debated, particularly among social scientists [98]. One of the problems with so-called objective risks is that risks change once they have been quantified. It is common that, when the incidence of an adverse effect in some population has been measured and communicated to them, that population will modify its behaviour towards the hazard or risk. This is readily illustrated in the context of people's attitudes towards road safety, and their behaviours as road users [98]. However, the theory can also be illustrated using food-related situations and behaviours. For example, it may be argued that if at some point in the future, fresh meat were to be routinely subjected to an irradiation treatment to eliminate pathogens at the point of production, food handlers and consumers might then be tempted to drop their guard about hygiene matters generally. Consequently, the incidence of meat related food poisoning might not improve as expected. Of course, it is not very likely that meat will be routinely irradiated in the UK in the near future, given the fact that consumers currently perceive that particular technology as highly hazardous and therefore, unacceptable for routine food use [101]. A slightly different food related example of consumers potentially resetting their risk thermostat [98] concerns anti-bacterial washing-up liquids [102]. Again, it might be argued that, by using such a product, some consumers might be lulled into a false sense of safety, which might actually lead them to neglect certain basic hygiene practices. In the commercial field, it is well recognised that for food handlers to wear disposable gloves does not guarantee satisfactory food hygiene, and that the wearing of such gloves can encourage unsafe practices.

In the context of a cost-benefit analysis, which involves the balancing of (likely) gains against (likely) losses, risk tends to be evaluated by the consumer as an element of cost. As a matter of principle, no risks are worth taken unless actual, or at least potential, choice outcomes include benefits to the consumer. Hence, where the genetic modification of foods is concerned, as benefits to consumers were not apparent when the technology was first introduced, it is not surprising that it was widely rejected [103]. Many food choice situations constitute gambles, i.e., choices between more or less uncertain outcomes. Choosing a product in the supermarket exemplifies the cost-benefit analyses routinely carried out by consumers. A shopper may thus weigh up the calorific, cholesterol and fibre contents of competing products, their likely contamination with pesticides, and so on [98]. When purchasing grain- or nut-based foods, they may be weighing up the likely presence of either fungicide residues (conventional production methods) or mycotoxins (chemical-free production methods). When purchasing salmon, they may be weighing up the presence of healthy fish oils against the possible presence of environmental contaminants [104]. Eventually, some purchase takes place, with many such purchases manifesting themselves in the economist's demand for a product (see Chapter 3). If food suppliers, and retailers in particular, perceive low demand to reflect consumers' avoidance of certain products, they will see this as a risk to their profitability. Complex risk assessments of this kind, which involve the continual redefining of risk by all stakeholders, lead to actions such as retailers removing unpopular additives, or genetically modified ingredients, from their own-label foods. However, information associated with consumption is rarely complete, especially information available to the consumer. First purchases of unfamiliar foods are particularly risky, although the level of risk varies. New products that differ in some small way from familiar ones, e.g., a new shape of dried pasta, are less risky, and easier to fit into existing household structures than radically new products, e.g., an unfamiliar meat. To reduce the buying decision risk for consumers, new products may initially be offered in small portions or as a free sample; detailed use instructions and swift complaint handling may be provided in support of the product, and endorsements may be used to promote it. When a food retailer introduces an own-label version of a leading manufacturer's brand, they may select so-called look-alike packaging, or a similar-sounding product name, to reduce risk, i.e., neophobia, for the consumer. However, some of these strategies are likely to backfire, as 3 % per cent of shoppers will basket a look-alike product by mistake [105].

Processes of risk perception are characterised by their complexity. A wide variety of beliefs and attitudes may be schematically linked together to yield drivers of individual purchasing or consumption decisions. Individuals' responses to risky situations may not always seem consistent or rational, e.g., indulging in chain smoking cigarettes on the one hand, but rejecting beef for fear of BSE. Risk characterisation means the (qualitative and quantitative) estimation, including attendant uncertainties, of the probability of occurrence and severity of known potential adverse effects in a population [99]. However, risk to the population is not the same as risk to the individual. Risk assessment at the level of, and directly by, the individual consumer takes place in a context specific to that individual. It takes into account their susceptibility to detrimental effects from a particular hazard, the place of the food in question in a whole-diet context, and cognitive factors, i.e., their general view of the world. How consumers respond to information about general risks depends on the amount of information available, how information is framed and what is the source of the information. Another important question concerns control, i.e., whether the individual is given the option to expose themselves, or to not expose themselves, to a particular risk. Perceptions of food related risk may be lifestyle-related, e.g., fat consumption or nu-

tritional deficiencies, product-related, e.g., microbiological and chemical contamination, or technological, e.g., genetic modification [106]. Public estimations
of some particular risk do not always coincide with official, or expert, risk assessments, and risks cannot always be accurately described or quantified. Individuals tend to fall prey to optimistic bias, which means that they may believe
themselves to have more control over a hazard than they perceive other people
to have over the same hazard [106]. This may be due, in part, to information not
being targeted sufficiently closely, and in part because sources of information
lack credibility in the eyes of the individual consumer. There is less optimistic
bias where technological risks are concerned, as these are perceived as being
controlled at the level of society, rather than that of the individual [106]. The
food industry tends to adapt rather than invent new technology. When medicine
innovated in the area of gene transfer, food science followed suit. The difference
in public acceptance of the technology in medicine and food may be due, in part,
to differences in perceived need or benefits, and in part because of medical applications being thought to be subject to more stringent controls [107].

Food safety is managed jointly by the various players in food supply chain,
since breakdown at one point of the chain, will often compromise the whole. Government may be tempted at times to manage a food safety risk by designing it out
of the system through various bans. An example of this was the ban on the sale of
beef on the bone at the height of the BSE affair, an intervention that was by no
means popular with all consumers, many of whom accused the government of a
nannying attitude towards the public [108]. Whilst this ban constituted a tempo-
rary measure, current regulatory approaches and official attitudes towards raw
milk, and particularly, to cheeses made from such milk, have been a bone of contention over a long period of time. To their devotees, the fundamental question
asked concerning foods that are, in objective terms, slightly more risky to eat or
drink than their more highly processed substitutes, amounts to "Why have we got
people telling us what we can eat and drink?" [109]. At this juncture, the meaning
of safety is worth a closer look. The dictionary offers several alternative definitions of safety, including protection from danger or risk and something being unlikely to cause harm or injury [97]. At the inquiry into BSE, the mother of a victim of vCJD said that governmental advice that beef was safe had suggested to her
that it was safe to eat, and not, as was being suggested now, that it was fairly, or
possibly, safe to eat [110]. Clearly, risk communication from official sources to the
individual consumer is in a poor state. In the wake of numerous major food safety
scares, those responsible for food safety at government level suggest that they
have become more cautious in their approaches to food safety management. To
this end, they have adopted what is known as the precautionary principle. This is
to be applied in cases where scientific evidence is insufficient, inconclusive or
uncertain, and where preliminary scientific evaluation indicates that there are
reasonable grounds for concern [111, 112]. This approach will only succeed if it
includes a move away from the current practice of providing blanket advice.
What is needed instead is a more straightforward approach, away from such
advice and towards hard-fact information.

In fact, people have a natural need for taking risks, i.e., for arousal (see Chapter 3), and fulfilling this need enables people to lead better lives. The alternative

to accepting risk is protracted boredom. Although the propensity to engage in risky behaviour varies, without any risk taking, without ever encountering and dealing with the unexpected, there is no learning. Trying a new food for the first time implies risk. Never trying a new food implies a lack of desire to learn, as well as a strong desire for comfort and tradition. Still, even within a fairly rigid, traditional diet, there will be elements of surprise, also associated with arousal, e.g., in terms of products currently available at a local market. Too much choice can mean too much risk, especially for older people, or others not used to selecting from large assortments of goods [113]. Risk is associated with variety, in the sense that variety-seekers will not be risk averse. In the context of personality research, risk is more than just uncertainty about outcomes; it is an expectation that a loss will occur [77]. Risk takers, also described as thrill seekers, have a higher than average need for stimulation and become bored easily [77]. One of the more extreme examples of deliberate risk taking in relation to food concerns a Japan delicacy, the blowfish, or fugu, some of who's organs contain a potent neurotoxin [114]. Despite the fact that fugu chefs are specially licensed, every year people die from consuming the fish [114]. Since information reduces risk, food labelling plays a potentially important role in the satisfactory outcome of consumption experiences involving unfamiliar products. Increased perceived personal control may reduce perceptions of personal risk [106]. On the other hand, as perceived threats increase, people may respond by being more careful [98], and this applies to the thoroughness with which they examine food labels. Food claims create some kind of order in an area of almost unlimited choice Claims serve as signals for different segments of the food market, e.g., as calorie-reduced consumption and ethical consumption. Branding foods reduces the risks for consumers by guaranteeing that products do not vary in their eating quality. In the unfamiliar surroundings of a foreign country, the tourist can reduce a whole range of risks associated with eating out by purchasing food from a McDonald's fast food outlet. Taking risks and exploring the unknown are preconditions of much active learning. Without the financial resources to experiment, the opportunity for learning is diminished. For example, poor consumers find it hard to change their families' diets partly because they cannot afford to experiment with, or waste, food. They are therefore likely to persist in feeding their children foods that they know will be accepted, rather than trying to introduce healthier foods [115].

8.4
Food Skills Development

The ability to learn forms a key aspect of survival, but beyond that, opportunities for learning are sought out due to people's intrinsic motivation to explore their environment. Learning is therefore a fundamental aspect of living, necessary for survival both in a physical and an intellectual sense. Learning is fundamentally enjoyable, especially in young children, who explicitly engage in the process through the medium of play. Learning is about change, and changes concern knowledge and understanding, skills and behaviours, and preferences and attitudes (see Chapter 3). Because of the central role food and eating play in people's lives, these serve as vehicles for more generalised learning. Throughout life, consumers develop behavioural rules, which enable them to cope with choice. Food consumers learn the balance of diet necessary to maintain a desired body shape, how much beer can be safely drunk during a drinking episode, and which foods can be eaten freely and without any detrimental effects. Such lessons may be learned relatively quickly if the foods and beverages that are available in the market do not change much, but when there is frequent change, effective learning becomes much more challenging. For example, how much of a new, low-fat spread may be used when it replaces butter, how many glasses of alcopop may be drunk when they replace beer? As choices become more complex, the impulses for deep learning increase, raising the opportunity for highly customised individual product use and eating patterns. On the other hand, some consumers respond to a high level of change and innovation in the food offer with confusion. If meaningful learning becomes impossible, consumers may respond by paying less attention to the relevant issues, which means that their involvement is reduced. Learning what to expect in a given situation makes life less risky and more comfortable, because uncertainty is decreased. Trial and error re-enforce learning, and the increasing sophistication gained in this way leads the individual to an appreciation of the rules of rational decision making, as exemplified by the rational consumer choice model of economics (see Chapter 3). Poorer consumers may be prevented from fully exploring food as a vehicle for experimentation, because for them the risk inherent in trial and error may be too high, i.e, they may feel that they cannot afford to risk wasting food.

Observational, or social, learning teaches an individual the norms adopted by their social reference group(s). By conforming to these norms, they themselves then contribute to the cohesion and permanence of such groups. Fashion cycles promote change and experimentation, but within the safe boundaries of group norms. Conditioned learning establishes associations between fundamentally separate events, including associations between consumption and reward generally, and reinforcement of a response to some stimulus in particular (operant conditioning). In evaluative conditioning, evaluations change after contingent pairing of an event with an already positive or negative event, e.g., conditioned taste aversions [116]. Social mediation appears to be involved in much of the evaluative conditioning that underpins the acquisition of food preferences. An unconditioned stimulus often associated with a new food is another person and their expressed attitudes towards that food [116]. Whether such a preference is sustained depends, in part, on the postingestive consequences associated with the new food. Satisfaction with some aspect of consumption means that the consumer wishes for a repetition. Insightful learning relies on timely and accurate feedback, and if this is not forthcoming, learning motivation may be stifled. As consumers cannot verify all of the quality attributes present in a particular food for themselves, they rely on communications with the food supply chain and regulatory agencies to complete this information. Communication is a two-way process, which involves the repeated encoding and decoding of information between players who trust each other. Miscommunication and overcommunication are common, but can be avoided by regularly checking the impact of information or educational materials on the target audience [117]. This includes that audience's awareness of, and interest in, an issue under consideration, as well as the suitability of the information disseminated in relation to the issue. Learning usually occurs in stages, with levels of sophistication in terms of product understanding increasing progressively. It is unlikely that, at any one time, all consumers are at the same stage of comprehension or involvement with a particular product or technology. Despite the increasing levels of public participation in higher education and education generally, government still typically adopts a top-down, rather than a communicative, approach to consumer education in relation to food. This may be explained, in part, by the focus of government on the well-being of the population as a whole, whereas the individual consumer will be focusing on what happens to them personally. From such disparate perspectives, issues such as BSE may therefore be judged as not much of a problem and, at the same time, a very serious problem.

One of the types of basic consumer value identified by Holbrook is play [118] (see Chapter 1). Play differs from the other types of value in that it assigns uncertainty as a positive feature of consumer value. It may therefore be argued that play and learning are closely related. According to Grayson, play-as-value requires an active consumer, who can attain this value either by following rules or by challenging them [119]. If the consumer adheres to the marketer's rules, satisfaction is predictable and the consumer's perceived competence is therefore reinforced; if the consumer challenges the rules, the opportunity for creativity, and deep learning, presents itself. Of course, consumers' disregarding of marketers' rules implies risk and, at worst, product abuse with food safety implications. For example, a consumer may decide to serve cold a food that is intended to be heated before use. In fact, food products are often formulated to accommodate abuse, at least where it can reasonably be predictable. Sauces in instant noodle snacks will contain certain functional starches, which will prevent lumping if the water used by the consumer to regenerate the product is not hot (as suggested in use instructions). Timely involvement of consumers in new product development is important to obtain an idea about how the product is likely to be used. This is especially useful where a new product challenges existing household activities, skills and kitchen hardware. A many-pronged approach may be needed, with inputs from, for example, cookery writers or broadcasters and suppliers of cooking equipment. The widespread promotion of Chinese stir-fried foods in the UK in the 1980s serves as an example. Entirely novel products are challenging, as they require extensive learning, including reassessment and modification of present household activities. Marginally new products will fit more comfortably into established household patterns, representing more gentle learning which builds on existing knowledge.

The concept of food skills covers a broad area of learning, with the main assumption being that a skilled food consumer is equipped to make fully rational choices within the constraints of a given food market. They will also have internalised all the rules pertaining to food selection within specific cultural boundaries. Food skills development, over the course of an individual lifespan, begins with the transmission of culture-wide food preferences to children. Social learning about what constitutes a food is particularly persistent, both because of its emotional (childhood) associations and because of it being continually practised, with consistent reinforcement by a variety of agents, in particular, family and friends. Cultures teach the use of names, or generalisations, for ranges of objects, e.g., "vegetables". This kind of schematic learning means that members of a class of objects may be accepted or rejected simply on the basis of that membership. In this situation, the label becomes a barrier to exploration, and ultimately, meaningful learning. However, adolescence is the age of independent experimentation, and of the challenging of received ideas, and with increasing independence, food is experimented with against the backdrop of ever increasing knowledge. The development of cognitive skills allows an individual to make increasingly wiser, more rational, decisions. Experience is gained by repeatedly engaging in specific consumption behaviours, and this means reinforcement of learning and the promotion of in-depth understanding of issues. Travel, reading, meeting people from other cultural groups all may provide an impetus for experimentation, leading to learning and possibly, behavioural changes and changes in preferences. The more successful experimentation, i.e., learning by doing, becomes, the more childhood neophobia regarding foods is likely to retreat.

Food skills are acquired through information and education. Specific product information does not contribute to skills development unless it can be integrated into wider frameworks, or schemas, of understanding. This includes knowledge of how the product compares with other products, how it can be used and what specific short- and longterm benefits will accrue to the user. Ideas about what constitutes important food skills also differ between individuals. A person may be highly skilled in survival techniques in the wild, having learned to make full use of the resources available within different natural, and perhaps hostile, environments. This individual will have an extensive knowledge about the edibility of wild plants. A second person might have no knowledge of wild food plants, but be highly skilled in interpreting nutrition labels on pre-packed ready meals in the context of healthy eating patterns. A third person may be unsure about how to cook dried pasta, never mind making pasta from scratch; but they may be very successful at selecting among ready pasta meals for maximum pleasure at minimum cost. These examples indicate that it is unlikely that there is a person anywhere who possesses all possible food skills. Skills are acquired within individual, motivational frameworks, and what is regarded as a skill by one person may be seen as irrelevant by the next. However, to the extent that the modern, rational, skilled food consumer needs to be able to differentiate between competing products, they must be advertising literate. In particular, they must be aware of the basic difference between ingredient, or process, information and advertising messages or claims; and they must be able to interpret advertising signals and claims correctly. In segmented markets, advertising is both necessary and informative, because product positioning has to be signalled somehow. Signalling is however achieved by a wide variety of means, of which labelling claims is only one. Other techniques include the physical display of foods, e.g., the placing of ambient-stable products in chilled supermarket cabinets to suggest, perhaps, freshness of the product and the use of high-quality ingredients in the recipe [120]. Signalling, by whatever means, has the potential to mislead; but for this to actually take place, there needs to be a consumer who will have themselves misled. In order to protect consumers, certain claims are prohibited either by food law or discouraged by advertising codes of practice. However, not all false claims are misleading, and such claims are common in humorous advertising. Advertising literacy, combined with access to full and meaningful product information, is therefore the only real protection to the consumer from unsatisfactory foods.

Food advertising, and quality signalling generally, utilises Pavlovian conditioning, i.e., the creation of associations between a new food and a familiar, positive stimulus. These stimuli also serve to draw attention to the product. Jingles, sensual imagery and verbal associations are examples of this mechanism. The name given to a food is particularly important in this. The arbitrary appropriation of names of established, usually foreign, dishes for new products is fairly commonplace, but might nevertheless be regarded as fraudulent [121]. As discussed previously, the name given to a food does matter to how the consumer perceives it. The attaching of customary names to "any recipes of someone's devising", with a disregard of the ingredients and methods of cooking which made these dishes famous [121], would certainly deceive the consumer familiar with the genuine, authentic article, if perhaps only for the first (and probably last) time they try the product. Unfortunately, the cavalier use of names likely to mislead consumers is not confined to the commercial context, as examples of the practice may be found in legally sanctioned food standards. Scotch Beef (as opposed to Scottish beef) is a case in point. For beef to be marketed under this designation, an animal will have had to reside in Scotland for at least three months. It is hardly surprising that this does not tally with the ordinary consumer's interpretation of the term Scotch [122]. And whilst it is true that the statutory specification for Scotch Beef is in the public domain and readily accessible via the internet, advertising leaflets put out by Quality Meats Scotland do not point this out. In fact, the slogan "raised the way you want it" is likely to reinforce the impression of the cattle having been raised, rather than just finished, in Scotland.

Where food producers and their ultimate customers meet face to face, communication between them is straightforward. Thus the traditional fishmonger's role is partly one of educating consumers, about the nature and provenance of the product, what it tastes like and possible ways of preparing it for consumption [48]. Compared with other small food traders, fishmongers' communication with their customers is especially critical, partly because of the relative unpredictability of supplies to the fishmonger, and partly because of the perishability of fish and the related issue of product freshness. The commercialisation of the food market, in particularly, the sophistication of the marketing function itself, are sometimes held responsible for consumers' loss of confidence in their own ability to provide simple, healthy dishes cheaply for themselves from scratch [28]. Dieticians now habitually deal with patients whose eating habits have become confused, and who do not know how to process the many diet-related messages directed towards them through the media [28]. Information about food is acquired from many different sources. Food supply chain players, who may supply such information include producers, manufacturers and retailers. Producers may belong to marketing groups, which will provide advertising as well as in-store recipe and information leaflets, e.g., Quality Meats Scotland. Manufacturers may do the same. However, food retailers are particularly active in providing information for their customers. Here, leaflets may cover general issues, e.g., healthy eating choices and food safety, as well as recipe leaflets. The most important vehicle for product information is the packaging of a food. This includes the graphics, in particular, on-pack food photography, and information panels for ingredients, nutrients, shelf life, use instructions, and so on. Food information and education are also provided through educational institutions, government agencies, the commercial media and health professionals. One of the most important, and most trusted, sources of food related information is obtained through the quality press. This includes investigative journalism, feature articles, newspapers' lifestyle sections and readers' letters. TV food and lifestyle shows, and radio programmes too, are regular suppliers of information on a variety of food related issues. However, the latest, and potentially most promising medium for food information and advice is the internet. All major food companies and professional organisations concerned with food can now be accessed via their websites, many already providing feedback facilities for site users. Recently, the UK Food Standards Agency has taken on the responsibility to represent the consumer's needs in so far as the food industry does not appear to be satisfying them, especially insofar as information asymmetries which disadvantage consumers exist in supply chains. This means, in particular, the need for detailed product and process information, and for information about good nutritional and hygiene practices. One of the issues that causes concern to government generally is the public's perceived ignorance of science [123], to which is attributed consumers' rejection of innovative food technologies, e.g., irradiation and genetic modification.

Where consumers experience difficulties in understanding and applying labelling information, e.g., nutrition labels in terms of a healthy diet, there may be several explanations. The problem may be due to cognitive limitations in translating the nutrient (label) format into a food (diet) format, and this may be aggravated by a general lack of understanding of nutrition science generally. On the other hand, the problem may be due to a lack of motivation to understand the information, which would lead to insufficient attention being given to the label. In either case, the style of presentation will affect the ease with which such information is processed [124], irrespective of the particular barriers impeding comprehension. In terms of dietary advice, the messages given also differ in their effectiveness. These range from advice to eat a varied diet (unhelpful – meaningless), to eating five portions of fruit and vegetables a day (slightly more helpful – but what is a portion?) to eating nothing except raw, ripe fruit [6] (easy to understand). Governmental advice expresses goals for the population overall (reduce fat consumption), and is therefore meaningless at the level of the individual consumer. Perhaps the simplest advice would be to eat as much fruit and vegetables as possible and as little refined/processed (rich in fats and sugars) food as possible; and at all times to observe the body's feedback to the food it is given. "Healthy Eating" ready meal ranges are popular, and may serve as simplified choice rules for consumers with certain diet-related concerns, e.g., body weight. However, such product ranges are rarely optimal in all aspects of healthy eating as defined by public health and nutrition policy. Instead, they tend to signal some difference in respect of some other ranges. Many people are strongly attached to their food habits which, after all, have been acquired over many years; others are more flexible. Thus the same type of cognitive learning in different individuals does not necessarily result in the same behavioural changes. The reduced-calorie, ready meal market illustrates the point, as it will only attract a proportion of people concerned about their weight. Others may take action by eating smaller portions of their normal foods, or by selectively eliminating high-calorie food items. A similar example is that of two hypothetical consumers, who learn about the detrimental effects of sugar consumption on health. Consumer A responds by eliminating sweetened foods and beverages from their diet, knowing that their preferences, e.g., for sweetened coffee, will eventually change as a result. Consumer B however is unwilling to change their eating habits and food preferences, opting instead to substitute all sugar with artificial sweeteners. On the face of it, both of these are valid, rational responses to a threat. However, a comparison of the two options taken by consumers A and B does raise questions about the actual skills of each as a consumer. For example, has the switch from sugar to artificial sweeteners made life less complicated for consumer B? Has it conferred genuine control to them? Was the wiser choice simply too difficult for consumer B to make, and if so, why?

Skills levels vary considerably between individual consumers, but so do motivations to appreciate the finer points of gastronomy. A taste for a wide range of flavours and textures is encouraged early in a child's life. Being fed home-prepared foods will make children more aware of nuances within familiar dishes and allows parent and child to discuss these; in contrast, formula foods taste the same every time [125] – as they should! Mastery of a skill usually involves progressive simplification, and perfection focuses on the essential. The person who is skilled in food preparation enjoys experimentation, and is able to indulge in this without fear of jeopardising the edibility of the dish they are creating. In many respects, the skilled domestic cook resembles the commercial chef. On any particular day, they will try to shop for some fresh, central meal ingredient, offered in peak conditioned on that day, but they will then be able to incorporate stock and larder items into their final dish. They will also make maximum use of freshly harvested, seasonal produce, recognising such products when they see them. A sure sign of a fresh Jersey potato is its flaky skin, which can be readily rubbed off; however, most consumers no longer understand this signal of freshness in a potato [126]. In fact, the widespread lack of understanding of the nature of many traditional foods and food ingredients is, at times, alarming. For example, the BBC's consumer programme "Watchdog" expressed surprise at finding gelatin in confectionery jellies and anchovies in Worcestershire Sauce [127]!

8.5
Food Acceptance and the Context of Consumption

Individual foods and food technologies vary in terms of their suitability for, or performance under, different circumstances. When the "Brendan", a replica of a 16th century boat, sailed across the Atlantic in 1976, its crew quickly discovered that their modern, dehydrated stores were highly susceptible to spoilage, in particular, as a result of sea water leakage [128]. Their medieval-style replacements, i.e., smoked sausages, smoked beef, salt pork, as well as hazelnuts, oat cereal and cheddar cheese, proved considerably more satisfactory. The oat cereal turned out to be good work food, and the smoked and salt meats survived being swamped by waves or soaked by rain [128]. In this example, the focus is strictly on the suitability of some technology in some context, as the ship's crew might have been equally satisfied with the dehydrated foods – had they remained edible! This example also serves to illustrate the difference between perceived and experienced appropriateness. The choice of the higher-technology foods, which are well established for use in mountaineering and space travel, was misjudged, as not enough attention had been paid to the likely effects of the watery environment of the journey.

There is a high degree of association between appropriateness and the expected liking or disliking of a food [129]. However, food preference and acceptance measures obtained through affective attitude measurement are relatively poor predictors of consumption [129]. Assuming that a consumer likes some food in principle, and that this food is in good condition when presented to them, the precise context of consumption can still cause the food to be rejected. Hedonic evaluation of foods in the sensory laboratory, perhaps in the course of a product development project, tends to take place without reference to the context in which the food would normally be consumed. Firstly, only a small amount of product will be eaten. In addition, the tasting environment will be kept deliberately free from any distractions. Finally, the assessor will not have access to other foods that might normally accompany the item being evaluated. A panellist required to carry out hedonic scoring on, say, half a dozen samples of smoked salmon, might identify and evaluate relatively small sensory differences between them. However, in practical terms, these differences could be insignificant, given the fact that smoked salmon is rarely eaten by itself – although, of course, it might be. With traditional accompaniments for smoked salmon designed to counteract its greasiness, e.g., lemon and onion, strong bread and vodka, as well as a stimulating eating environment, the small differences in preference identified in the sensory laboratory are likely to dissolve. In fact it seems safe to assume that, when consumer panels carry out food acceptability tests within such neutral environments, each member carries in their mind some reference point regarding more or less appropriate consumption contexts for the product. For the marketer, the task is to determine the nature of such idealised settings. If consumers have as yet little, or no, experience of a food product, as is typical of many new market introductions, consumer education based on in-depth marketer-consumer communications becomes a critical factor in the success of the product. Letting consumers take new products home, in order to evaluate them, will provide valuable feedback to the product development and marketing teams by turning the prospective consumer into an active participant in the process.

Appropriateness means being presented with the right food at the right time for the right purpose, and in the right way. Contextual variables have a major impact on food acceptance, not only during eating and drinking, but prior to this, i.e., in the acquisition phase. Key contextual factors in both these aspects of consumption include finding the foods one expected to find; the way in which these foods are presented, including lighting effects; the general pleasantness of the surroundings, including background music; and the presence of more or less congenial fellow consumers. Product information too is a potential modifier, which can affect the hedonic ratings of a food and of food attributes [130]. And of course, it is very much in the nature of advertising to set the frame within which consumption is to take place. Hence Lucozade, which began life as a drink for the sick has recently been redefined as a fashionable, high-energy drink; and breakfast cereals have been repositioned as "anytime" snacks. Perceptions of appropriateness are not static, so that yesterday's waste fish, discarded by fishermen, can become today's delicacy, as is currently happening in Scotland with prawns, monkfish and hake [131]. As with eating and drinking, food retailing may be designed to satisfy a need for speed, reliability, predictability and convenience, in which case a systems approach is the most suitable. Alternatively, retailing approaches may respond to some need for a more thoughtful, involved, personal, and less hurried shopping experience, where there is room for some level of unpredictability. The offer itself is, of course, important. In purely functional terms, small chocolate Easter eggs might be considered a convenient shape and size for a year-round snack. However, the seasonal and festive associations evoked by the product would irritate many shoppers were the product to be found on offer in the shops over the Christmas period. Another issue concerns the way in which retailers display meat. Many food shoppers prefer to minimise the association between the meat or fish they buy and some animal that was once alive. These customers would favour visually sanitised displays of meat. At the same time, consumers with more realistic attitudes towards meat eating would value the quality cues inherent in the display of whole fish, carcasses and carcass sections, and so on. Food choice invariably takes place with reference to some consumption value to be realised, and therefore to some use situation. Food categories reflecting common uses and contexts include: prestige foods; celebratory foods; power meals and business lunches; staple foods, a culture's main source of calories; comfort foods to lift mood or reduce stress; social foods; snackfoods and finger foods; performance foods; nourishing foods; square meals; and body-image foods. Food use has also been categorised as "utilitarian", "casual", "satiating" or "social", and appropriateness as "really like", "for teenagers", "unhappy", "as a main dish", "at parties" and "inexpensive" [132].

Important contextual variables in the eating event itself include other foods and beverages served with a food, and variables in the physical and social environment [133]. Perspectives on what food goes with what other food are often culturally determined, e.g., the mixing of sweet and savoury items, of meat with fruit, and the serving of pickles as an accompaniment of British cheeses. In the context of multi-course, acceptability is dominated by the acceptability of the main dish [133]. When manufactured snack foods and fruit were studied in term of their perceived suitability for snacking uses, major perceptual differences were found to exist between the two categories [64]. Manufactured foods were considered to be, at the same time, more convenient in use and more suitable for indulgence and comfort eating. Convenience aspects included storability, pre-dictability of eating quality and absence of waste and mess. Cheese is a versatile product, and has an important role in UK diets. A study of the role of the qual-ity attributes of cheese in appropriateness found a number of significant use factors, of which textural and melting properties were the most important [134]. The food matching properties of several categories of drink are also of interest, with rules of thumb about choosing a wine to partner a particular food widely shared. One such rule posits that sourness in food diminishes sourness in wine, so that wines suitable for complementing sour foods (e.g., oysters with lemon juice, salads with dressing, dishes with tomato sauce, cheeses) should themselves possess a certain degree of "acidity" [135]. Whilst wine is the most familiar topic in terms of food-and-drink partnering, there are foods for which wine is thought to be unsuitable. Beer might be suitable to accompany some of these. Foods generally perceived as beer foods include sausages in any form, cold pork pies, and hot (Thai) noodles [136]. Whilst it is customary in the West today to consume a beverage with one's food, alcoholic beverages are not always suitable. When commercially available apple juice was investigated to establish sensory attributes determining suitability as a meal accompaniment, only clear juices were perceived as appropriate [47]. Clear juices were described as thirst-quench-ing, and visually, they were closely associated with wine. In contrast, cloudy apple juice was associated with orange juice, the UK's most popular breakfast beverage, and it was basically perceived as a healthy (liquid) snack. Some years ago, UK consumers tended to align themselves with a particular alcoholic drink which, they felt, encapsulated their personality. In particular, there were wine drinkers, beer drinkers and whisky drinkers. Today it is more common for people to adjust their drinking to specific social (and food choice) contexts, demonstrating greater knowledge and sophistication [137]. Historically, the consumption of beverages, and in particular, contexts of consumption, have often served as media for social interactions. For example, the fashion in Europe for taking three new and exotic drinks, i.e., tea, coffee and chocolate, helped to revolutionise social customs. It enabled men and women of the upper classes to consort without impropriety [138]. As discussed in Chapter 2, in many cultures certain foods are traditionally associated with quasi-medical uses; this is still strongly the case in countries such as China [67].

The appropriateness of individual foods for specific use contexts can be ex-plored from a number of different perspectives. These include the nature of the meal or eating episode, its venue and how the food is assembled, either in terms of individual dishes or in terms of the courses that constitute a meal. Influences exerted by fellow diners are particularly important. Informal eating is tied up with fewer, less heightened expectations than formal, or special, occasions. If the diners have made a special effort in dressing up to attend a formal dinner, they will expect some level of finesse in the way that food is presented to them. Whilst finger-friendly foods, such as breaded, deep-fried "scampi in the basket", or a plate of langoustines in the shell, are fun, shellfish served at a formal dinner will be expected to have been dressed. Chic diners will expect finesse in the food, rather than rusticity. To them, the expense and status of the food needs to reflect their own status. A key aspect of formal dining concerns conventions about the order in which dishes are brought to the table, and the way in which each course harmonises with, and frames, all the other courses. The above scampi and langoustines symbolise fast foods and slow foods, respectively. Slow food implies both discerning eating and appreciation of the social functions of a meal. It means cooking with care and searching out the best. Fast foods are distinguished from slow foods primarily by the individual's level of involvement with them. However, this involvement is situation specific. Meal occasion also refers to the time of day. For example, certain foods are traditionally associated with breakfast, others with lunch or dinner, yet others are thought of as all-day snacks. Such boundaries are becoming increasingly flexible, with "all-day cooked breakfasts" typically available in roadside catering establishments. Rare delicacies are not normally desired after fatiguing labour [69]. Food service establishments will also aim to cater for the non-food needs of their customers. The décor and ambience, and the style of service play an important role in this. Customer needs may mean fun and boisterous entertainment at one end of the scale, and a gentle, relaxing environment at the other – as still represented by the traditional English tea room. In all cases, the circumstances of the serving of the food will be designed to create, or promote, some desired mood, into which all the guests presents are invited to tap. In certain situations, the cook actively contributes to this, e.g., the sushi chef, who converses with the customer, assembling dishes to individual requirements; the virtuoso pizza chef, who stretches the pizza dough by the spinning it in the air; and the traditional breakfast cook, who prepares cooked breakfasts in full view of the hotel guests.

International differences in the demand for certain basic commodities can often be traced to differences in food use customs. Hence, the high level of milk consumption in the UK corresponds to the consumption of breakfast cereals. Countries also differ in how they traditionally prepare, present and combine staple foods, such as bread, cheese and sliced meats. The type of bread favoured within a particular culture reflects not only differences in taste, but also how it is to be used. German bread, rye breads in particular, tend to be of a sturdy structure, which makes them easy to slice. The bread also has a robust taste. The slices are then combined with thinly sliced toppings, e.g. cheese or cold meats, or with spreads, including spreadable sausages. In contrast, the popular English "ploughman's lunch" consists of a hunk of (crusty but soft) bread, which is served with a suitable chunk of cheese, which may be quite crumbly. Cold meats in Germany are invariably eaten as open sandwiches or in rolls. In Britain, they are associated with salads, and in Italy and Spain with appetisers, where they may be eaten on their own. Another difference concerns German and British consumers' perceptions of potato crisps, which are popular snacks in both countries. In Germany, crisps are mainly considered as party nibbles, whereas in the UK they are general snack foods. Pizza-type flavour is one of many varieties of crisps successfully marketed in the UK, but the concept was not acceptable to German consumers. The reason was that they associated pizza with bread, and hence, bread-flavoured potatoes made not sense to them [139]. UK consumers clearly take flavour claims less literally; after all, they have seen hedgehog-flavoured crisps come and go. Similarly, in the late 1990s, chocolate flavoured vegetables, supposedly designed by the retailer Iceland to change the eating habits of British children, came and went – fast. The concept of inappropriateness is readily illustrated in connection with airline food. The ultimate aim here is to provide food that is widely acceptable in cultural, as well as sensory terms. Hence, smelly foods such as garlic and most fish will be inappropriate, as will delicate foods that do not stand up well to re-heating. Chicken is often thought of as the inoffensive food par excellence [138]. However in today's food market, this could only be said with confidence if the chicken could be demonstrated to have been reared to what every meat eater would consider to be humane standards.

9 "Good" and "Bad" Foods

9.1
Introduction

With all that has been said so far, is there any sense in categorising foods as either good foods or bad foods? Does not the mere fact that something is identified as a food, and that there is an established market for it, also make it a good food? The intense, public debate conducted in the UK throughout the 1980s and 1990s on what constitutes good and bad food supplied major impulses to write this book. To date, this debate has been characterised by strong antagonisms, and very little conflict resolution. One reason for this may be that underlying assumptions and frames are rarely explicit, and are often highly personal. One of the UK's foremost protagonists and defenders of what he refers to as good food is the journalist Derek Cooper. His definition of good food is food that is "natural, pure, nourishing and, above all, full of taste and flavour" [140]. To Cooper, modern food technology, and the terminology associated with it, are baffling [140]. Traditional and home-made foods are best, with "home style" tolerated as an industrial substitute [140]. Cooper expresses concern about the intensive, mass production of foods, because he sees this as leading to food being deprived of its traditional cultural role in society, with farms functioning purely as production units, and animals as meat machines [141]. Henrietta Green, author of the Food Lovers' Guide to Britain and organiser of a variety of "good food" events, worries about farmers, growers and processors who are willing to compromise their standards, e.g., by substituting cheaper alternatives, or by speeding up processes [142]. Common bad-food terminology includes junk food and, in the special case of genetically modified foods, Frankenstein food. From this perspective, the term "processed food" is almost always used with negative connotations, although traditional processes, however severe, are invariably seen as producing good foods. The industry perspective on the good-food-vs.-bad-food debate is typically one of frustration with the perceived stupidity of the non-technologist and non-scientist, who simply refuses to understand that there simply are no bad foods, only bad diets [143]. This argument appears not to apportion any role to the activities of food formulation, or marketing, in terms of the development of particular food consumption patterns among consumers, that may be detrimental to their health, to the environment, and so on.

It is certainly justified to turn the focus of the debate away from the product (food) and onto the consumer (food habits). Given any particular food or beverage, say, rich chocolate cake or wine, whilst consumer A will choose to indulge themselves, which is good for their mood (and maybe their immune system), consumer B will overindulge, which speaks for itself. The indulgence-related terminology implies that for consumer A, rich chocolate cake is an exceptional treat rather than a regular dietary habit, and that wine, although it may be consumed habitually, is consumed in moderation by them. Clearly, the individual (consumer) realises virtue by developing and practising good habits. However, the question remains whether specific foods and marketing approaches play some role in the way in which individual food habits are formed. After all, an important aspect of food marketing is to persuade potential consumers of the benefits to them of specific products. Taken out of context, the terms good and bad do not transmit much meaning, other than signalling the presence of positive or negative attitudes about something in a food. Yet, in relation to food, the terms are common currency, and even children in the UK are quite able to recite those foods that are "good for you" (vegetables) and foods that are "bad for you" (chips). One question concerns who is to derive which benefits from a particular food, i.e, in whose eyes is it a good food, and in what terms? Hard-boiled eggs are a very good food as far as the UK government is concerned. If people sacrifice their love of soft-boiled eggs, the government will suffer no further bad publicity due to salmonellosis from eggs, because the bacteria will not survive the process. But it is a bad egg day for UK consumers who, for centuries, have dunked their (bread) "soldiers" into their soft-boiled eggs.

Clearly, old, stale and spoiled foods, or dangerously contaminated foods are bad foods; but then they are not really foods. Also, depending on the type of contamination, they are not bad for everybody. So, does bad signify foods that are somehow dangerous, more so because the danger is only realised over extended times scale? Equally, does good signify foods that are simple, seasonal and fresh, and local, and is their goodness due, in part to the trust that people have in such foods. Processed foods are processed foods irrespective of whether they are cooked at home, manufactured to so-called traditional recipes or whether they are manufactured using modern ingredients. If the latter are often salty and fatty, so are the former; after all, cooking, and the industrial processes developed from it, exist to render a food more palatable, at least in the view of those to whom it is targeted. Trust may play some role in the not uncommon perception of so-called processed foods as inferior. If something is mass produced at distant locations, opportunities for deception and adulterations may be perceived to increase, e.g., pumping water into hams and chickens, hiding carcass waste and fat in processed meats such as sausages, and so on. The hamburger is a case in point. Whilst nutritionists may view favourably a meal consisting of a lean-meat hamburger served in a bun with salad vegetables, the reality of the fast food hamburger seems to suggest an entirely different sort of product. Yet to modern children, Big Macs and chicken nuggets are "real" foods, rather than foods laboriously prepared at home by their mothers. They are real because they are what everybody likes and eats. Vegetables are both good (perceived health benefits) and bad (enjoyment), with watery, slimy, smelly, overcooked institutional cabbage wreaking much havoc with children's perceptions of foods. Vegetables are foods children are made to eat against their own inclinations. There are many different frames for assessing goodness and badness in a food, as many as there are consumption benefits to be derived, all of which have to be seen within the context of specific eating pattern. They include time frames, i.e., whether an act has long term (health), medium term (body shape) or short term (sensory enjoyment, alertness, stamina) benefits. Unless a food is judged with reference to some standard, sensory frames are unhelpful simply because tastes are subjective and differ markedly between individuals. Bad food habits are a particular issue with children's food consumption. Children are deeply involved in learning food preferences for the rest of their lives. The chapter explores four specific aspects of goodness and badness in foods. The first of these concerns the concept of healthy, or wholesome, foods, the context in which "good/bad for you" is frequently encountered. This is followed by emotional aspects of foods, authenticity and related concepts, and finally, good foods in the ethical sense.

9.2
Healthy Foods

An awareness of the possible positive and negative health consequences of diet is one of a range of factors that influence individual food choices, and the adoption of individual dietary routines. As health is increasingly becoming an issue for both public policy and for individual preoccupation [144], people are expected, and themselves expect, to manage personal health through self-control based on information from government. However, national campaigns to reduce the incidence of obesity and diet-related diseases, e.g, heart disease and cancers, have only met with limited success [145]. This is despite the placing on the market of many ostensibly healthier, "low and lite", versions of traditional processed foods by most of the major food manufacturers and retailers. These so-called healthy options include products with reduced calorie, fat, sugar and salt levels. An obvious example is the market for fat spreads, which is segmented not only in terms of total fat contents, but also in terms of the contents of undesirable saturated and trans-unsaturated fatty acid [146]. Healthy options also include basic foods. For example, selective breeding has achieved a 35% fat reduction in British pigs since the late 1960s [147]. This type of commercial approach towards promoting healthier eating habits mirrors healthy-eating advice from health educators, i.e., that consumers should substitute healthier versions of familiar foods. This might mean advice to bake with wholemeal flour and sultanas rather than white flour and sugar [28]. The underlying assumption characterising this approach is that many small steps to cut down on highly refined foods or food ingredients will in time achieve the ultimate goal, e.g., a slimmer, healthier body. Others dismiss such conventional approaches as compromise cookery, which will never lead to optimum nourishment, well-being and disease resistance [6]. The latest trend in marketing foods with possible health benefits is the development of functional foods [145] (see Chapter 6). Here in contrast to the "low and lite" approach, health-promoting food components are actively designed into products, and health benefits are positively framed. Health promotion models have been challenged by those calling for a more social understanding of health and illness and the recognition of material limitations to choice [145]. This includes both lack of access by those on low incomes and over-simplification of risks.

If food is conceptualised as a tool, diet is the process of applying it. And since all excess is, by definition, bad, individual foods can be used for either good ends or bad. Both the frequency of use of individual dietary items and the dietary patterns into which they fit, are important factors in this. However, there are certain important differences. For example, the advice with regard to fruit and vegetable consumption generally is that people can eat as many of these as they like and stay healthy. In fact, there are certain, well-known examples of people who have taken this advice to its logical conclusion, i.e., their diets consist of nothing else, except perhaps the odd nut, and who insist that their physical and mental health are excellent as a result [6, 148]. Yet, conventional dietary advice would fight shy of such extremes, and warn of nutrient deficiencies that must surely accompany them. Given these constraints, the opposite of fruit and vegetables as desirable edibles would be artificial sweeteners. Whilst consumption rules about fruit and vegetables are easy to absorb, rules about artificial sweeteners are complicated. In fact, not only are there limits for safe consumption, there are, in addition, rules about which products may contain them. Thus, food and drink intended for children aged three years or under are not allowed to contain artificial sweeteners, yet the pack design of many soft drinks containing them would appeal to young children [149]. Consumption rules about artificial sweeteners are very hard to learn, and, given the range of products containing them and that might be components of individual diets, consumption levels are, in fact, impossible to manage. Different life stages require different foods and eating patterns for optimum performance [28]. Dietary observances during pregnancy are an especially important example of this because, during pregnancy, women are more susceptible to certain pathogens. In addition, certain foods and food ingredients, which may be beneficial in terms of the wider population, are unsuitable for some consumers, such as people with food intolerances and allergies. Food intolerances, which are defined as non-psychological, reproducible unpleasant reactions to a specific food or ingredient, occur in around 1–2% of adults and 5–7% of children [150]. Coeliac disease, a bowel disease, is the main form of wheat intolerance and occurs when gluten is present in the diets of susceptible consumes [150]. Lactose intolerance, which is dose related, is the most common adverse reaction to cows' milk, and is due to an enzyme (lactase) deficiency [150]. A small proportion of food intolerances is classified as food allergies, i.e., intolerances involving an abnormal immunological reaction [150]. Among allergies, peanut allergy has assumed a particularly high public profile in recent years, largely due to its potentially lethal consequences. Although the prevalence of peanut allergy is not known, it is believed that a small number of people suffer severe reactions within minutes of exposure, which can result in anaphylactic shock [150]. Metabolic disorders, such as diabetes, do not necessarily require specific foods to be excluded from the sufferer's diet altogether; instead, they may need to be rationed. The management of eating disorders, such as anorexia nervosa, bulimia nervosa and compulsive overeating, reaches beyond mere dietary recommendations, as the causes of these disorders are largely psychological [151]. With up to 2% of school children currently suffering from anorexia, one expert view that has been expressed is that children are becoming anxious and confused about food because their parents are [152].

However, the greatest perceived food-related threat to children's health, and to the health of the population, is a fondness for so-called junk foods. The debate about whether or not such a thing as a junk food actually exists is somewhat spurious. It is usually clear from the context that eating junk foods refers to bad patterns of eating, with an excessive reliance on foods which, if eaten in moderation, would not be thought of as jeopardising health. Such junk foods are typically characterised by high levels of nutritionally controversial ingredients, e.g., salt, sugar and fat, in particular, saturated and *trans*-unsaturated fatty acids. Junk foods, in common with many other cooked foods, typically do not bring with them the enzymes that help digest raw fruit and vegetables, i.e., they may be conceptualised as devitalised [153]. As well as being low in nutrient density, junk foods typically do not require much chewing [153]. Lack of chewing is likely to be accompanied by a low level of attention to digestion generally, including saliva production and attendance to signals of satiety [153]. The term fast food reflects this lack of attention on the act of eating, as well as the ready availability of these foods. Junk food habits are increasingly being linked to physiological imbalances in children, in particular, asthma [154] and rickets [155]. In the short term, the overconsumption of junk foods may lead to indigestion and lethargy. The presence of stress during eating episodes can similarly lead to symptoms related to poor digestion so that, where fast foods are concerned, it may be the manner of consumption, as much as the food itself, which may be to blame for any negative postingestive consequences. However, stress at the table in the course of a slow meal would have a similar effect [28]. A so-called wholesome food is a food that is generally conducive to good health and well-being, promoting an alert mind, freedom from disease and longevity. Eating wholesome foods, like eating junk foods, refers to an individual's food habits. However, although no individual food can be classified as complete junk, there are some foods that are regarded as particularly wholesome, the classic example being mother's milk in respect of the nursing infant. In the context of adult foods, a superfood usually designates an easily digestible food that is a rich sources of key nutrients, i.e., EFA, IAA, vitamins and essential minerals. Sprouted grains and legumes are examples of foods which provide large amounts of available, easily digested nutrients [153]. Meat is generally recommended as a source of the full spectrum of IAA. However, there are also vegetarian sources of complete protein, e.g., buckwheat, lentils, quinoa and mung beans [153]. The increasingly fashionable raw food diet, with its focus on live food [156], is taken to its ultimate logic in a diet which permits only wild plants, i.e., plants that will grow healthily without the need for human intervention [148]. No single food could be absolutely ideal simply because different individuals, and different situations, will require different amounts of dietary fuel.

Individual foods that have received an unfavourable press in recent times, and whose health benefits are controversial, include eggs, red meat, oily fish and salt. Cooked and processed foods high in protein, fat and refined carbohydrates are under suspicion because of their tendency to generate acidity within the body [153]. This can lead to acidosis, an excessively acid condition of the body fluids and tissues [97]. Fruit on the other hand, although tasting sour in the mouth, are metabolised in the body to release alkaline compounds [153]. A possible negative scenario associated with raw fruit and vegetable consumption concerns biological contamination, in particular, with parasites. Eggs have been controversial not just in terms of the high incidence of pathogens, in particular, salmonellae (see Chapter 6). They have also come under the spotlight because of their relatively high cholesterol contents, and because of the link between blood cholesterol levels and heart disease (see Chapter 4). However, current thinking regarding this issue is that the impact of one egg on the average person's serum cholesterol level is trivial [157]. In addition, it has been shown that dietary cholesterol has less of an effect on serum cholesterol than had been assumed, with a major role attributed to SaFA instead [157]. The recent debate about safe levels of consumption of red meat has been heated [158], and the existence of such a debate is, in itself, of some interest. A number of research studies have established links between meat consumption and various cancers [158]. However, whether the problem is meat as such, or the processing techniques typically employed for meat, e.g., frying and grilling, is unresolved. Oily fish, and farmed Scottish salmon in particular, has been the subject of debates weighing up the beneficial effects of fish oil against the detrimental effects of contamination of that same oil with environmental pollutants, e.g., PCB and dioxins [159]. Salty food may be responsible for the death of infants [160]. However, in adult diets the subject of salt consumption is more controversial, as salt restriction does not show an effect on blood pressure in all subjects [161]. As a general rule, and in accordance with the whole foods principle, intrinsic food ingredients are favoured above the same substances after they have been isolated from their native environment, e.g., intrinsic vs. extrinsic sugar, and intrinsic (fresh fruit etc.) vs. extrinsic (bran) fibre.

The traditional view of a balanced diet was a mix from three so-called nutrient groups: proteins foods for growth and repair, fats and carbohydrates for energy, and rich sources of specific vitamins and minerals [162]. Although relatively easy to follow, this scheme proved too crude [162]. Today in the UK and northern Europe, the general diet is said to be too low in starchy foods, and in fruit and vegetables, too high in fat, particularly saturated fat, and too high in sugar and salt [162]. The high saturated fat intake explains the high average levels of cholesterol in British adults and is linked with increased risk of coronary heart disease [162]. More than 35% of the adult population of the UK are overweight, reflecting the overeating and inactivity that characterises certain modern lifestyles [162]. One of the traditional rules of thumb for vegetarians to obtain the full spectrum of IAA from a dish has been to combine pulses with grains. However, current advice is that it is not necessary to combine foods at one meal for the sole purpose of achieving a complete protein [153]. Eating well means combining foods in such a way that a feeling of vitality results in the short term, and in the longer term, a healthy body. There is a school of thought that certain food groups do not combine well. For examples, concentrated animal proteins, foods rich in complex carbohydrates, and oils are said not to combine well with fruits because these foods take a long time to digest and slow down the passage of the fruit, causing it to undergo fermentation in the digestive tract

[153]. A good, healthy meal is a meal that leaves the eater feeling energetic and clear-headed, whereas a bad meal might leave them lethargic, or even bloated. Surprisingly, meals dominated by starchy foods are highly favoured in conventional nutrition education, even when such meals are explicitly recognised as soporific [28]. Those preferring fruit and vegetable based meals clearly avoid such an outcome. One professional view of a balanced diet envisages a combination of fresh fruit, vegetables and pulses, whole grains, lean protein, dairy products and water, but offers no suggestions on proportion [28]. As can be seen, this further muddles the message by mixing foods with nutrients. After all, protein is present in fruit, vegetables, pulses and whole grains, as well as many dairy products.

Yet the concept of the balanced diet is difficult to grasp, and it is therefore important that nutrition educators provide workable rules of thumb and commitment devices that people can use in everyday life. One useful public policy initiative that does this is the promotion of the five-a-day rule, i.e., the recommendation that everyone ought to consume at least five portions of fruit or vegetables every day [163]. This rule is easy to grasp conceptually, as the message it gives is simple. Whether individuals find it easy to follow it in practice is another matter. However, anyone who increases their fruit and vegetable consumption significantly will find that their overall dietary habits will gradually be affected. This is simply because fruit and vegetable are relatively filling. Assuming that the individual attends to satiety feedback, they are likely to reduce their intake of the more energy dense foods they would have favoured previously, simply in order to accommodate the additional volume of fruit and vegetables. The recommendation to follow a varied diet is often made in the context of choosing from four main food groups: starchy foods; dairy products; meat, poultry, fish and alternatives; and vegetables and fruit [164]. The idea here is that starchy foods should form the main part of most meals, with foods from other groups being added [164]. In the USA, a coalition of over 3000 medical doctors has requested the authorities to eliminate meat and dairy products from the four basic food groups [153]. A so-called varied diet is often said to take care of all nutrient requirements, thus obviating the need for dietary supplements. There are two problems with this particular advice. Firstly, it is too vague to be implemented with any confidence; secondly, too much variety, especially within individual eating events, tends to override SSS mechanisms and lead to overeating.

Although many people follow some aspects of current healthy eating advice, few find such advice inspirational [145]. Instead, people tend to find out through their own experience which consumption patterns suit them, with short-term priorities generally dominating [145]. Many use calorie-counting and portion-control approaches to control food intake. The retailers' healthy option schemes too tend to incorporate one or other of these approaches. Yet, there is no real shared understanding of what constitutes a portion. As fast food snacks are increasingly used as meal substitutes, portion sizes may increase [165], so that the boundary between snacks and meals becomes blurred. To some commentators, the so-called Americanisation of food markets, i.e., trends towards larger portions, constitutes yet another encouragement for people to overeat [165]. In some cases, the presentation of a food makes clear whether a food is to be regarded as a treat or a regular dietary item. A small, after-dinner mint will clearly be seen as an example of the former; on the other hand, the message communicated by a Mars bar, or similar count-line confectionery item, is that here is a snack. Diet and exercise promotion messages may be seen as tiresome, e.g., in the context of the treatment of diabetes [166]. Health promotion campaigns are seldom assessed for their effectiveness, and the notion that such promotions may be misconceived and damaging is even less familiar [167]. In order for the promotion of specific foods or food groups to be targeted accurately, consumers' perceptions of them have to be ascertained. Thus, identifying psychological rules that guide inclusion of fruit and vegetables in the diet may facilitate behaviour change strategies to encourage their consumption [168]. Old established food habits may be difficult for the individual to relinquish. Such habits are frequently associated with a wide range of feelings and emotions, which makes it all the harder to break them in favour of healthier eating patterns.

9.3
Emotional Food Value

The Italian born UK chef and author Antonio Carluccio is well known for his passion for his native cuisine. His book "Passion for Pasta" starts with the author reminiscing about his childhood [169]. Carluccio's father having been a stationmaster, every day his mother would ask the boy to check whether the trains were running on time, so that she could time her pasta perfectly. In "A Passion for Mushrooms", Carluccio evokes the pleasure of hunting for mushroom in the early-morning atmosphere of woods and hills; anticipation, i.e., of a large haul, of some especially desirable specimens, or nothing at all, all heighten that pleasure [170]. Although certain food ingredients can have direct physiological effects on mood (see Chapters 4 and 7), the above examples illustrate that the emotional value of a food to an individual is largely psychologically based. This value derives from positive associations, in particular, emotionally charged associations with persons and events, but also more generally with places and activities. The sight and flavour of produce may become associated with the seasons, e.g., soft fruit in early summer and wild mushrooms in autumn, thus evoking those seasons. However, among such food-related associations, nostalgic associations with childhood, family, culture and nation are especially powerful. A celebrated literary example of a food evoking deeply nostalgic childhood memories is that of Marcel Proust's petites madeleines. When the, by now bed-ridden, narrator is presented at a certain point with these "squat, plump little cakes which look as though they had been moulded in the fluted valve of a scallop shell" – they instantly recapture for him his early, close relationship with his mother as "an exquisite pleasure had invaded my senses, something isolated, detached, with no suggestion of its origin" [171]. By extension, to the admirer of Proust, the term petite madeleine, or the sight of a petite madeleine, will serve to evoke the writer's work and its particular mood. Similar food-related associations are familiar to many people. The author's father often returned home from work with his lunch sandwiches still uneaten. Although these might show signs of curling at the edges, their journey through some imaginary forest had transformed them into highly desirable (to his children) "Hasenbrot" (bread for the hares). During periods lived in exile, whether enforced or voluntary, foods that have strong associations with childhood or family take on an added significance, as they assist in summoning up loved persons and places and, in more general terms, the culture of one's birth and perhaps, a lost lifestyle [172]. Emigrants often take their food culture with them and adapt it to the new local conditions in which they find themselves. Hence, the bratwurst makers of Wisconsin trace their recipes and skills back to the Germans who settled close to Lake Michigan in the mid-19th century [173]. During that same period, the Silesian settlers of the Barossa valley in Australia exploited the natural resources of their new country to allow them to preserve their own food culture. From these beginnings, pickled cucumbers, egg noodles, sauerkraut, smoked meats, sausages, pickled pork, breads and Streuselkuchen survive in the Barossa to this day [174]. Today, these local specialities are highly valued as heirlooms, as well as being exploited in the context of cultural tourism. As for Italian food, its popularity in many parts of the world is said to be due, in some measure, to the fact that there are more people of Italian origin living abroad than in Italy itself [175].

Homesickness can often be assuaged by sufferers treating themselves to – or even with – foods that are strongly representative of home. Thus the British expatriate community in New York enthusiastically embraced the city's first British-style fish and chip shop, the authenticity of which extends to the food being wrapped in British newspapers [176]. In a related scenario, a tourist in San Francisco, on being identified as a Briton by his waitress, was offered Sarson's malt vinegar with his chips; and "because of this double recognition of my nationality and my culture, I remember her for life" [177]. Mark Twain toured Europe in 1878 and, loathing the food, phantasised about what he would eat when he got home, e.g., buckwheat cakes with maple syrup, hot bread, fried chicken, soft-shell crabs, Boston baked beans, hominy, squash, and "a mighty porterhouse steak on inch and a half thick, hot and sputtering from the grill; dusted with fragrant pepper; enriched with melting bits of butter of the most impeachable freshness and genuineness …" [178]. Today's global food service chains have removed some of the risks associated with foreign travel experienced by Twain. Regional food preferences exist in all countries and are tied in with cultural identity. While any food that co-defines a particular culture, region or nation is imbued with emotional associations, this is especially true of celebratory and festive foods, ceremonial and ritual foods, and foods defined by religious dietary codes. The role of Cajun foods in the ethnic identity of the Cajuns has been researched in some detail [179]. Special food events, e.g., crawfish boils and boucheries blend work and play, thus highlighting both Cajun food skills and joie de vivre [179]. At these events, ethnic differences between Cajuns and outsiders are clearly evident [179]. Crawfish, boiled crab and boudin all are difficult for outsiders to manipulate and eat. Outsiders may also find it difficult to tolerate the pepper content of Cajun foods generally, and may even categorise certain foods as inedible or repulsive [179]. For people leaving Scotland, the traditional Scotch pie, potato scones and haggis are among the foods that they are likely to miss most [180]. The Scotch pie is one of the nation's most popular snacks, and has long been associated with people watching football matches [181]. The

Burns Supper, on or about 25th January, commemorates the birth of Scotland's national poet, Robert Burns, in 1759. It is the annual ritual wherever Scots gather around the world, and has haggis as the central food [182]. The annual Thanksgiving Dinner celebrated by US citizens is another example of a ritualised food event with patriotic associations, as is the Chinese New Year.

Rituals, especially rites of passage, provide tangible markers of culture, and support the continuity of a culture. The historical links of culture are explicit in occasions such as the Burns Supper, but in certain cultures, such links are more direct. For example, in China, ancestor worship lies at the root of religion, and until recently, has underpinned traditional family life for thousands of years [67]. Thus, when Hom visits his ancestral home for the first time, certain food rituals are performed as a mark of respect to the ancestors [67]. In the classic British wedding cake, form, rather than edibility or nutrition, are what matters; it is the cutting rather than the eating of the cake which provides cultural significance [183]. Food can be used to make general affirmations, and this is particularly true in the context of religious dietary law. Dietary exclusivity such as that of the Hebrews and, to a lesser extent, the Muslims are examples of this [184]. Foods can become reflections of spiritual inclinations and a vehicle for the development of a unified mind [185]. Thus, the word shōjin is composed of the Japanese characters for "spirit" and "progress", and originally had the meaning of zeal in progressing along the path to salvation [69]. In the context of the yogic diet, three kinds of foods are traditionally held to influence human personality. These are pure (sattvic), stimulating (rajasic), and impure (tamasic) foods. Each of these food groups has a corresponding state of consciousness, i.e., gross, intermediate, and spiritual for tamasic, rajasic and sattvic foods, respectively [186].

Food and drink are often given in the context of hospitality, as a contribution to a social meal, or as a general gift. In each case, emotional meaning is carefully encoded. Giving food to her child traditionally reaffirms the mothering role. Giving food to strangers allows the host to signal their understanding of the relationship between themselves and their guest. The writer Gertrude Stein's servant Hélène was said to understand perfectly the niceties of making up menus. "If you wished to honour a guest you offered him an omelette soufflé with an elaborate sauce; if you were indifferent to this, an omelette with mushrooms of fines herbes; but if you wished to be insulting you made fried egg. With the meat course, a fillet of beef with Madeira sauce came first, then a leg of saddle of mutton, and last a chicken ..." [187]. The giving of food can also have romantic associations. The traditional Irish confectionery yellowman is a golden yellow, hard, brittle, boiled toffee. It has a particular romantic association with the Lammas Fair in Ballycastle, Country Antrim, an annual event where young boys and girls were traditionally able to meet each other [188]. Romantic food gifts and meals in the UK are a tradition on Valentine's Day, although no particular foods are associated with this. There is some debate about the nature of so-called aphrodisiac foods. Some of these, such as oysters and asparagus, may evoke sexual associations simply because of their visual and other sensory properties. It is not only generic foods, with associations of home cooking that allow emotional associations to be established. The same applies to commercial products and brands. Character merchandising, e.g., the use of Disney and other popular characters to promote foods strongly taps into children' feelings.

People use comfort foods as a means of relaxation, and to disperse states of anxiety, or general low feeling. Comfort eating may result from work- or relationship-related stress, and serve to boost mood. Such eating may be a solitary behaviour, and using food persistently in this context may result in overeating. Cold winter weather and darkness typically cause people to reach for comfort foods in an attempt to lift their spirits. As comfort foods are often associated with childhood, they vary between different cultures. However, comforting dishes are often hot and fatty. In Britain, steamed puddings and custard are popular comfort foods, and when the weather is cold, people want gravy, sauces and hot dinners rather than salads [189]. Dupleix refers to her selection of comfort foods as pyjama foods, and includes "sloppy soups and stews, scones dripping with butter, gentle risotto, baked pasta, stir-fried noodles, scrambled eggs, cheesy gratins and steamed puddings" [190]. Meatballs are another favourite with her [191]. Italians traditionally go for gnocchi, a dish of homely, soft and light dumplings [192]. Following the recent terrorist attack on their city, New Yorkers are taking comfort in cupcakes and apply pie, whilst spurning their usual gourmet restaurant dinners [193]. The so-called nursery foods served in certain London gentlemen's clubs also represent comfort foods, transporting the (possibly) high achiever on a journey back to a warm and protected childhood. The opposite of a comfort food would be an unfamiliar, delicate, haute cuisine dish served in a formal setting. Comfort foods fulfil their role because they provide reassuring familiarity. Apart from their emotional associations, and their fat and sugar contents, it is noticeable that comfort foods share certain textural characteristics. Meat balls do not contain bones or gristle, and gnocchi are soft and light, as are puddings and cakes. In contrast to most fruits and salads, comfort foods require little chewing and oral effort. Emotional associations with foods may be negative, as well as positive. Memories of school meals tend to be negative and often colour British adults' perceptions of foods such as semolina and cabbage. Some children are made to feel guilty about not eating their mother's food or about eating too much, something that can lead to eating disorders and food aversions [28]. Another downside of the connection between foods and emotions is the fact that this can become a barrier to desirable dietary change – either because the individual finds it difficult to change, or because change would upset those that had considered it their duty to pass down food traditions future generations.

9.4
Purity Value in Food

A practising UK dietician contrasts "real, fresh, unadulterated food" with "processed and ready-made meals", explaining that the nutrients in real food are packaged "in the way that nature intended", and that the body would not have to deal with additives that it was not designed to digest [28]. A celebrity cook describes a vegetable bouillon powder as containing no chemicals and endorses the product as entirely natural [194]. A government committee grapples with providing advice to the food industry on the use of the terms fresh, natural, pure, traditional, original, authentic, home-made and farmhouse in the marketing of its products [195]. The committee bases its work on a collective feeling that consumers are currently significantly misled by advertising claims incorporating such terms, and appears to equate stricter rules on their application with giving consumers access to the information they need [195]. In food marketing, such terms may be used more or less interchangeably, and for the simple purpose of signalling some positive quality attribute or other in a product which is sold competitively alongside similar ones. The context in which such terms are used is therefore highly relevant. It needs to be considered carefully by the consumer weighing up his or her options if they are to extract the appropriate meaning from them. This is not to say that the food industry is entirely free from unscrupulous operators, who will cynically and deliberately mislead their customers. Marketers tend to choose punchy product descriptions for industrially produced foods in terms familiar to consumers, whilst glossing over the reality of industrial food production [195]. Consumers may respond highly negatively when confronted with unexpected ingredients in foods, e.g., the presence of non-meat proteins in sausages [59].

Pure spring water, drunk at source, may convey the ultimate image of purity, i.e, an untouched, virginal state. A wild fruit picked and eaten in an ancient or primeval forest conjures up similar images. Beyond these unlikely scenarios, things become more complicated. Purity can mean, on the one hand, removal of undesirable substances from a raw material, or it can mean leaving a food whole. Refining food can therefore be either very good or very bad, depending on who is talking. For example, for Hindus one of the purest, and most sacred, of human foods is ghee, a clarified butter which plays a prominent role in Hindu mythology [138]. Thus any inferior food cooked in ghee instantly becomes "pakka", i.e., edible even by a Brahmin [138]. Clean and unclean foods are also part of the Mosaic dietary regulation of the Hebrews [184], still familiar in contemporary Jewish eating, and the choice of the right (kosher) foods in that context. Similarly, to Muslims, pork is impure, animal blood means pollution, and alcohol consumption is an abomination [184]. Thus, the concept of purity is familiar in the context of religious dietary law. Where food additives and flavouring substances are concerned, purity is again a key quality attribute. Only so-called food-grade chemicals are permissible in order to ensure freedom from harmful contaminants. Purity may also be important in more traditional food ingredients, e.g., starch and sugar fractions, in order to be able to accurately predict their behaviour in recipes and processes. On the other hand, refined foods, in particular, sugar, flour and oils, have received a poor press in recent years. Leaving aside the issue of functionality, refined foods are often seen as nutritionally inferior to their whole food equivalents. Refined foods tend to be more stable towards spoilage and infestation. However, one of the key features of what are considered to be good foods is that these foods go bad [196]. Refined, isolated nutrients are often seen as inferior to the same nutrients presented in the context of a whole food [197]. The holistic approach leads to the conclusion that foods that are naturally rich in vitamins are to be preferred over vitamin supplements. Synthetic food chemicals may also not entirely duplicate their natural counterparts, as there may be 50:50 chirality (handedness) in synthetic materials, whereas in natural materials, only one species (hand) is present [198]. This may pose risks to certain consumers, in terms of their exhibiting symptoms of food intolerance [198].

Reshaped meat may also be conceptualised as a food that has been refined, not literally, but in the sense that processing has allowed consumers to distance themselves from unwelcome thoughts about animal welfare and slaughter [138]. Examples are chicken chunks, sometimes heavily breaded to round out shapes, turning them into abstract forms [138]. Shaped meats and fish products may be perceived as being "made up of scraps", and "not very healthy to begin with, and are frequently oozing with additives and preservatives" [28]. People may also camouflage the basic identity and nature of food raw materials through the liberal application of sauces and seasonings that are fashionable at any particular time [69]. In advertising terms, the word pure is often used to denote single ingredient foods and beverages, e.g., pure orange juice or pure Maine maple syrup. In one particular instance, "Pure" has been used as a product brand name for a fat spread [199]. The words pure and sunflower appeared next to each other on the packaging, whilst the ingredients list included oils other than that from sunflowers. Such practices on the part of food suppliers, which can easily cause consumers to suffer ill effects, as happened in this case, are reprehensible and should not be tolerated. As a first step, the Food Standards Agency's Food Advisory Committee has recommended that the term pure should not be included in any brand or fancy name [195].

A closely related debate concerns the perception of foods as either natural or unnatural. This too may be looked at from a variety of points of view. The food naturalness scale may be visualised as being anchored at one end in the pure, wild foods of the primeval forest described above, and at the other in foods containing synthetic and artificial (not, so far, detected in nature) flavourings. Between these poles might lie organically produced foods, followed by the raw products of conventional farming, then physically processed foods, and finally, chemically processed or preserved foods. Both purity and naturalness are often associated with simplicity, with some original state, both in relation to food production and to food processing. Clearly, of the foods commonly eaten in the West, few have not been managed and developed over many decades, even centuries, so that most farmed foods differ greatly from their wild ancestors. Despite this fact, farmed foods are widely considered to be natural. Long established processes for foods, such as bread and cheese making, similarly tend to be perceived as natural. In these examples, naturalness is associated with the familiar and commonplace, with small-scale, intelligible processes and with slow change. Natural foods are foods that people feel comfortable about, primarily because they have a long history of human dietary use. By contrast, unnatural foods would be those foods that appear strange and unfamiliar, with uncertain consequences following consumption. Genetically modified and irradiated foods fulfil these criteria. Food law recognises a consumer's need for purity in certain foods by prohibiting the use of additives in basic foods such as honey and butter. The term natural is well established and understood in the context of the markets for yoghurt, fromage frais and cottage cheese, where it is used interchangeably with the term plain, to indicate that the product is unflavoured [195]. A combined Institute of Food Science and Technology (UK) and Association of Public Analysts (IFST-APA) working group felt that the extensive use of the word natural in food labelling and advertising arises from a public misconception that natural implies safe, healthy and nutritious [200]. In the USA, where hormone injection into food animals is legal, public revulsion occasionally manifests itself against that practice [138]. Hormone injection in chicken presents a special problem because the attraction of chicken involves connotations of mildness, innocence and healing power [138]. Unlike hormone injections, the use of antibiotic growth promoters is common in the conventional farming industry within the European Union, and like hormone injections, it is often viewed as a food-polluting practice. This perception may be further reinforced due to the association of the routine application of antibiotics with the emergence of human drug resistant microorganisms [201]. Natural and unnatural practices are predominantly identified in connection with the rearing of food animals. Thus, caged hens live unnatural lives. They are in a position neither to scratch the ground, flap their wings nor to express a natural pecking order [138]. Yet only a small proportion of consumers refrain from purchasing the products from such systems. For the rest, animal abuse for the benefit of humans appears to be natural. To the routine purchaser of organic produce, visible soil is a sign of naturalness, whilst invisible pesticide residues are pollutants, all the more undesirable because of their invisibility. Visible dirt can be washed away, but one cannot be sure about what happens to invisible dirt. Hydroponic growing systems, e.g., for lettuce, can be exploited to reduce the physical distance between producers and consumers, allowing a freshly picked product to be sold. However, these soil-less systems may be considered unnatural by some because of the lack of the visible growing medium [138]. There is some concern, in the organic food market, about the true nature of natural flavouring substances and the potential of their label declaration for deceiving consumers. These are therefore being phased out by the Arbeitsgemeinschaft ökologischer Landbau in Germany [202].

When confronted with a term such as pure and natural, but especially fresh, the consumer's immediate response ought to be to ask what is meant by fresh. Depending on the precise context, the term fresh can have a variety of meanings. One does not have to be too clever to realise that absolute freshness is one thing modern food supply chains deny consumers [138]. However, mussels and oysters are sold alive, hence fresh, and a bunch of carrots or an orange, still attached to their foliage, signal a certain degree of freshness. IFST-APA define as fresh "the condition of a short shelf-life perishable unprocessed food prior to perceptible evidence of physical, chemical or microbiological change" [200]. The term fresh is normally applied to unprocessed foods, e.g., fresh eggs and fresh meat, showing that they are in their original state [200]. Fresh is also used in apparently contradictory terms, e.g., fresh pasteurised cream, to distinguish it from more highly processed types of cream [200]. In the context of fruit and vegetables, fresh implies products that have not been frozen, dried, canned or preserved in vinegar and other liquids or sugar [195]. Fresh milk is understood to mean a product with a limited shelf life of 10–12 d and which requires chilled storage [195]. Fresh in relation to food generally has also been defined as not old or spoiled, as recently harvested, obtained or made, as in its raw state, in good condition, refreshing (as an oral sensation), excitingly different, and so on [195]. The term fresh thus has a wide range of application. When people argue the benefits of eating fresh food, bearing in mind the practicalities of modern food provisioning, they tend to mean cooking raw materials fresh, rather than picking them shortly before use [68]. However, the fish for Bouillabaisse should be "more than fresh, it should be caught and cooked the same day" [187]. It is worth noting that in the context of meat, fish and fermented foods, such as cheese, there is no universal rule for the desirability of freshness. Meat may need to be hung for tenderness and depth of flavour, Dover sole should be aged for the same reason, eggs and Brazil nuts cannot be shelled cleanly if too fresh, and as far as fermented foods are concerned, ageing is fundamental to their very existence. Some classic dishes rely on leftovers as ingredients, and may be very popular. In Scotland for example, stovies, a slow-cooked dish made with left-over potatoes, gravy and meat, usually the remains of the Sunday roast, are such a dish [203].

Definitions broadly relating to the area of authenticity are particularly complex. The Food Safety Act 1990 prohibits the sale of any food which is not of the nature, substance or quality demanded by the purchaser [195], and in the absence of direct communication between producers and consumers, food names, descriptions and labels serve to transmit important information to the latter. Names of established foods have long been appropriated both by domestic and commercial cooks for their own creations. Thus in 1962, Elizabeth David lamented the decline of proper mayonnaise, which was being displaced by mass produced imitators (including salad cream) with distinctly different ingredients and sensory properties compared with those of the traditional, home-made product [121]. Nevertheless, there will always be many versions of certain traditional recipes, reflecting individual and household preferences. Questions as to the proper name of foods arise frequently, e.g.: Should yoghurt without the presence of an active culture be allowed to be called yoghurt? Should farmed fish have to be labelled as such? Should so-called novelty cheeses, with additions such as apricots, pickled onions or whisky, be allowed to be called cheeses [204]? Unless a food name has been legally protected, authenticity is a problematic concept, as it is largely subjective. IFST-APA define genuine and authentic as terms to distinguish ingredients which might otherwise be synthetic (e.g., vanilla ice cream made with genuine vanilla), or as terms that may establish the origin or type of a food (e.g., genuine Manzanilla olives, genuine Italian olive oil) [200]. Supermarkets cannot be expected to provide genuinely fresh foods [138], in particular, freshly dug or picked produce, nor, clearly, home-made foods. Clarke supports organic food, because "it is real food, and it tastes like it" [28]. However, she merely expresses a personal preference. The food market shows clearly that people do like highly processed foods, including salad cream [205] and salt and vinegar flavoured crisps, the latter being described by Cooper as "the logical and inevitable end result of the sterile philosophy of over-processing" [206]. Authentic foods are foods that are correctly described and labelled [207]. Misdescription can take many forms, from the undeclared addition of water or cheaper materials (adulteration) to the wrong quantitative declaration of ingredients or the incorrect declaration of origin (animal species, plant variety or geographical) [207]. However, false label descriptions are not the only case of consumers being deceived. Food suppliers also need to be aware of, and fulfil, consumer expectations that have been raised implicitly. When celebrity chef Gary Rhodes was found to be serving factory produced, frozen chips in his Edinburgh restaurant, customers were horrified [208]. Various campaigns have tried to stem the tide of mass produced, and increasingly standardised, foods. The first of these may have been the Campaign for Real Ale (CAMRA), which supports the British public house as a focus of community life and campaigns for greater appreciation of traditional beers, ciders and perries as part of British national heritage [209]. With traditional cheeses increasingly under threat, the concept of real cheese has also made an appearance [210]. Real bread, real beef and real pork refer to traditional systems of production which are believed to have yielded better tasting products than those which replaced them. Real foods are said to result from taking time, in sourcing, growing and preparation – in expressing respect and gratitude. From a different perspective, especially that of children, real and proper food may be food that everybody else eats and likes. In this way, chicken nuggets and fish fingers may seem more real than the organic roast chicken, or the freshly made chicken risotto, proudly presented to them by a conscientious parent. The expression real food may also be used in the context of simple, hearty, to-be-shared food, as opposed to the "tortured little bonsai vegetables" of nouvelle cuisine [211]. Terms such as real and pure can only be understood in specific contexts. It is not the use of these terms per se that should be deplored, but the general lack of ready public access to the definitions and specifications that underpin their use in each case.

9.5
Ethical Value in Foods

In common language usage, the term ethics refers to sets of principles, or rules, according to which actions are either sanctioned or forbidden [212], e.g., the ethics of meat production. From the standpoint of philosophy, ethics is concerned, more fundamentally, with normative speculation about the nature of the good life and the ultimate worth of the goals that people set for themselves [212]. Throughout recorded history, the great philosophers have added to the public debate about the rights and wrongs of people's actions. For example, in utilitarianism, moral agents are said to act so as to produce the greatest amount of pleasure for all concerned with a particular issue [212]. Food market ethics are segmented to reflect the ethics of groups of consumers, e.g., organic foods and vegetarian foods. There are various quality assurance schemes that focus specifically on ethical consumer value, e.g., the Fairtrade Mark, administered by the Fairtrade Foundation, a UK Registered Charity [213]. This mark provides guarantees on the welfare of, as well as financial rewards to, food workers in developing countries, who are typically involved in the production of commodities such as coffee, tea, cocoa and bananas [213]. Whilst some consumers will develop personal ethical rules about the types of food they are prepared to consume, most will not. Even some of the most controversial food production systems, i.e., eggs from caged hens, do not generally attach social stigma to the consumption of their products. In fact, for anyone who ever uses processed convenience foods, or consumes food away from home, it is almost impossible to consistently avoid such products. It is estimated that only 10% of UK food consumers take an active interest in how their food is produced [214]. The rest either abdicate responsibility entirely (48%) or, although potentially interested, are not sufficiently involved to search for the relevant information (42%) [214]. Minimum standards of public morality are expressed in a society's laws, which are designed to protect individuals from what is either injurious or offensive [215]. Market mechanisms allow added value to be delivered where personal standards are dissatisfied by the minimum set by the law. Ethical food choices may be said to result from the individual consumer's preparedness to sacrifice some of their consumption value due to consideration of others, and in the case of food, these others may well be animals. Whilst ethics appears as one of eight types of consumer value hypothesised by Holbrook [118], it may be troubling to some to think of ethics as a consumer value that can be obtained in market place exchanges [216]. Perhaps, what is really gained in ethical consumption is consonance among the individual's beliefs, attitudes and behaviours, providing a sense of integrity and virtue. Despite the apparent apathy of the majority of UK consumers, many people have been sufficiently affected by a decade of headlines about food scares, environmental standards and the crisis in farming, to seriously examine their past food consumption habits. This recent history of food production has led to a widespread perception that damage is being done to the land and to human health, and that many animals used for food are made to suffer unnecessarily [217]. Such perceptions would cause the individual consumer to become more involved in their consumption practices and choices, leading them to accept a share of responsibilities in the nature of food supply chains. Subscribing to a vegetable box scheme, perhaps in conjunction with some community supported agricultural project, offers high levels of participation and involvement.

There are several types of ethical concerns in relation to food. The so-called green consumer represents the environmental aspect of ethical food consumption. People holding moral views about environmental issues are committed to an environmental ethic consisting of one or more principles [218]. An example of a basic principle in such an ethic would be the view that the extinction of species as a result of human actions is a bad thing [218]. Human-centred environmental ethics place the impact on humans at the centre of environmental policies [218]. Care of the natural environment affects consumers directly. The pollution of land, rivers and oceans with harmful chemical or biological agents implies potential pollution of food obtained from them. This in turn poses threats to human health. However, environmental concerns may be wider ranging, including consideration of the interests of other species. In fact, classical utilitarianism includes animal suffering in its calculations of maximum benefit to stakeholders [218]. Broadly based environmentalism is concerned with the sustainability of current practices in terms of the inheritance left to future generations. Central to this is the debate about the responsible use of non-renewable resources. Food packaging waste, the cost of food transport (food miles), and the decline of food species through over-exploitation, e.g., cod [219], exemplify this wider environmental debate. The survival of tropical rain forests also enters this debate, e.g., in the sustainable management of such forests for tropical crops such as Brazil nuts. Recently, the environmental debate has entered quasi-religious realms, with the question of whether humans should allow themselves to "play god" by transferring genes between species [220]. Whilst from the religious perspective, genetic modification is not necessarily considered an ethical issue, the patenting of crops by commercial companies may well be more generally understood as such [221]. This is an example of the concept of the social responsibility born by companies involved in the food market.

The second major area of ethical consideration in relation to food therefore concerns what might be summarised as respect for life. This includes the needs of other people, both producers and consumers of foods, and the needs of farm animals and other animals used for human food. Consumers in developing countries generally do not enjoy the standards of protection familiar in the West. Nevertheless, the practices of global food companies do come under scrutiny there too. A well-known example is that of the inappropriate marketing of Nestlé breast milk substitutes, which has been under more or less continuous surveillance by consumer pressure groups for many years [222]. Among unethical practices in the West, mismarketing and the withholding of vital information are major issues. This is exemplified by the recent history of the unadvertised flooding of the market with genetically modified foods. It is thought that one of the key factors in the eventual, widespread rejection of the technology may have been a disregard for public values by big business [220]. Wholesale fortification or, depending on the individual point of view, contamination of food, denies choice. This applies to the surreptitious introduction of genetically modified soy into a wide range of processed convenience foods. It may also be seen to be relevant to issues such as the fluoridation of water and fortification of bread with folate. Both are examples where autonomy is denied the individual for the sake of, arguably, a greater public good. The issue of social responsibility has arisen in a number of different contexts. Retailers have been accused of not only cashing in on, but watering down the impact of religious festivals, e.g., by starting to sell Easter eggs in January [223]. According to the retailers, this practice had been demand driven [223]. Companies have also been accused of endangering consumers' lives by imbedding small toys in confectionery targeted at children [224], and of developing alcoholic drinks to appeal to under-age drinkers [225]. Food retailers have had to tackle the issue of dealing with large quantities of, still palatable, food waste in a socially acceptable manner [226].

However, the ethical issue in food supply chains which is, undoubtedly, the most emotionally charged, is that of the factory farming of animals, and of animal welfare and slaughter. The welfare of an animal may be defined as being determined by its capacity to avoid suffering and sustain fitness [227]. The slaughter of animals for food, as an ethical issue in its own right, has assumed added importance as it has become clear that the consumption of animal protein is not required for good human health, and that in fact, too much meat is being consumed in these terms. Many consumers choose to ignore, as far as possible, the connection between living animals, the process of slaughter and the meat on

their plate. Meat consumers in Scotland have been shown to perceive meat quality attributes as belonging into distinct categories [49]. There are those attributes that a consumer will address at the point of purchase, in particular, price
and visible cues to eating quality [49]. Consideration of certain other attributes
is deliberately suppressed at this point, in particular, attributes related to the
quality of life of an animal [49]. In consumers' minds, the moral benefits of meat
consumption to the family, especially if children are present, vastly outweigh any
moral costs to an animal [49]. Although animal welfare is a concern in general
terms, it is seen as one relating to citizenship rather than specifically relating to
consumption, and therefore delegated to the sphere of Government [49].

What of those consumers who do care and who do engage in active and moral
citizenship both in and outside consumption? Many are able to rationalise meat
consumption and reconcile it with their beliefs about the rights of animals. It is
possible to identify benefits of meat eating from the point of view of the animal.
The individual farm animal, destined for human consumption, is alive only because it will one day be slaughtered. Well housed and well fed, in good health and
able to engage in its inborn behaviours, as far as we can understand, the animal
values its life. Unaware of its ultimate fate, it has both a life and a comfortable living. In reality, the situation is more complicated, and the costs and benefits cannot be readily assessed. Many farm animals do not lead comfortable lives, and
death for many is miserable, whether at slaughter or as a result of inherent mor-
tality rates of specific rearing systems. Certain production systems are even designed to produce a sick animal, e.g., veal calves fed on iron deficient diets, which
leave them anaemic, and force-fed geese and ducks (for pâté de fois gras) with
pathologically enlarged livers. Not thinking about the fate of the animals negates
the opportunity to show some respect to the animal that is being sacrificed.
Whilst one may today view practices such as Halal slaughter as primitive rituals,
these rituals do at least imply respect for the life being taken [227]. Lack of respect for food animals is also implicit in their categorisation as livestock and in
the increasing selectivity of consumers in terms of which part of an animal they
are willing to consume – the rest going to waste. Consumer boycotts of particular products, companies or even countries are moral acts and can be instrumental in bringing about social change, forcing companies to demonstrate
greater acceptance of their social responsibilities. Lobbying and consumer
choices do have an effect on the market, although progress may be slow. In 1997,
Marks and Spencer announced that it was de-listing the eggs of caged hens
[228]. Four years later this policy was extended to eggs used in the retailer's food
products [229]. Tesco, one of its competitors, started labelling "eggs from caged
hens", although the statement was barely legible [230]. By the year 2009, the use
of battery cages for hens will be prohibited throughout the EU [231]. Today, for
the all-round ethical consumer, and for the consumer who would like to appear
that way, organic foods may be the market sector of choice. To some extent, the
organic label represents a substitute for local knowledge. It re-establishes the
lost link between producer interests and consumer interests. This is especially
true as far as the pioneers of organic production are concerned [232]. Evidently,
the producer-consumer bond will be closer where the producer follows a certain
path from conviction than if he or she merely responds to market demand.

Given the current growth in demand for organically produced foods [83], concerns are already being raised about the integrity of the label in the future. Because all mass production may be seen as putting soil quality at risk, committed organic growers may in future have to find new terminology [233].

Conclusion

Part 3 of the book has provided the focus on individual consumers and their interactions with individual foods. It has attempted to draw together some of the social scientific issues first raised in Part 1 and certain aspects of the physical, chemical and biological nature of foods discussed in Part 2. It has been demonstrated that even food attributes generally considered to be objective, e.g., aspects of food safety, contain subjective components. These may be perceptual in nature (e.g., risk-seeking vs. risk-averse personalities), or they may be related to individual differences in physiological function. The sensory attributes of foods are measurable in more or less objective terms. At the same time, what such attributes mean to the individual is closely bound up with individual psychological and physiological factors. In this final part of the book, what is eaten, in what quantities, when and why, has become clearer. Individual foods differ in their satiating capacity, and hunger does not necessarily reflect a physiological need. Eating usually takes place in anticipation of hunger, or because certain private or social eating routines have been set up. Once a consumer understands all this, they will be able to design and implement beneficial dietary patterns for themselves, including food intake reduction programmes, with increased confidence.

References

1. Epstein AN. Prospectus: Thirst and Salt Appetite. In: Stricker EM (ed). *Handbook of Behavioral Neurobiology. 10. Neurobiology of Food and Fluid Intake.* New York: Plenum Press, pp 489–512, 1990
2. Friedman MI. Making Sense Out of Calories In: Stricker EM (ed). *Handbook of Behavioral Neurobiology. 10. Neurobiology of Food and Fluid Intake.* New York: Plenum Press, pp 513–529, 1990
3. Blundell JE, Tremblay A. Appetite Control and Energy (Fuel) Balance. *Nutritional Research Reviews* 1995; 8: 225–242
4. Montgomery D. One in two adults 'face future of obesity' in epidemic threat. *The Scotsman*, 26th May 2000
5. Stuart J. How dieting became a national obsession. *The Independent*, 6th September 2001
6. Wandmaker H. *Rohkost statt Feuerkost.* München: Wilhelm Goldmann Verlag, 1999
7. Rozin PN, Schulkin J. Food Selection. In: Stricker EM (ed). *Handbook of Behavioral Neurobiology. 10. Neurobiology of Food and Fluid Intake.* New York: Plenum Press, pp 297–328, 1990
8. Lieberman DA. *Learning. Behaviour and Cognition.* 2nd edition. Pacific Grove, CA: Brooks/Cole, 1992
9. Carlson NR. *Physiology of Behavior.* 7th edition. Boston: Allyn and Bacon, 2001
10. Benton D. *Food for Thought.* London: Penguin Books, 1996
11. Anon. The Autonomic Nervous System. Retrieved on 13th September 2000 from http://faculty.washington.edu/chudler/auto.html

12. Stricker EM. Homeostatic Origins of Ingestive Behavior. In: Stricker EM (ed). *Handbook of Behavioral Neurobiology. 10. Neurobiology of Food and Fluid Intake.* New York: Plenum Press, pp 45–60, 1990

13. Stellar E. Brain and Behavior. In: Stricker EM (ed). *Handbook of Behavioral Neurobiology. 10. Neurobiology of Food and Fluid Intake.* New York: Plenum Press, pp 3–22, 1990

14. Kissileff HR. Physiological Controls of Single Meals (Eating Episodes). In: Meiselman HL (ed). *Dimensions of the Meal. The Science, Culture, Business and Art of Eating.* Gaithersburg: Aspen, pp 63–91, 2000

15. Collier R. *Wie neugeboren durch Darmreinigung.* 6th edition. München: Gräfe und Unzer, 1998

16. Grill HJ, Kaplan JM. Caudal Brainstem Participates in the Distributed Neural Control of Feeding. In: Stricker EM (ed). *Handbook of Behavioral Neurobiology. 10. Neurobiology of Food and Fluid Intake.* New York: Plenum Press, pp 125–149, 1990

17. Deutsch JA. Food Intake: Gastric Factors. In: Stricker EM (ed). *Handbook of Behavioral Neurobiology. 10. Neurobiology of Food and Fluid Intake.* New York: Plenum Press, pp 151–182, 1990

18. Rolls BJ, Rolls ET, Rowe EA, Sweeney K. Sensory Specific Satiety. *Physiology and Behavior* 1981; 27: 137–142

19. Gibson EL, Wardle J. Effect of contingent hunger state on development of appetite for a novel fruit snack. *Appetite* 2001; 37: 91–101

20. Knibb RC, Smith DM, Booth DA, Armstrong AM, Platts RG, Macdonald A, Booth IW. No unique role for nausea attributed to eating a food in the recalled acquisition of sensory aversion for that food. *Appetite* 2001; 36: 225–234

21. Rogers PJ. Food choice, mood and mental performance: some examples and some mechanisms. In: Meiselman HL, MacFie HJH (ed). *Food Choice, Acceptance and Consumption.* London: Blackie Academic & Professional, pp 319–345, 1996

22. Bender AE, Bender DA. *Oxford Dictionary of Food and Nutrition.* Oxford: Oxford University Press, 1995

23. Friedman MI. Making Sense Out of Calories. In: Stricker EM (ed). *Handbook of Behavioral Neurobiology. 10. Neurobiology of Food and Fluid Intake.* New York: Plenum Press, pp 513–529, 1990

24. Stricker EM, Verbalis JG. Sodium Appetite. In: Stricker EM (ed). *Handbook of Behavioral Neurobiology. 10. Neurobiology of Food and Fluid Intake.* New York: Plenum Press, pp 387–419, 1990

25. Blass EM. Ontogeny of Ingestive Behavior. In: Fluharty SJ, Morrison AR, Sprague JM, Stellar E (ed). *Progress in Psychobiology and Physiological Psychology. Volume 16.* San Diego: Academic Press, pp 1–51, 1995

26. Woods SC. Insulin and the Brain: A Mutual Dependency. In: Fluharty SJ, Morrison AR, Sprague JM, Stellar E (ed). *Progress in Psychobiology and Physiological Psychology. Volume 16.* San Diego: Academic Press, pp 53–81, 1995

27. Stubbs RJ. Nutritional implications of ingesting modified foods. In: Henry CJK, Heppell NJ (ed). *Nutritional Aspects of Food Processing and Ingredients.* Gaithersburg: Aspen Publishers, pp 84–111, 1998

28. Clarke J. *Body Foods for Life.* London: Seven Dials, 1999

29. Cardello AV. The role of the human senses in food acceptance. In: Meiselman HL, MacFie HJH. *Food Choice, Acceptance and Consumption.* London: Blackie Academic & Professional, pp 1–82, 1996

30. Goldstein EB. *Sensation and Perception.* 5th edition. Pacific Grove, CA: Brooks/Cole, 1999

31. Monell Chemical Senses Centre. Nutrition and Appetite. Retrieved on 13 September 2000 from http://www.monell.org/nutrition.htm

32. Galef BG, Sherry DF. Mother's milk: A medium for transmission of cues reflecting the flavor of mother's diet. *Journal of Comparative and Physiological Psychology* 1973; 83(3): 374–378

33. Bartoshuk LM. The biological basis of food perception and acceptance. *Food Quality and Preference* 1993; 4: 21–32

34. Bartoshuk L. Separate worlds of taste. *Psychology Today* 1980; September: 48–56, 63
35. Lawless HT, Hartono C, Hernandez S. Thresholds and suprathreshold intensity functions for capsaicin in oil and aqueous based carriers. *Journal of Sensory Studies* 2000; 15: 437–447
36. McEwan JA, Colwill JS. The sensory assessment of the thirst-quenching characteristics of drinks. *Food Quality and Preference* 1996; 7: 101–111
37. Stevenson RJ. Is sweetness taste enhancement cognitively impenetrable? Effects of exposure, training and knowledge. *Appetite* 2001; 36: 241–242
38. McBride RL, Johnson RL. Perception of sugar-acid mixtures in lemon juice drink. *International Journal of Food Science and Technology* 1987; 22: 399–408
39. Shepherd R, Wharf SG, Farleigh CA. The effect of a surface coating of table salt of varying grain size on perceived saltiness and liking for paté. *International Journal of Food Science and Technology* 1989; 24: 333–340
40. Pliner P, Rozin P. The Psychology of the Meal. In: Meiselman HL (ed). *Dimensions of the Meal. The Science, Culture, Business and Art of Eating.* Gaithersburg: Aspen, pp 19–46, 2000
41. Marsh J. Cooking for children. *AGA Magazine* 2001; 7(22): 23
42. Karmel A. If it looks good, they'll eat it. *The Times Weekend*, 5th February 2000
43. Dimbleby J, McEvedy A. Our first meal. *The Times Magazine*, 27th May 2000
44. Francis FJ. Quality as influenced by color. *Food Quality and Preference* 1995; 6: 149–155
45. Tom G, Barnett T, Lew W, Selmants J. Cueing the consumer: The role of salient cues in consumer perception. *The Journal of Consumer Marketing* 1987; 4(2): 23–27
46. Hutchings JB. *Food Colour and Appearance.* London: Blackie Academic & Professional, 1994
47. Schröder MJA, Jack FR. Adding value to commodities through product differentiation: the example of apple juice. In: Booth D, Sobal J (ed). *Abstracts: The 9th annual multidisciplinary Conference on Food Choice. Appetite* 2000; 35: 193–214
48. Diamond P. *Covent Garden Fish Book.* London: Kyle Cathie, 1992
49. McEachern MG, Schröder MJA. The role of livestock production ethics in consumer values towards meat. *Journal of Agricultural and Environmental Ethics* 2002; 15: 221–237
50. Carluccio A. *Passion for Pasta.* London: BBC Books, 1993
51. Jaffrey M. *Far Eastern Cookery.* London: BBC Books, 1989
52. Hom K. *Chinese Cookery.* London: BBC Books, 1984
53. Psilakis M, Psilakis N. *Cretan Cooking.* Heraklion: Karmanor, 2000
54. Davidson A. *The Oxford Companion to Food.* Oxford: Oxford University Press, 1999
55. Schafheitle JM. Meal Design: A Dialogue with Four Acclaimed Chefs. In: In: Meiselman HL (ed). *Dimensions of the Meal. The Science, Culture, Business and Art of Eating.* Gaithersburg: Aspen, pp 270–310, 2000
56. Hicks D. The Importance of Colour to the Food Manufacturer. In: Counsell JN (ed). *Natural Colours for Food and Other Uses.* London: Applied Science Publishers, 1981
57. Elliot R. *The Bean Book.* Glasgow: Fontana/Collins, 1979
58. Shaida M. *The Legendary Cuisine of Persia.* London: Penguin Books, 1994
59. Schröder MJA, Horsburgh K. Communicating food quality to consumers. *Journal of Consumer Studies and Home Economics* 1997; 21: 131–139
60. Schutz HG, Wahl OL. Consumer perception of the relative importance of appearance, flavor and texture to food acceptance. In: Solms J, Hall RL (ed). *Criteria Perception of Food Acceptance. Conference Proceedings.* Zürich: Forster, pp 97–116, 1981
61. Muñoz AM, Civille GV. Factors affecting perception and acceptance of food texture by American consumers. *Food Reviews International* 1987; 3(3): 285–322
62. Szczesniak AS, Kahn EL. Consumer Awareness of and Attitudes to Food Texture. I. Adults. *Journal of Texture Studies* 1971; 2: 280–295
63. Baxter IA, Jack FR, Schröder MJA. The use of repertory grid method to elicit perceptual data from primary school children. *Food Quality and Preference* 1998; 9(1/2): 73–80
64. Jack FR, O'Neill J, Piacentini MG, Schröder MJA. Perception of fruit as a snack: A comparison with manufactured snack foods. *Food Quality and Preference* 1997; 8(3): 175–182

65. Birch LL, Orlet Fisher J, Grimm-Thomas K. The development of children's eating habits. In: Meiselman HL, MacFie HJH. *Food Choice, Acceptance and Consumption*. London: Blackie Academic & Professional, pp 161–206, 1996

66. Kern DL, McPhee L, Fisher J, Johnson S, Birch LL. The Postingestive Consequences of Fat Condition Preferences for Flavors Associated With High Dietary Fat. *Physiology & Behavior* 1993; 54: 71–76

67. Hom K. *The Taste of China*. London: Pavilion Books, 1996

68. Morningstar A. *Ayurvedic Cooking for Westerners*. Twin Lakes, Wisconsin: Lotus Press, 1995

69. Yoneda S, Hoshino K. *Zen Vegetarian Cooking*. Yokyo: Kodansha International, 1998

70. Prudhomme P. *Louisiana Kitchen*. New York: William Morrow, 1984

71. Gopnik A. *Paris to the Moon*. London: Vintage, 2001

72. Graham P. *Mourjou. The Life and Food of an Auvergne Village*. London: Penguin Books, 1999

73. Borssén J. *Eat the Heat*. Berkeley: Ten Speed Press, 1997

74. Sage A. The French turn up their noses at smelly cheese. *The Times*, 20th November 1999

75. Anon. A very different cup of tea. *London Evening Standard*, 25th March 1999

76. Dawson A, Hutchins R. *The Truth Will Out. Consumer Attitudes to the UK Food & Grocery Industry*. Watford: IGD Business Publications, 1999

77. Engel JF, Blackwell RD, Miniard PW. *Consumer Behaviour. International Edition*. 8th edition. Forth Worth: The Dryden Press, 1995

78. Dade P. Interesting Times – The New Consumer: Curse or Opportunity? *British Food Journal* 1988; 90(3): 105–110

79. National Statistics. Matheson J, Summerfield C (ed). *Social Trends*. No 31. London: The Stationery Office, 2001

80. Schröder MJA. FSA in Scotland: the consumer's viewpoint. *Food Science and Technology Today* 2001; 15(1): 54–58

81. Straughan R. Freedom of Choice. Principles and Practice. In: National Consumer Council (ed). *Your Food: Whose Choice?* London: HMSO, pp 135–156, 1992

82. Griffiths J (ed). *Key Note. UK Food Market. 1999 Market Review*. 11th edition. Hampton, Middlesex: Key Note, 1999

83. House of Commons. Select Committee on Agriculture. *Organic Farming*. London: The Stationery Office, 2001

84. Anon. Food fight. The FSA is poorly placed to pronounce on organic products. *The Times*, 2nd September 2000

85. Wheatcroft P. Iceland shivers as its customers get cold feet over the organic revolution. *The Times*, 23rd January 2001

86. Cope N. Tesco aims to boost organic food sales to £2bn by 2006. *The Independent*, 2nd November 2001

87. Dawson A. *Consumer Watch*. June. Watford: IGD Business Publications, 2000

88. Spencer M. 'Omigod, I'm a Smug Organic!'. *London Evening Standard. ES Magazine*, 29th June 2001

89. Judd T. Celebrity chefs 'stop people entertaining'. *The Independent*, 14th December 2001

90. Carter M. Delicious, did you buy it yourself? *The Times*, 14th August 1999

91. Dawson A. *Consumer Watch*. August. Watford: IGD Business Publications, 2000

92. MAFF. *Working Together for the Food Chain: Views from the Food Chain Group*. London: HMSO, 1999

93. Maxted-Frost T. *The Organic Baby Book*. London: Green Books, 2001

94. Barfe L (ed). *The Under-16 s Market June 1998*. Hampton, Middlesex: Keynote, 1998

95. Sherwin A. Monsters come out from under the bed. *The Times*, 16th February 2002

96. Foxall GR, Goldsmith RE, Brown S. *Consumer Psychology for Marketing*. 2nd edition. London: International Thomson Business Press, 1998

97. Pearsall J (ed). *The New Oxford Dictionary of English*. Oxford: Oxford University Press, 1998

98. Adams J. *Risk*. London: UCL Press, 1995

99. Commonwealth of Australia. *Food: a growth industry.* Canberra: Food Regulation Review Committee, 1998
100. Royal Society. *Risk assessment: a study group report.* London: Royal Society, 1983
101. Robertson V. Hard struggle to sell the safety of irradiation. *The Scotsman,* 15th March 1999
102. Arthur C. Don't be a sponge. There's nothing to worry about. *The Independent,* 26th September 1997
103. Cuthbertson WFJ. Why does the consumer object to genetic manipulation? *Food Science and Technology Today* 1998; 12(1): 61
104. Stuttaford T. Balancing risks against benefits. *The Times,* 4th January 2001
105. Walker N. Supermarket brands packaged like the top sellers are under scrutiny. *The Independent,* 31st October 1995
106. Frewer LJ, Howard C, Shepherd R. Consumer perceptions of food risks. *Food Science and Technology Today* 1995; 9(4): 212–216
107. Magnusson M, Koivisto Hursti U-K. Swedish consumers' attitudes towards genetic engineering and genetically modified foods. In: Booth D, Sobal J (ed). *Abstracts: The 9th annual multidisciplinary Conference on Food Choice. Appetite* 2000; 35: 193–214
108. Anon. Make it safe, but a menu means choice. *The Independent,* 15th January 1998
109. Conran T, Harding B, Errington H (readers letters). Give us the cheese, dirt, bugs and all. *The Sunday Times,* 21st March 1999
110. Burke DC. Making British food safe. *Food Science and Technology Today* 1999; 13(1): 12–18
111. European Commission. *IP/00/96. Commission Adopts Communication on Precautionary Principle.* Brussels, 2nd February 2000
112. Food Standards Agency. *Draft 21/08/00. The Food Standards Agency's Approach to Risk.* London: The Food Standards Agency, 2000
113. Lawrence A. 'Who needs 50 different types of cheese?' *The Scotsman,* 24th April 2000
114. Werb J. Food to die for. *The Scotsman,* 26th May 2000
115. Wavell S. Fat of the land. *Sunday Times,* 2nd November 1997
116. Rozin P. The socio-cultural context of eating and food choice. In: Meiselman HL, MacFie HJH. *Food Choice, Acceptance and Consumption.* London: Blackie Academic & Professional, pp 83–104, 1996
117. Lucas F, Woodhead S. So you think you communicate? *Food Science and Technology Today* 1999; 13(1): 2–4
118. Holbrook MB. Introduction to consumer value. In: Holbrook MB. *Consumer Value. A framework for analysis and research.* London: Routledge, pp 1–28, 1999
119. Grayson K. The dangers and opportunities of playful consumption. In: Holbrook MB. *Consumer Value. A framework for analysis and research.* London: Routledge, pp 105–125, 1999
120. Meyrick N. Chill out! *Food Processing* 2001; August: 19
121. David E. *An Omelette and a Glass of Wine. The True Emulsion.* London: Penguin Books, 1986
122. Schröder MJA, McEachern MG. ISO 9001 as an audit frame for integrated quality managememtn in meat supply chains: the example of Scottish beef. *Managerial Auditing Journal* 2002; 17(1/2): 79–85
123. Tudge C. A brief history of misunderstanding. *The Independent,* 5th October 2001
124. Byrd-Bredbenner C, Wong A, Cottee P. Consumer understanding of US and EU nutrition labels. *British Food Journal* 2000; 102(8): 615–629
125. Karmel A. Some babies like it fresh. *The Times,* 4th March 2000
126. Beckett F. Time to get fresh with your potato. *The Times,* 27th May 2000
127. Anon. Walkers crisps update. *BBC Watchdog.* Retrieved on 20th October 2000 from http://bbc.co.uk/watchdog/reports/food/wcrisp.shtml
128. Severin T. *The Brendan Voyage.* London: Abacus, 1996
129. Cardello AV, Schutz H, Snow C, Lesher L. Predictors of food acceptance, consumption and satisfaction in specific eating situations. *Food Quality and Preference* 2000; 11: 201–216

130. Kähkönen P, Tuorila H, Rita H. How information enhances acceptability of a low-fat spread. *Food Quality and Preference* 1996; 7(2): 87–94
131. Muir K. Fraught & Social. *The Times Magazine*, 8th September 2001
132. Schutz HG. Multivariate Analyses and the Measurement of Consumer Attitudes and Perceptions. *Food Technology* 1988; November: 141–144, 156
133. Meiselman HL. The contextual basis for food acceptance, food choice and food intake: the food, the situation and the individual. In: Meiselman HL, MacFie HJH. *Food Choice, Acceptance and Consumption*. London: Blackie Academic & Professional, pp 238–263, 1996
134. Jack FR, Piggott JR, Paterson A. Use and appropriateness in cheese choice, and an evaluation of attributes influencing appropriateness. *Food Quality and Preference* 1994; 5: 281–290
135. Natt D. White Wine and Food. The Acidity Factor. *Wine Times* No. 72, Summer 1994
136. Dupleix J. The Times Cook. *The Times*, 22nd September 2001
137. Fracassini C. Whatever you're having yourself. *The Scotsman*, 16th February 2000
138. Visser M. *Much Depends on Dinner*. London: Penguin Books, 1989
139. Sander T, Scharf A. Does exposure really reduce neophobia? In: Booth D, Sobal J (ed). *Abstracts: The 9th annual multidisciplinary Conference on Food Choice. Appetite* 2000; 35: 193–214
140. Cooper D. Good food, good ingredients. Retrieved on 7th November 1999 from http://www.saga.co.uk/publishing/foodnwine/cooperjan.html
141. Cooper D. Foreword. In: Green H. *Henrietta Green's Food Lovers' Guide to Britain 1996–1997*. London: BBC Books
142. Green H. *Henrietta Green's Food Lovers' Guide to Britain 1996–1997*. London: BBC Books
143. Woollen A. Junk food hysteria. *Food Processing* 2001; January: 8
144. Williams J. 'We never eat like this at home': food on holiday. In: Caplan P (ed). *Food, Health and Identity*. London: Routledge, pp 151–171, 1997
145. Keane A. Too hard to swallow? The palatability of healthy eating advice. In: Caplan P (ed). *Food, Health and Identity*. London: Routledge, pp 172–192, 1997
146. Anon. A fine spread. *Which Magazine* 2001; June: 24–27
147. Higgs JD. Leaner meat: an overview of the compositional changes in red meat over the last 20 years and how these have been achieved. *Food Science and Technology Today* 2000; 14(1) :22–26
148. Konz F. *Der Gross Gesundheits-Konz*. München: Universitas Verlag, 1995
149. Anon. A fruit drinks healthy? *Which Magazine* 2000; March: 8–11
150. British Nutrition Foundation. *Food Allergy and Intolerance*. Retrieved on 18th December 2000 from http://www.nutrition.org.uk/Publications/briefingpapers/allergy.htm
151. Neill F. Hearts and Bones. *The Times Magazine*, 22nd September 2001
152. Treneman A. Let them eat Jaffa cakes. *The Independent*, 23rd December 1997
153. Romano R. *Dining in the Raw*. New York: Kensington Books, 1992
154. Norton C. Junk food diet blamed for asthma in children. *The Independent*, 22nd August 2000
155. Norton C. Junk food blamed for the return of rickets. *The Independent*, 6th October 2000
156. Werb J. The Raw and the Beautiful. *The Scotsman*, 22nd May 2000
157. Meister K. Eggs: Not as Bad as They're Cracked Up to Be. *ACSH Priorities* 1996; 8(3). Retrieved on 14th July 2000 from http://www.acsh.org/publications/priorities/0903/eggs.html
158. Laurance J. Even a burger a day can increase your risk of cancer. *The Independent*, 26th September 1997
159. Food Standards Agency. *Agency advises consumers fish is an important part of a healthy balanced diet*. Press release 2001/01. 3rd January 2001
160. BBC Watchdog. *Salt*. Watchdog Healthcheck 2nd August 1999.
161. IFST. *Salt*. IFST Position Statement. *Food Science and Technology Today* 1999; 13(4): 221–225
162. James P, Ralph A. What is a healthy diet? Retrieved on 18th January 2000 from http://www.rri.sari.ac.uk/ar/diet.htm

163. Sharp I (ed). *At Least Five a Day. Strategies to increase vegetable and fruit consumption.* National Heart Forum. London: The Stationery Office, 1997
164. HEA. *Enjoy Healthy Eating.* London: Health Education Authority, 1991
165. Sanghera S. Food industry's big obsession could yet be a gross mistake. *Financial Times,* 28th/29th August 1999
166. Cohn S. Being told what to eat: conversations in a Diabetes Day Centre. In: Caplan P (ed). *Food, Health and Identity.* London: Routledge, pp 193–212, 1997
167. Watts G. Cases in need of evaluations. *The Times Higher,* 2nd July 1999
168. Drenowski A. From Asparagus to Zucchini: Mapping Cognitive Space for Vegetable Names. *Journal of the American College of Nutrition* 1996; 15(2): 147–153
169. Carluccio A. *Passion for Pasta.* London: BBC Books, 1993
170. Carluccio A. *A Passion for Mushrooms.* London: Pavilion Books, 1989
171. Proust M. *Remembrance of Things Past.* Harmondsworth, Middlesex: Penguin Books, 1983
172. Roden C. Middle Eastern Food. In: Levy P (ed). *The Penguin Book of Food and Drink.* London: Viking, pp 322–332, 1996
173. Holledge R. Everything you've always wanted to know about... Sausages. *The Times,* 10th February 2001
174. Ioannou N. *Barossa Journeys.* Kent Town, SA: Paringa Press, 1997
175. Carluccio A. Mamma mia – it's so delicious. *The Times,* 13th November 1999
176. Bone J. Americans put their trust in cod. *The Times,* 4th November 2000
177. Acton J. English, French or Belgian chips, sir? *The Times,* 19th February 2000
178. Twain M. *A Tramp Abroad,* 1894. (cited in 184)
179. Gutierrez CP. *Cajun Foodways.* University Press of Mississippi, 1992
180. Gray A. Search for upper crust is all pie in the sky. *The Scotsman,* 29th October 1999
181. Walker A. Scots' favourite snack to get image bakeover. *The Scotsman,* 23rd November 1999
182. Marshall N. *Chambers Companion to the Burns Supper.* Edinburgh: W&R Chambers, 1992
183. Charsley S. Marriages, weddings and their cakes. In: Caplan P (ed). *Food, Health and Identity.* London: Routledge, pp 50–70, 1997
184. Tannahill R. *Food in History.* London: Penguin Books, 1988
185. Pitchford P. *Healing with Whole Foods.* Berkeley: North Atlantic Books, 1993
186. Hewitt J. *The Complete Yoga Book.* London: Rider, 1991
187. Toklas AB. *The Alice B. Toklas Cook Book.* Harmondsworth, Middlesex: Penguin Books, 1961
188. Cowan C, Sexton R. Ireland's traditional foods. *Food Science and Technology Today* 2000; 14(1): 27–28
189. Fletcher I. Steamed pud and custard banishes the blues. *The Scotsman,* 21st January 2000
190. Dupleix J. The Times Cook. *The Times,* 5th January 2002
191. Dupleix J. The Times Cook. *The Times,* 13th January 2001
192. Beckett F. Acquiring the knack for gnocchi. *The Times,* 15th September 2001
193. Bone J. New Yorkers take comfort in cupcakes and apple pie. *The Times,* 17th November 2001
194. Denholm A. Delia cooks up a sales storm. *The Scotsman,* 10th January 2000
195. Food Advisory Committee. *FAC Review of the use of the terms Fresh, Pure, Natural etc. in Food Labelling 2001.* Hayes, Middlesex: Food Standards Agency, 2001
196. Walker C, Cannon G. *The Food Scandal.* London: Century Publishing, 1985
197. Holford P. *The Optimum Nutrition Bible.* London: Judy Piatkus (Publishers), 1997
198. Strudwick DJ. Wrong-handed chemicals. *Country Garden & Smallholding* 1998; (2): Letters
199. Anon. Pure confusion. *Which Magazine* 1998; January: 4
200. IFST/APA. *Definitions and Interpretations of Some Words and Terms in Relation to Food Products and Processes.* Institute of Food Science & Technology (UK), 2001
201. O'Brien D. Farmers 'could create incurable epidemic'. *The Scotsman,* 19th August 1999
202. Anon. Naturkost. Verbot für "natürliche" Aromen. *Schrot & Korn* 2001; 3: 8

203. Main L. *Scottish Food: A Consumers Perception. Honours Project.* Queen Margaret University College, Edinburgh, 2000
204. Hyman C. Cheese with extra bits. *The Times*, 29th April 2000
205. Robinson A. Anne Robinson *The Times*, 16th October 1999
206. Cooper D. *Snail Eggs & Samphire.* London: MacMillan, 2000
207. Food Standards Agency. *Research Programme Annual Report 2001*
208. Ronay E. Chilly reception for Rhodes's chips. *The Scotsman*, 14th February 2000
209. Campaign for Real Ale. Retrieved on 12th June 2000 from http://www.camra.org.uk/site/camrainfo/camrainfo.htm
210. Freeman S. *The Real Cheese Companion.* London: Little, Brown and Company, 1998
211. McGlone J. That old tradition of love and taste. *The Scotsman*, 6th January 1999
212. Stroll A, Popkin R. *Introduction to Philosophy.* New York: Holt, Rinehart and Winston, 1961
213. Anon. *The Fairtrade Foundation: Further Details.* Retrieved on 1st November 1998 from http://www.gn.apc.org/fairtrade/foundatn.htm
214. IGD. *The Changing Consumer.* Watford: Institute of Grocery Distribution, 1999
215. Lee S. *Law and Morals.* Oxford: Oxford University Press, 1986
216. Smith NC. Ethics and the Typology of Consumer Value. In: Holbrook MB. *Consumer Value. A framework for analysis and research.* London: Routledge, pp 147–158, 1999
217. Fearnley-Whittingstall H. Foreword. In: Friends of the Earth. *The real food book.* London: Friends of the Earth, 1999
218. Elliot R. Environmental ethics. In: Singer P (ed). *A Companion to Ethics.* Oxford: Blackwell, pp 284–293, 1993
219. WWF-UK Life without fish 'n' chips? It's on the horizon. *WWF-UK News*, 20th July 2000
220. Bruce D. Whether it be cloning or GM foods, the kirk has been on the case. *The Scotsman*, 6th May 2000
221. ActionAid campaigns against food patenting. *The Guardian*, 28th June 2000
222. Yamey G. Nestlé violates international marketing code, says audit. *British Medical Journal* 2000; 321: 8
223. Anon. January start to Easter attacked. *The Herald*, 15th January 1999
224. Cooper G. Nestlé chocolate balls lose Disney magic. *The Independent*, 3rd October 1997
225. Thompson T. Young 'preyed on' by drinks industry. *The Scotsman*, 17th January 2000
226. Pring A. Retailing's Social Conscience. *The Grocer* 1997; 27th September: 44–47
227. Webster J. *Animal Welfare. A Cool Eye Towards Eden.* Oxford: Blackwell Science, 1995
228. Anon. M&S wins praise as first supermarket to remove battery eggs from shelves, *The Independent*, 6th October 1997
229. Peachey P. Marks & Spencer vows to use only free-range eggs. *The Independent*, 24th July 2001
230. Anon. Large eggs, small print. *Food Magazine* 1998; 40: 20
231. Compassion in World Farming. European Parliament votes for Ban on Battery Cages. CIWF Celebrates! *Press Release PR5951*, 28th January 1999
232. Cousins D. Slotting the Organic Link into Position. *Farmers Weekly* 1995; 25th August: 75–76
233. Shaw D. Let's find another word for 'organic' food. *The Times*, 14th October 2000

Conclusion

Over the past decade-and-a-half or so, the debate about food quality, and its constituent parts, has intensified in the UK. The detail of this debate owes much to the fact, that quality is a practical, business issue. It is recognised that value must be created for consumers, and that quality must be managed. There are two key components to the management of quality, i.e., the management of innovation and consumer value (quality of design) and the management of product and process conformity. This book focuses largely on the quality of design, i.e., on the fit between the nature of some raw material, a company's capabilities, and a consumer's requirements and buying motives. Food marketers commonly exploit the quality concept in order to draw consumers' attention to their products. The need to draw attention to individual offerings exists in all consumer-oriented markets, in which the consumer's input constitutes one half of the market mechanism. The consumer demands a product, the producer offers it, and finally the consumer evaluates it. Quality in any product relates to individual requirements, which will have been identified in respect of some target market. In the absence of a consumer capable of perceiving quality, there is none. Hence in advertising terms, quality and similar words act as signals rather than as carriers of specific meaning. Specific foods, and food consumption patterns, can be adopted by a consumer to express certain aspects of his or her lifestyle, e.g., green consumerism. Producers are similarly able to express their business philosophy through the products that they offer, through the products that they do not offer, and in their actions in respect of society.

If a processed food is commercially successful, by definition, the processing will have added value to the food. However, that value may be perceived only by the target market for that food. In many cases, processing value is related to some aspect of convenience, in particular, extended shelf life. Various food technologies are available to achieve extended shelf life, e.g., chilling (with and without MAP), freezing, drying and chemical preservation. Here, one type of value is invariably gained at the cost of the loss of another, in particular, and obviously, freshness. However, in the case of MAP technology, perceived freshness may remain high. Processed foods are also likely to contain food additives. It is for the individual consumer to weigh up the costs and benefits of each situation as it affects them personally. Not processing a food can also add value to it, again in the eyes of the target market. In this case, it may be the sophistication of the distribution logistics that enables added value to be created. For example, developments in logistics have opened up the UK market for fresh sandwiches and tree-ripened mangoes. There are two mechanisms for adding perceived value to a food. The added-value concept typically calls to mind some food to which certain desirable ingredients have been physically added, e.g., added seeds on a loaf of bread, or added calcium or vitamins in a juice drink. This added-feature approach is common in highly processed foods. Alternatively, value can be added to a basic commodity by revealing some aspect(s) of its intrinsic value, e.g., the natural presence of flavonoids and other bioactive compounds in apples. This is an issue of marketing and product positioning. All foods need to be positioned, i.e, visibly targeted towards their intended market. It is often possible to revive the fortunes of some long-established food, which is approaching the final stages of its life cycle, by repositioning it. This means revealing some type of value that it represents, but that has not previously been highlighted. Consumer value refers to the consumption, or buying, motive for some product and is therefore determined, in part, by the measurable attributes of that products. Motives are also influenced by a consumer's cultural background. Cultural mechanisms serve to maintain certain patterns of consumption. On the other hand, fashion provides a group mechanism for challenging and revising obsolete behavioural patterns.

Objective and subjective aspects of quality are equally important. In food, quality is closely linked with product attributes such as nutrient content, sensory performance, and the absence of contaminants. Objective food quality also increasingly refers to production and manufacturing processes, even though these do not always translate into measurable product attributes. For many modern consumers, the way in which a food is produced, e.g., how a meat animal is raised and what it is fed, plays an important part in its overall palatability. In the wake of the UK's BSE crisis, events in the early parts of food chains are assuming increasing importance. Yet each consumer views food attributes through his or her own personal set of evaluative lenses, each lens focusing on a particular type of consumer value, e.g., safety, impact on health, sensory indulgence, overall palatability, and so on. Irrelevant product attributes, or those with low importance for a consumer's satisfaction with a product, are thus removed from evaluations. Objective food attributes are translated into food quality attributes when they impact a user's satisfaction in some way. This is where the subjective component of the quality concept is first encountered.

The three types of consumer value represented most prominently in modern foods are commonly cited as indulgence, convenience and health. However, the analysis of consumer value in respect of any consumer product can be taken much further, revealing greater detail and differentiation between motivations. Morris B. Holbrook's recently published typology of consumer value allows, for the first time, the identification of consumer value to be carried out systematically [1]. Basic types of value, such as esteem, status, play and ethics here derive from intersections of more fundamental motivations, e.g., self-oriented and other-oriented, active and reactive, and extrinsic and intrinsic. It would be useful to refine this typology further, bearing in mind the segmentation of the consumer market for foods. At present, rather a large proportion of food-related types of consumer value in Holbrook's typology is likely to assemble under the rubric "efficiency". For example, where in the typology do comfort foods fit in?

Analysed coldly, comfort eating is a means to an end, the end being an improvement in mood. However, the actual activity of comfort eating tends to be rather less conscious than is suggested by the concept of efficiency.

The rational consumer represents one of the pillars of economic theory (see Chapter 3). With the increasing levels of obesity in the populations of Western, industrialised countries, blame is often laid at the door of the food industry for supplying not only the wrong kinds of foods, but the wrong kinds of product information. What manufacturers may produce for the food label, if they are making a nutrition-related claim, is a nutrient breakdown of their products. What consumers may find difficult is to integrate nutrient knowledge with knowledge of what constitutes a healthy diet, and a memory of what has already been consumed on a particular day or in a particular week. Under these circumstances, maintaining optimum nutrition and body weight may seem a tall order. This is why individuals tend to devise rules, and develop routines, in connection with their food intake. However, assuming that an individual understands the information available and is capable of manipulating it effectively, the question may be asked: Would such a person be considered to be rational if they persisted with what would be regarded as unhealthy dietary habits? The answer could be: Yes. Scottish lorry drivers prefer fried foods to salads whilst working, because they enjoy eating them more. But there is a second, more telling motivation for the driver's preferred choice, i.e, perceived food safety. Fried foods are seen as less likely to lead to food poisoning than do cold foods [2]. For a lorry driver, who finds him- or herself on the road, a long way from home, a stomach upset is a highly undesirable condition. In contrast, the long-term health consequences of a fat-rich diet do not seem very immediate. In making explicit the types of value a consumer realises through some act of consumption, fresh light may be cast on the concept of rational consumption. Many commentators now see the rational food consumer as one who strives, through their food choices and eating habits, to optimise health, not just for the moment but well into old age. However, health is only one amongst many types of consumption value, and old age is not a vivid image that younger consumers, in particular, routinely call up about themselves. All likely gains and losses associated with specific behaviours are evaluated by the individual consumer in the context of their short-, medium- and long-term effects.

In the current debate about the public understanding of science, because consumers are liable to obstruct the introduction of radically new food technologies, they are often portrayed as irrational. This irrationality is then attributed to the public's ignorance of scientific issues generally. The argument may then be advanced that scientists, e.g., geneticists, should simply be trusted, not just because of their superior knowledge, but also because they are consumers too. There are two key issues in consumers' understanding of food science, or any other issue, i.e., cognitive processing limitations and lack of attention to an issue because of insufficient motivation. If consumers are confronted with new technologies that, by their very nature, carry risks or uncertainties, they will not automatically be motivated to learn about any potential benefits of the technology. They may however be motivated to find out about the potential disadvantages of the technology, especially if it has been introduced without the appropriate level of public debate, and without safeguarding choice to the individual. Scientific rationality may also be questioned. After all, a scientist working on the genetic modification of organisms has a strong, vested interest in the technology. Their livelihood may depend on it. This is likely to influence their perceptions, and their ability to argue and act rationally in relation to the issue at hand. Beliefs about food quality are often based on scant evidence, because information is often incomplete. Beliefs do not represent straight inputs into perceptions. Instead they are modified by the individual's current emotional state and by the context of the consumption situation.

Consumers should be seen as active participants in food markets. They do not just choose among alternatives that are offered to them. Each consumer also actively builds products into his or her household function. Although individual consumers are responsible for how much they eat, and of what, marketing may override individuals' personal or social inhibitions. For example, at a Scandinavian smorgasbord, all the food that is to be served is laid out on one large, central table, from which diners help themselves. No formal rationing or portion control takes place. In principle, those present can take as much food as they find it physically possible to eat. Yet this rarely happens. A smorgasbord, by definition, has many participants, and personal greed is not socially sanctioned. In contrast, an "eat-all-you-can", fixed-price pizza menu (see Chapter 3) deliberately creates a social expectation that people will overeat. Similarly, large confectionery bars are an invitation to the consumer to overindulge on a type of food that should perhaps be considered only as a treat. Unlike bars of chocolate, even the type of packaging used in these bars makes it difficult for the product to be consumed in smaller portions. Government policy can also be found guilty of encouraging overeating. Varied diets are often recommended to ensure adequate nutrient intakes. However, the more varied a meal, the more likely it is for the diner to overeat, due to SSS mechanisms being circumvented.

Although all the major food retailers offer a wide range of product lines, that offer has to be selective. No supermarket could be expected to provide thousands of different varieties of apple, nor would this be desirable, as shopping would turn into an impossible task. These varieties of apple exist however, and they are available from somewhere. The active consumer, if sufficiently motivated, will always find ways of getting hold of niche products. In the UK today, there are few foods that a consumer cannot obtain if he or she truly wishes to do so. Some critics of the modern food system accuse food companies of trying to maximise their sales, objecting to them promoting their products too vigorously. If current food consumption patterns really are harmful to public health, perhaps it is the consumer who should be helped to better understand the issues surrounding consumer value, and who should be taught how to maximise food value for themselves. This is a task primarily for the education system. Consumers must also be given ready access to the detailed product information they require, in particular, factual evidence in support of food marketing claims. This information must be presented in an appropriate format, and at an appropriate level. This means that issues, in particular scientific issues, need to be presented at several access levels, so that individuals can take charge of their own learning. This may be an issue for future legislation.

References

1. Holbrook MB. Introduction to consumer value. In: Holbrook MD (ed). *Consumer Value. A framework for analysis and research.* London: Routledge, pp 1–28, 1999
2. Jack FR, Piacentini MG, Schröder MJA. Perception and Role of Fruit in The Workday Diets of Scottish Lorry Drivers. *Appetite* 1998; 30: 139–149

Name Index

Subject Index

acetic acid 147, 151, 174
acid 105, 120, 129, 147–148, 159, 168, 170,
182, 226, 229
acidity regulator 168
acidosis 120, 277
ACNFP 30, 44
actin 108–109
actomyosin 109
adaptation 72, 228–229
added-value food 23, 27, 29, 51, 157, 164,
253, 289, 301–302
additive 37, 47, 49, 51, 97–100, 104, 107, 157,
159–161, 167–183, 193, 197, 259, 285, 301
adenosine triphosphate 126, 229
adulteration 167, 193, 196, 274, 283
advertising 11, 24, 32, 36, 47–49, 58, 62, 66,
74, 81, 87, 213, 250, 264–265, 269, 284–285
Aeromonas species 188
aesthetics 19, 21, 67, 84–85, 237, 251
affect 75, 137–138, 220
agar 104, 107
alcohol 135, 149, 159, 182, 227
algae 100, 107, 116, 124, 131, 186, 192–193
alginate 104, 107, 169–170
alkaloid 135, 230, 244
aloe vera 186
aluminium 121, 178
amino acid 107–108, 121–123, 125, 129,
148, 151, 183, 185–186, 217–218
indispensable 51, 103, 119–121, 123,
197, 212, 217, 277–278
ammonia 231, 243
amniotic fluid 229
α-amylase 165
amylose 102, 105–106, 171
anaphylactic shock 192, 276
anaemia 127
analgesia 218
animal breed 52, 275
animal secretions 102
animal welfare 4, 10, 19–21, 50, 255–256,
285, 289–291

annatto 142, 178
anthocyanin 132, 142, 179
antibiotic 99, 133, 175, 190, 194, 286
anti-caking agent 168
anticipation 219, 234–235
anti-inflammatory agent 133
anti-foaming agent 168
antimicrobial agent 163, 174, 186, 258
antioxidant 124–125, 128–129, 132–135,
161, 167–168, 173–176, 182, 186, 197
anti-thrombic agent 133
anti-tumour agent 132
aphrodisiac 214, 241, 282
apigenin 132–133
appearance 38, 48, 51, 98–99, 101, 137–143,
146, 150, 162, 168, 171, 176–179, 212, 221,
231–239, 243–244, 270
appetite 50, 83, 135, 138, 211–212, 215,
220–221, 223, 228, 232–234, 237–238, 243,
245, 247
salt 212, 222, 225, 243
apple 102, 131–133, 140–141, 144–145, 153,
191, 302
arabinose 104–105
arachidonic acid 122, 130
arousal 75, 211, 216–220, 225, 242, 260–261
arsenic 121, 195
artificiality 47
ascorbic acid 124, 174–175
asparagus 102, 127
Aspartame 148, 179, 189, 226
aspartic acid 123
Aspergillus species 191
aspirational target 67
association 48, 87, 144, 148, 228, 231, 236,
241, 243, 247, 263, 265, 268–269, 280,
282–283
animal-meat 269, 290
colour-flavour 235, 237
social 83, 85, 88, 212
astaxanthin 143, 179
asthma 277

Printing: Mercedes-Druck, Berlin
Binding: Stein+Lehmann, Berlin